COMPUTATIONAL OPTIMIZATION of Systems Governed by Partial Differential Equations

Computational Science & Engineering

The SIAM series on Computational Science and Engineering publishes research monographs, advanced undergraduate- or graduate-level textbooks, and other volumes of interest to an interdisciplinary CS&E community of computational mathematicians, computer scientists, scientists, and engineers. The series includes both introductory volumes aimed at a broad audience of mathematically motivated readers interested in understanding methods and applications within computational science and engineering and monographs reporting on the most recent developments in the field. The series also includes volumes addressed to specific groups of professionals whose work relies extensively on computational science and engineering.

SIAM created the CS&E series to support access to the rapid and far-ranging advances in computer modeling and simulation of complex problems in science and engineering, to promote the interdisciplinary culture required to meet these large-scale challenges, and to provide the means to the next generation of computational scientists and engineers.

Series Volumes

Borzì, Alfio and Schulz, Volker, *Computational Optimization of Systems Governed by Partial Differential Equations*

Ascher, Uri M. and Greif, Chen, *A First Course in Numerical Methods*

Layton, William, *Introduction to the Numerical Analysis of Incompressible Viscous Flows*

Ascher, Uri M., *Numerical Methods for Evolutionary Differential Equations*

Zohdi, T. I., *An Introduction to Modeling and Simulation of Particulate Flows*

Biegler, Lorenz T., Ghattas, Omar, Heinkenschloss, Matthias, Keyes, David, and van Bloemen Waanders, Bart, Editors, *Real-Time PDE-Constrained Optimization*

Chen, Zhangxin, Huan, Guanren, and Ma, Yuanle, *Computational Methods for Multiphase Flows in Porous Media*

Shapira, Yair, *Solving PDEs in C++: Numerical Methods in a Unified Object-Oriented Approach*

COMPUTATIONAL OPTIMIZATION
of Systems Governed by Partial Differential Equations

Alfio Borzì
Universität Würzburg
Würzburg, Germany

Volker Schulz
University of Trier
Trier, Germany

Society for Industrial and Applied Mathematics
Philadelphia

10 9 8 7 6 5 4 3 2 1

Library of Congress Cataloging-in-Publication Data
Borzì, Alfio.
Computational optimization of systems governed by partial differential equations / Alfio Borzì, Volker Schulz.
p. cm. – (Computational science and engineering series)
Includes bibliographical references and index.
ISBN 978-1-611972-04-7 (pbk.)
1. Mathematical optimization. 2. Differential equations, Partial. I. Schulz, Volker, 1965- II. Society for Industrial and Applied Mathematics. III. Title.
QA402.5.B655 2011
515'.353–dc23

2011033263

Dedicated to our wives,
Melanie and Petra

Contents

List of Figures

List of Tables

List of Algorithms

Preface

This book provides an introduction to some modern computational techniques for optimization problems governed by partial differential equations (PDEs). The optimization framework used is that of optimization on functional spaces based on the Lagrange formalism. The computational techniques discussed in this book represent recent developments resulting mainly from the combination of modern techniques for the numerical solution of PDEs and of sophisticated optimization schemes. Based on these two main features, the book provides a bridge between continuous optimization and PDE modeling focusing on the numerical solution of the corresponding problems.

Although the topics considered in this book are advanced, the methods and problems are presented in much detail such that the book may serve as a textbook for graduate students, requiring previous knowledge of the reader on the numerical solution of PDEs and on basic principles of continuous optimization. The book should also be suitable as an introduction for researchers in the field of scientific computing with PDEs to work in the field of optimization and for researchers in optimization to consider methodologies coming from the field of numerical PDEs. It should also allow scientists of natural sciences and engineering to formulate and solve optimization problems appropriate for their research purposes.

This book is timely considering the increasing interest in the field of PDE optimization, as can be seen from the many recent conferences and workshops worldwide and from the enormous increase in publications in this field. The focus of the book is on methodological aspects, and therefore applications are less emphasized. Problems with elliptic, parabolic, and hyperbolic PDEs and coupled systems of PDEs are considered.

It is clear that the purpose of this book is too ambitious to succeed completely. However, it fills a gap between more theoretically oriented texts on PDE constrained optimization and algorithmic oriented books. Furthermore, the intention to cover many emerging topics in numerical PDE optimization obviously results in less background material for which many references are provided. Nevertheless, enough details are devoted to each topic to understand the formulation of the problems and of the proposed solution procedures such that this book could be used as an introductory textbook for some very recent research topics of PDE optimization. It is obvious that the choice of these topics is biased by the research interests and experience of the authors, and this fact explains some focusing on specific topics rather than others.

This book is organized as follows. After a brief introduction, we start discussing the theoretical framework of PDE optimization. We focus on the characterization of the solution of optimization problems by means of the so-called optimality systems. These are coupled systems of PDEs and differential or scalar equalities or inequalities called the optimality conditions. It should be remarked that these systems have very special structures posing new challenges for the scientific computing methodologies. This challenge starts with the appropriate discretization of optimality systems that is the subject of one of the chapters of this book. Once the PDE optimization problem is discretized, a solution-optimization procedure can be started. One chapter is devoted to single-grid optimization schemes that should refer to the optimization procedures that do not exploit the possibility to represent the given problem on multiple scales of resolution. On the contrary, a chapter on multigrid methods illustrates recent advances that combine classical optimization schemes with the multigrid strategy. Another chapter is dedicated to recent methodologies that combine PDE optimization with statistical approaches in order to solve PDE optimization problems with uncertainty. Finally, a chapter of applications is presented with the purpose of outlining important emerging application topics such as shape optimization, quantum control, and time-dependent electromagnetic inverse problems. Summarizing, we have the following chapters:

1. Introduction
2. Optimality Conditions
3. Discretization of Optimality Systems
4. Single-Grid Optimization
5. Multigrid Methods
6. PDE Optimization with Uncertainty
7. Applications

We would like to thank many friends who encouraged this work and contributed through remarks, suggestions, and scientific collaboration. In particular, we would like to acknowledge the support and collaboration of (in alphabetical order) Mario Annunziato, Hans Georg Bock, Giuseppe Borzì, Nicolas Gauger, Omar Ghattas, Sergio Gonzalez Andrade, Matthias Heinkenschloss, Roland Herzog, Ulrich Hohenester, Kazufumi Ito, Do Young Kwak, Karl Kunisch, Roberta Mancini, Stephen Nash, Hans Josef Pesch, Ekkehard Sachs, Julien Salomon, Stephan Schmidt, Claudia Schillings, Georg Stadler, Fredi Tröltzsch, Michelle Vallejos, Stefan Volkwein, and Greg von Winckel.

We would like to thank Ms. Elizabeth Greenspan of SIAM Publications very much for her kind and generous assistance in publishing this book. Furthermore, we would like to thank an unknown referee, who helped to improve the final version of this book significantly.

We hope that this book will become a useful tool to everyday research and teaching work and that it comes to further editions because "Perfection is achieved, not when there is nothing more to add, but when there is nothing left to take away" (Antoine de Saint-Exupéry).

Chapter 1

Introduction

Optimization with partial differential equations (PDEs) is a modern field of research in applied mathematics, starting in the early 1970s with the work of Lions [235], although it has a very important predecessor and companion in the research on optimal control of ordinary differential systems that started with the works of the Bernoulli brothers in 1700; see [274, 278].

As in the optimal control case, it is the idea of being able to influence PDE application systems that is attracting enormous research efforts toward PDE optimization. In fact, it is the ultimate aim of any application problem to manipulate systems in a desired way.

The purpose of PDE optimization appears manifold and ranges from the need to control application systems to that of how to optimally change or estimate features of real-world systems modeled by PDEs. An important class of problems in optimization results from optimal control applications. These consist of an evolutionary or equilibrium system including a control mechanism and a functional modeling the purpose of the control. Other important classes of optimization problems are shape design, topology, and parameter optimization. Optimization is also an essential tool for solving many inverse problems. In all these cases we are concerned with systems that are modeled by a set of PDEs. Therefore we are in the framework of infinite-dimensional optimization problems where one distinguishes between state and optimization variables.

Because of the nature of PDE problems, solving the related optimization problems requires realization of optimization strategies having increasing complexity and the ability to solve large-scale problems in an accurate and computationally efficient way. These needs are revolutionizing the scenario of the optimization discipline since many classical methodologies appear overwhelmed by the size of the problems to be solved and by the type of additional requirements that the solution must satisfy. On the other hand, classical PDE solvers cannot be straightforwardly applied to solve optimality systems since new differential structures and a new type of coupling appear.

In this book, we present some recent results and outline present developments in the field of the numerical solution of PDE optimization problems. On the one hand, we give an introduction to well-known optimization methodologies that have been extended to the present case. We discuss these methodologies with less emphasis because they are also the subject of well-written books on optimization. On the other hand, we discuss with greater detail some methodologies that originate from the field of numerical PDE and are less known in the optimization community. Furthermore, we illustrate problems and methodologies like PDE optimization with uncertainty, shape optimization, quantum optimal control, and space-time electromagnetic inverse scattering problems that are emerging topics in PDE optimization.

In Chapter 2, we provide an introduction to optimization problems with PDE constraints using the terminology and notation that is usual in the PDE optimization community. The notions of reduced objective functional, gradient, Hessian and Lagrangian functions, optimality systems, etc., are introduced. A detailed derivation of optimality systems characterizing representative PDE optimization problems is illustrated.

In Chapter 3, we illustrate basic concepts concerning the discretization of optimality systems focusing on finite differences. Accuracy of optimal solutions is discussed considering an elliptic and a parabolic control problem. Further, we discuss the case of higher-order discretization and the case of optimal control problems governed by integral equations.

In Chapter 4, we recall some well-known unconstrained optimization methods. Our purpose is to emphasize some aspects of the implementation and application of these methods in the case of PDE optimization. We discuss nonlinear conjugate gradient schemes, Newton-type methods, and other black-box schemes. Further, we illustrate the main features of semismooth Newton methods and SQP schemes. The issue of preconditioning of KKT systems is also addressed.

In Chapter 5, multigrid methodologies are discussed. We roughly distinguish between the direct multigrid approach where the optimization problem is implemented within the hierarchy of grid levels, the use of multigrid schemes as inner solvers within an optimization scheme, and solution strategies where the multigrid method defines the outer loop.

In Chapter 6, we illustrate recent approaches to computing solutions to PDE optimization problems which are robust with respect to the stochasticity of the application framework. We discuss the treatment of uncertainty in PDE constrained optimization in the case where the coefficients of the PDE models are subject to random perturbations. Further, we discuss the problem of aerodynamic design under geometric uncertainty. A Bayesian approach to quantify uncertainty in PDE optimization problems is also illustrated.

In Chapter 7, we discuss novel applications of PDE optimization with representative models of mathematical physics, that is, shape optimization with Navier–Stokes equations, optimal control of quantum systems governed by the Schrödinger equation, and inverse scattering with Maxwell equations.

A long but certainly incomplete list of references completes this book.

Chapter 2

Optimality Conditions

2.1 Introduction

This introductory chapter is intended to set the stage for the optimization and discretization methods to come in the subsequent chapters. On the one hand, it presents the necessary notation and recalls well-known optimality conditions in differential optimization where we borrow a lot of ideas from [339]. On the other hand, it tries to provide also a certain point of view on optimization such that the subsequent discussions can be understood more easily. In particular, the point of view of the choice of functional spaces and corresponding scalar products is of importance for the efficiency as well as efficacy of the optimization methods to be employed. This topic is tightly intertwined with spectral Hessian analysis as well as preconditioning of one-shot methods.

2.2 Optimality Conditions

In this book, we will consider optimization problems of the form

$$\min_{y,u} J(y,u), \tag{2.1}$$

$$c(y,u) = 0, \tag{2.2}$$

$$u \in U_{ad}. \tag{2.3}$$

We assume that the constraint model equation $c(y,u) = 0$ uniquely defines the state y if the decision variable u is given, which means that c_y is invertible. This particular separation of the variables in the constrained optimization problem is usually called the separability framework. The objective $J(.,.)$ is a scalar-valued function. All functions are supposed to be as smooth as required. The decision variable is constrained to some admissible set U_{ad}. For ease of presentation, we deliberately exclude state constraints $y \in Y(u)$ at the moment and refer to Section 5.7.1 for practical consequences. In this chapter, we give a brief overview on the theoretical framework of existence of solutions and optimality conditions. We will always try to frame the results in the finite-dimensional setting as well as in the function space setting. The function space setting is of importance if one tries to solve optimization with increasing space and/or time discretization refinement.

The finite-dimensional setting—although a simplified consequence of the general function space setting—has its own justification, since it is easier to understand and may thus help to obtain a better intuitive understanding of the more general concepts.

Let us first take the point of view of black-box methods; that is, let us exploit the implicit function theorem and consider the state variable y as a function depending on u, that is, $y = y(u)$ such that $c(y(u), u) = 0$. That means the constrained optimization problem is equivalently formulated as

$$\min_{u \in U_{ad}} \hat{J}(u),$$

where $\hat{J}(u) := J(y(u), u)$ is the so-called reduced objective.

Now, we state the existence of an optimal solution in a finite-dimensional setting. We consider a generic function $f : U_{ad} \to \mathbb{R}$ having in mind $\hat{J}$ as a formal particular instance.

Theorem 2.1. *If $U_{ad} \subset \mathbb{R}^n$, for some $n \in \mathbb{N}, n < \infty$, the function $f : U_{ad} \to \mathbb{R}$ is continuous and bounded from below, and the level set*

$$L_t := \{u \in U_{ad} \mid f(u) \leq t\}$$

is a nonempty, closed, and bounded set for some $t < \infty$, then there exists an optimal solution $\hat{u} \in U_{ad}$.

Proof. Since f is bounded from below, there exists

$$j := \inf_{u \in U_{ad}} f(u).$$

Therefore, we can define a sequence $\{u_m\}_{m=1}^{\infty}$ with $f(u_m) \to j$ $(m \to \infty)$. For all m greater than some number $M < \infty$, all u_m have to lie in L_t with $t > j$. Since L_t is closed and bounded and therefore compact, we can select a converging subsequence $\{u_{m_k}\}_{k=1}^{\infty}$ such that $u_{m_k} \to \hat{u}$ $(k \to \infty)$. Since f is continuous, we know that $\hat{u}$ is a minimum, because

$$j = \lim_{k \to \infty} f(u_{m_k}) = f(\hat{u}). \quad \square$$

The general case, where U_{ad} is a subset of a function space, is more complicated, since closeness and boundedness no longer yield compactness. However, the situation is only slightly more complicated if $U_{ad} \subset H$, where H is a Hilbert space, that is, a complete linear space with a scalar product $(.,.)_H$. Then, the only property in addition to the finite-dimensional case is convexity to ensure existence of a solution.

Definition 2.2. *A function $f : U_{ad} \to \mathbb{R}$ is called convex if*

$$f(tu + (1-t)v) \leq t\, f(u) + (1-t) f(v) \quad \forall u, v \in U_{ad}, \forall t \in (0,1).$$

It is called strictly convex, if the inequality is strict (i.e., $<$ instead of $\leq$).

Theorem 2.3. *If $U_{ad} \subset H$, for some Hilbert space H, the function $f : U_{ad} \to \mathbb{R}$ is continuous, convex, and bounded from below, and the level set*

$$L_t := \{u \in U_{ad} \mid f(u) \leq t\}$$

is a nonempty, closed, and bounded set for some $t < \infty$, *then there exists an optimal solution* $\hat{u} \in U_{ad}$.

Proof. See [339] or other books. □

If one aims at computing an optimal solution numerically, one should hope for some argument for uniqueness of the optimal solution in addition to existence—or at least local uniqueness is a prerequisite to make numerical computations manageable. In both cases, the finite-dimensional as well as the function space case, we observe that strict convexity of f ensures uniqueness of the optimal solution.

Theorem 2.4. *If* f *is strictly convex and it possesses a minimum, this minimum is unique, provided* U_{ad} *is a convex set.*

Proof. Let us assume the existence of two separate minima $u \neq v \in U_{ad}$ with $f(u) = j = f(v)$. Then we find that the vector $\frac{1}{2}(u+v)$ yields an even lower objective value

$$f\left(\frac{1}{2}(u+v)\right) < \frac{1}{2}(f(u)+f(v)) = j$$

because of strict convexity of f. This is in clear contrast to j being the infimum. □

Strict convexity overall in U_{ad} is a rather strong requirement. But if we have convexity in some region and can construct a minimizing sequence as above, which stays in that region, we can conclude at least local uniqueness, which is enough for numerical purposes. The minimizing sequence can be constructed by employing a variant of the steepest descent method. In order to be able to formulate this method, we need to define the gradient of a scalar function.

Definition 2.5.

(a) *For a differential function* $f : \mathbb{R}^n \to \mathbb{R}$, *we define the Euclidean gradient as the vector*

$$\nabla f = \begin{pmatrix} \frac{\partial f}{\partial u_1} \\ \vdots \\ \frac{\partial f}{\partial u_n} \end{pmatrix}.$$

(b) *For a (Fréchet-)differentiable function* $f : H \to \mathbb{R}$, *where* H *is a Hilbert space, we define the gradient as the Riesz-representation of the derivative* f', *such that*

$$(\nabla f, v)_H = f'(v) \quad \forall v \in H,$$

where f' *is defined as the linear operator*

$$f'(u) : v \mapsto \left.\frac{d}{dt}\right|_{t=0} f(u+tv) \quad \forall v \in H.$$

With this definition, it is not possible to talk about the gradient without mentioning the specific scalar product employed to represent the derivative. Also, we do not have to deal with dual spaces, which can be consistently avoided by the usage of Riesz representations.

Let us study some examples in order to get used to this definition of the gradient.

Example 2.6. (a) First we study $H = \mathbb{R}^n$ endowed with the scalar product $(x,y) = x^\top Ay$, where A is a symmetric and positive definite matrix. The derivative of a differentiable function $f : H \to \mathbb{R}$ is defined as the mapping

$$f'(u) : v \mapsto \sum_{i=1}^{n} \frac{\partial f(u)}{\partial u_i} v_i.$$

Now

$$f'(u)v = \sum_{i=1}^{n} \frac{\partial f(u)}{\partial u_i} v_i = \left(A^{-1} \begin{pmatrix} \frac{\partial f(u)}{\partial u_1} \\ \vdots \\ \frac{\partial f(u)}{\partial u_n} \end{pmatrix}, v \right) \qquad \forall v \in H.$$

Therefore

$$\nabla f(u) = A^{-1} \begin{pmatrix} \frac{\partial f(u)}{\partial u_1} \\ \vdots \\ \frac{\partial f(u)}{\partial u_n} \end{pmatrix}.$$

If we choose $A = I$, we obtain the Euclidean gradient as defined in Definition 2.5(a). If we choose it differently, we obtain a different vector as the gradient. As we will see below, a steepest descent method based on this definition of the gradient profits from a good choice of the metric A. Basically, A acts as a scaling of the variables.

(b) Let us now consider the function space $H = L^2([0,1])$ endowed with the scalar product $(x,y) := \int_0^1 x(t)y(t)\omega(t)dt$ with some weight function $\omega \in L^\infty([0,1], \mathbb{R}_+)$. We consider the function $f : H \to \mathbb{R}$ defined by

$$f(u) := \int_0^1 \phi(u(t))dt$$

for some given twice differentiable function $\phi : \mathbb{R} \to \mathbb{R}$ with bounded second derivative. This is the integral over the so-called Nemytskii operator, whose first derivative exists in $L^2([0,1])$ but whose second derivative is a subject of subtle discussion in [339]. Now, the directional derivative is

$$f'(u)v = \left.\frac{d}{d\alpha}\right|_{\alpha=0} f(u + \alpha v) = \left.\frac{d}{d\alpha}\right|_{\alpha=0} \int_0^1 \phi(u(t) + \alpha v(t))dt = \int_0^1 \frac{1}{\omega(t)} \phi'(u(t))v(t)\omega(t)dt.$$

Therefore

$$\nabla f(u) = \frac{1}{\omega} \phi'(u).$$

Note that $\nabla f(u)$ itself is a function so that the long version of this L^2 gradient is as follows

$$\nabla f(u)(t) = \frac{1}{\omega(t)} \phi'(u(t)).$$

Here, the weight function ω defines the metric in H.

(c) Now, let us assume that we are only interested in gradient representations in the subspace

$$H = \{u \in H^1([0,1]) \mid u(0) = 0,\, u(1) = 0\} \subset L^2([0,1])$$

endowed with the scalar product

$$(x,y) = \varepsilon \int_0^1 x(t)y(t)dt + \int_0^1 \dot{x}(t)\dot{y}(t)dt\,,\ \varepsilon > 0.$$

If we want to determine the gradient of the function f in example (b), where we express the derivative of f in terms of the scalar product of H, we have to find a function g such that for all $v \in L^2([0,1])$

$$\int_0^1 \phi'(u(t))v(t)dt = \varepsilon \int_0^1 g(t)v(t)dt + \int_0^1 \dot{g}(t)\dot{v}(t)dt = \int_0^1 (\varepsilon g(t) - \ddot{g}(t))v(t)dt,$$

which means we have to solve the second-order ODE

$$-\ddot{g}(t) = \phi'(u(t)) - \varepsilon g(t)$$

with boundary conditions $g(0) = 0$, $g(1) = 0$. Thus the H^1 gradient given by

$$\nabla f(u) = g$$

is a smoother function than $\phi'(u)$. Here, we see that the choice of the scalar product can have a smoothing effect on the gradient. The effect that we observe here is sometimes called Sobolev smoothing, which can be naturally explained as expressing the gradient in a particular scalar product. ■

Since now the definition of the gradient depends on the choice of the scalar product, we have to reassure ourselves of well-known properties of the gradient.

Theorem 2.7. $-\nabla f(u_0)$ *is orthogonal to the level set of* f *in* u_0 *and it defines the direction of steepest descent of* f *if we consider the scalar product used for the definition of the gradient.*

Proof. The level set is locally characterized by curves $\gamma : (-\varepsilon;\varepsilon) \ni t \mapsto \gamma(t) \in H$ with $f(\gamma(t)) \equiv f(u)$ for all t and $\gamma(0) = u_0$. Orthogonality pertains to the tangent space which is spanned by the velocity vectors $\dot{\gamma}(0)$ of all of these curves. Therefore, we obtain

$$0 = \left.\frac{d}{dt}\right|_{t=0} f(\gamma(t)) = (\nabla f(u_0), \dot{\gamma}(0)).$$

The steepest descent direction can be found as the solution of

$$\min_h f(u_0) + f'(u_0)h,$$
$$(h,h) = 1.$$

We abbreviate $g := \nabla f(u_0)/\|\nabla f(u_0)\|$. Each vector h in the sphere $S := \{h \mid (h,h) = 1\}$ can be uniquely represented as

$$h = \alpha g + \beta g^{\perp}, \quad \alpha^2 + \beta^2 = 1,$$

where $g^{\perp}$ is some vector in H with $(g, g^{\perp}) = 0$ and $(g^{\perp}, g^{\perp}) = 1$. Now

$$f'(u_0)h = (\nabla f(u_0), h) = \alpha \|\nabla f(u_0)\|$$

is minimized at the minimum possible value for α, which is $\alpha = -1$, which means also $\beta = 0$. This concludes the proof. $\square$

We can also use the gradient, defined as a Riesz representation, in order to characterize solutions to optimization problems by necessary conditions.

Lemma 2.8 (necessary condition of first order). *If $\hat{u} = \arg\min_{u \in U(\hat{u})} f(u)$ for some open neighborhood $U(\hat{u})$ of $\hat{u}$, then $\nabla f(\hat{u}) = 0$.*

Proof. Let us consider the mapping $\varphi : (-\varepsilon; \varepsilon) \ni t \mapsto f(\hat{u} + th)$ for an arbitrary perturbation vector h. Then, $\varphi(0)$ is the obvious minimum of that function and from elementary calculus, we obtain

$$0 = \left.\frac{d}{dt}\right|_{t=0} \varphi(t) = (\nabla f(\hat{u}), h) \quad \forall h \in U.$$

In particular, we conclude that

$$0 = (\nabla f(\hat{u}), \nabla f(\hat{u}))$$

and therefore

$$\nabla f(\hat{u}) = 0. \quad \square$$

Analogously to the definition of the gradient, we define the Hessian operator in the following way.

Definition 2.9. *We assume that U is a real Hilbert space. Let $f : U \to \mathbb{R}$ be twice differentiable in $u \in U$ with the second derivative as a symmetric bilinear form denoted by*

$$D^2 f(u)[h_1, h_2] := \left.\frac{d}{dt_1}\right|_{t_1=0} \left.\frac{d}{dt_2}\right|_{t_2=0} f(u + t_1 h_1 + t_2 h_2) \quad \forall h_1, h_2 \in U.$$

Then for each $h_1 \in U$ there exists a vector $v(h_1) \in U$ which is the Riesz representation of the linear form $D^2 f(u)[h_1, .]$ such that

$$D^2 f(u)[h_1, h_2] = (v(h_1), h_2) \quad \forall h_2 \in U.$$

We call the linear mapping

$$\begin{array}{rcl} \text{Hess}\, f(u) : U & \to & U \\ h_1 & \mapsto & v(h_1) \end{array}$$

the Hessian operator at u.

Obviously, the operator is linear and self-adjoint. Let us see how the Hessian operators in Example 2.6 look.

Example 2.10. We investigate the Hessian operators of Example 2.6(a)–(c).

(a) We study $H = \mathbb{R}^n$ endowed with the scalar product $(x,y) = x^\top Ay$, where A is a symmetric and positive definite matrix. We are interested in the Hessian operator of a twice differentiable function $f : H \to \mathbb{R}$. Its Hessian operator is

$$\text{Hess}\, f(u) = A^{-1} \begin{bmatrix} \frac{\partial^2 f(u)}{\partial u_1^2} & \cdots & \frac{\partial^2 f(u)}{\partial u_1 \partial u_n} \\ \vdots & & \vdots \\ \frac{\partial^2 f(u)}{\partial u_n \partial u_1} & \cdots & \frac{\partial^2 f(u)}{\partial u_n \partial u_n} \end{bmatrix},$$

since

$$D^2 f(u)[h_1,h_2] = h_1^\top \begin{bmatrix} \frac{\partial^2 f(u)}{\partial u_1^2} & \cdots & \frac{\partial^2 f(u)}{\partial u_1 \partial u_n} \\ \vdots & & \vdots \\ \frac{\partial^2 f(u)}{\partial u_n \partial u_1} & \cdots & \frac{\partial^2 f(u)}{\partial u_n \partial u_n} \end{bmatrix} h_2 = (\text{Hess}\, f(u)h_1, h_2).$$

(b) Now, we consider $H = L^2([0,1])$ endowed with the scalar product $(x,y) := \int_0^1 x(t)\,y(t)\,\omega(t)dt$ and the function $f : H \to \mathbb{R}$ defined by

$$f(u) := \int_0^1 \phi(u(t))dt.$$

As pointed out in Section 4.9.4 of [339], this integrated Nemytskii operator is not twice differentiable in $L^2([0,1])$, but rather in $L^\infty([0,1])$. However, the second derivative can be represented as a linear operator $L^2([0,1]) \to L^2([0,1])$, which can be obtained by a formal derivation and which we also call the Hessian. It is defined by

$$\text{Hess}\, f(u)h = \frac{1}{\omega}\phi''(u)h$$

because

$$D^2 f(u)[h_1,h_2] = \int_0^1 \frac{1}{\omega(t)}\phi''(u(t))h_1(t)h_2(t)dt = (\text{Hess}\, f(u)h_1, h_2).$$

More details on the necessity to work with two norms, when dealing with second derivatives, can be found in [339].

(c) We investigate the same objective as in (b) but in a different space

$$H = \{u \in H^1([0,1]) \,|\, u(0) = 0\, u(1) = 0\}$$

endowed with the (Sobolev-smoothing) scalar product

$$(x,y) = \varepsilon \int_0^1 x(t)y(t)dt + \int_0^1 \dot{x}(t)\dot{y}(t)dt,\ \varepsilon > 0.$$

The Hessian in this scalar product is

$$\text{Hess}\, f(u) = \left(\varepsilon I - \frac{d^2}{dt^2}\right)^{-1} \phi''(u).$$

In many shape optimization problems, $\phi''(u)$ is spectrally equivalent to $(\varepsilon I - d^2/dt^2)$—c.f. Example 2.17. Then, we observe that the Hessian operator as defined in (c) is spectrally equivalent to the identity, which gives us an excellent performance of a gradient descent method. In those cases Sobolev smoothing is advantageous. ∎

As a first consequence of these definitions, we can derive a Taylor series expansion.

Theorem 2.11. *We assume for the function $f : U \to \mathbb{R}$ defined on a Hilbert space H, where the subset $U \subset H$ is open and convex and f is twice differentiable, the property*

$$\|\text{Hess}\, f(u) - \text{Hess}\, f(v)\| \leq L\|u - v\| \quad \forall u, v \in U$$

with a constant $L < \infty$. Then, we achieve the estimation

$$\left| f(u) - f(v) + (\nabla f(u), v - u) + \frac{1}{2}(\text{Hess}\, f(u)(v - u), v - u) \right| \leq \frac{L}{6}\|u - v\|^3.$$

Proof. Let us consider the mapping $\varphi : [0,1] \ni t \mapsto f(\hat{u} + t(v - u)) \in H$. We note that for all differentiable functions and in particular for φ it yields

$$\int_0^1 \int_0^t \varphi''(s) - \varphi''(0) ds\, dt = \varphi(1) - \varphi(0) - \varphi'(0) - \frac{1}{2}\varphi''(0).$$

Since

$$\varphi(1) = f(v), \quad \varphi(0) = f(u), \quad \varphi'(0) = (\nabla f(u), v - u), \quad \varphi''(0) = (\text{Hess}\, f(u)(v - u), v - u),$$

we observe

$$\left| f(u) - f(v) + (\nabla f(u), v - u) + \frac{1}{2}(\text{Hess}\, f(u)(v - u), v - u) \right| = \int_0^1 \int_0^t |\varphi''(s) - \varphi''(0)| ds\, dt$$

$$= \int_0^1 \int_0^t |((\text{Hess}\, f(u + s(v - u)) - \text{Hess}\, f(u))(v - u), v - u)|\, ds\, dt$$

$$\leq \int_0^1 \int_0^t \|\text{Hess}\, f(u + s(v - u)) - \text{Hess}\, f(v)\|\, \|u - v\|^2 ds\, dt$$

$$\leq \int_0^1 \int_0^t sL\|u - v\|^3 ds\, dt = \frac{L}{6}\|u - v\|^3. \quad \square$$

Now, we can exploit the Taylor expansion of Theorem 2.11 for necessary and sufficient optimality conditions.

Theorem 2.12. *Under the assumptions of Theorem 2.11 we obtain the following:*

(a) *If $\hat{u}$ is an optimal solution, then* Hess $f(\hat{u}) \geq 0$, *i.e.,* (Hess $f(\hat{u})h, h) \geq 0$ *for all $h \in H$.*

(b) *If $\hat{u}$ satisfies $\nabla f(\hat{u}) = 0$, and* Hess $f(\hat{u})$ *is coercive, i.e.,* (Hess $f(\hat{u})h, h) \geq c\|h\|^2$, *for all $h \in H$ and for some $c > 0$, then $\hat{u}$ is a local minimum, provided* $\text{Hess}_f(u)$ *satisfies a Lipschitz condition as in Theorem* 2.11.

Proof.

(a) Let us consider the mapping $\varphi : (-\varepsilon;\varepsilon) \ni t \mapsto f(\hat{u}+th)$ for an arbitrary perturbation vector h. Then, $\varphi(0)$ is the minimum, $\nabla f(\hat{u}) = 0$, and φ is convex, i.e., *Hess* $f(\hat{u}) \geq 0$.

(b) Choose a neighborhood U of $\hat{u}$ small enough so that $\|u - \hat{u}\| < 6c/L$ for all $u \in U$. Then for $h := u - \hat{u}$

$$f(u) - f(\hat{u}) \geq (\nabla f(\hat{u}),h) + (\mathit{Hess}\, f(\hat{u})h,h) - \frac{L}{6}\|h\|^3 \geq \left(c - \frac{L}{6}\|h\|\right)\|h\|^2 > 0. \quad \square$$

From the Taylor expansion, we observe at some point u close to $\hat{u}$

$$0 = \nabla f(\hat{u}) = \nabla f(u) + \mathit{Hess}\, f(u)(\hat{u} - u) + \mathcal{O}(\|\hat{u} - u\|^2)$$

and therefore

$$\hat{u} = u - \mathit{Hess}\, f(u)^{-1}\nabla f(u) + \mathcal{O}(\|\hat{u} - u\|^2).$$

That means that a Newton method of the form

$$u^{k+1} = u^k - \mathit{Hess}\, f(u^k)^{-1}\nabla f(u^k)$$

is locally quadratically convergent. Since often we can only afford a variant of the linearly converging steepest descent method

$$u^{k+1} = u^k - \nabla f(u^k),$$

in practice it would be highly profitable to choose the scalar product in H so that *Hess* $f(\hat{u})$ is close to identity, because that choice improves the convergence properties of the steepest descent method. At least, one should include the basic characteristics of the second derivative in the scalar product, e.g., the order of the operator.

Remark. Again, one should note that in general optimization problems the assumption of differentiability as in Theorem 2.11 is in question, even in simple cases. Also, the Hilbert space setting may not be applicable. Then, more refined techniques as in [339, 194, 207] for guaranteeing optimality conditions are to be applied which for many problems are still under investigation. Those issues are delicate, difficult, and important but beyond the scope of this book.

Now we come back to the constrained problem (2.1)–(2.3) and employ the reduced formulation

$$\min_u \hat{J}(u) := J(y(u),u).$$

The implicit function theorem gives us the expression

$$\hat{J}'(u) = \frac{\partial J(y,u)}{\partial u} - \frac{\partial J(y,u)}{\partial y}\left(\frac{\partial c(y,u)}{\partial y}\right)^{-1}\frac{\partial c(y,u)}{\partial u},$$

where y is fixed at $y = y(u)$. Note that the partial derivatives are meant only with respect to the explicit occurrence of the respective variable. The implicit dependency of $y = y(u)$ is taken care of by the formula itself. Since $\partial J(y,u)/\partial y$ is only one (dual) vector in contrast to the whole operator $\partial c(y,u)/\partial u$, it is computationally more convenient to compute this derivative in the order indicated by the brackets:

$$\hat{J}'(u) = \frac{\partial J(y,u)}{\partial u} - \left[\frac{\partial J(y,u)}{\partial y}\left(\frac{\partial c(y,u)}{\partial y}\right)^{-1}\right]\frac{\partial c(y,u)}{\partial u}. \tag{2.4}$$

This leads to the so-called adjoint approach to the gradient computation. As the name indicates, it is associated with adjoint operators, which will enable us to write the expression within the bracket in the form of a linear system of equations. In order to introduce this concept in a concise way, we make some simplifying assumptions on the vector spaces involved.

The PDE model $c(y,u) = 0$ is defined by a mapping

$$c : Y \times U \to Z,$$

where we assume that U is a Hilbert space. For the sake of clarity, we also assume for Y and Z that they are Hilbert spaces, where we denote by $(.,.)_U$, $(.,.)_Y$, and $(.,.)_Z$ the respective scalar products. The derivatives

$$\frac{\partial J(y,u)}{\partial y} \in Y^* \quad \text{and} \quad p^* := \frac{\partial J(y,u)}{\partial y}\left(\frac{\partial c(y,u)}{\partial y}\right)^{-1} \in Z^*$$

are elements of the dual spaces which can be pulled back to the primal space by the Riesz representation theorem. By use of adjoint operators (also denoted by *) corresponding to the scalar products of the Hilbert spaces, such that

$$\left(p, \left(\frac{\partial c(y,u)}{\partial u}\right)u\right)_Z = \left(\left(\frac{\partial c(y,u)}{\partial u}\right)^* p, u\right)_U \quad \forall p \in Z, u \in U$$

and

$$\left(p, \left(\frac{\partial c(y,u)}{\partial y}\right)y\right)_Z = \left(\left(\frac{\partial c(y,u)}{\partial y}\right)^* p, y\right)_Y \quad \forall p \in Z, y \in Y,$$

we can write the gradient in the form

$$\nabla \hat{J}(u) = \nabla_u J(y,u) + \left(\frac{\partial c(y,u)}{\partial u}\right)^* p, \tag{2.5}$$

where p solves

$$\left(\frac{\partial c(y,u)}{\partial y}\right)^* p = -\nabla_y J(y,u). \tag{2.6}$$

The engineering literature usually prefers the following derivation of this expression: Since $y(u)$ satisfies $c(y(u),u) = 0$, we can add it to the objective function in the following way, by using the scalar product in Z

$$\hat{J}(u) = J(y(u),u) = J(y(u),u) + (p, c(y(u),u))_Z$$

for any $p \in Z$. The gradient now is by chain rule

$$\begin{aligned}\nabla \hat{J}(u) &= \nabla\left[J(y(u),u)+(p,c(y(u),u))_Z\right]\\ &= \nabla_u J + \left(\frac{\partial y}{\partial u}\right)^* \nabla_y J + \left(\frac{\partial c}{\partial u}\right)^* p + \left(\frac{\partial y}{\partial u}\right)^* \left(\frac{\partial c}{\partial y}\right)^* p\\ &= \nabla_u J + \left(\frac{\partial c}{\partial u}\right)^* p + \left(\frac{\partial y}{\partial u}\right)^* \left[\nabla_y J + \left(\frac{\partial c}{\partial y}\right)^* p\right].\end{aligned}$$

If we define p by $\nabla_y J + (\partial c/\partial y)^* p = 0$, we can get rid of the expression in the brackets and arrive at the formula we have already obtained above

$$\nabla \hat{J}(u) = \nabla_y J + \left(\frac{\partial c}{\partial u}\right)^* p.$$

The expression of the gradient of $\hat{J}$ as derived above brings us to the necessary conditions of first order for equality constrained problems in the separability framework.

Theorem 2.13. *Define for the equality constrained problem* (2.1)–(2.3), *where* $U_{ad} = U$, *the Lagrangian as*

$$L(y,u,p) := J(y,u)+(p,c(y,u))_Z.$$

Then the necessary conditions for optimality are

$$\begin{aligned}0 &= \nabla_y L(y,u,p) = \nabla_y J + \left(\frac{\partial c}{\partial y}\right)^* p,\\ 0 &= \nabla_u L(y,u,p) = \nabla_u J + \left(\frac{\partial c}{\partial u}\right)^* p,\\ 0 &= \nabla_p L(y,u,p) = c(y,u).\end{aligned}$$

Proof. This follows immediately from the condition $\nabla \hat{J}(u) = 0$ and the derivation of a computational expression for $\nabla \hat{J}(u)$ above. □

Remark 2.14. *The choice of the sign of the adjoint variable in the definition of the Lagrangian is arbitrary and chosen only for aesthetic reasons. One can as well define*

$$L(y,u,p) := J(y,u)-(p,c(y,u))_Z$$

with obvious sign switches in all corresponding formulae.

Necessary and sufficient optimality conditions of second order follow from Theorem 2.12 by the observation that

$$\mathit{Hess}\,\hat{J}(u) = \begin{bmatrix}-(\partial c/\partial y)^{-1}(\partial c/\partial u)\\ I_U\end{bmatrix}^* \begin{bmatrix}\mathit{Hess}_{yy}\,L & \mathit{Hess}_{yu}\,L\\ \mathit{Hess}_{uy}\,L & \mathit{Hess}_{uu}\,L\end{bmatrix}\begin{bmatrix}-(\partial c/\partial y)^{-1}(\partial c/\partial u)\\ I_U\end{bmatrix}, \quad (2.7)$$

which can be seen from writing the gradient in the form

$$\nabla \hat{J}(u) = \begin{bmatrix}-(\partial c/\partial y)^{-1}(\partial c/\partial u)\\ I_U\end{bmatrix}^* \begin{pmatrix}\nabla_y L(y,u,p)\\ \nabla_u L(y,u,p)\end{pmatrix},$$

where p is defined by (2.6).

Theorem 2.15.

(a) *If* $(\hat{y},\hat{u})$ *is an optimal solution for the equality constrained problem* (2.1)–(2.3), *where* $U_{ad}=U$, *and* $\hat{p}$ *is defined by*

$$\left(\frac{\partial c(\hat{y},\hat{u})}{\partial y}\right)^{*}\hat{p}=\nabla_y J(\hat{y},\hat{u}),$$

then the operator

$$\text{Hess}\, L(\hat{y},\hat{u},\hat{p})=\begin{bmatrix}\text{Hess}_{yy}\, L(\hat{y},\hat{u},\hat{p}) & \text{Hess}_{yu}\, L(\hat{y},\hat{u},\hat{p})\\ \text{Hess}_{uy}\, L(\hat{y},\hat{u},\hat{p}) & \text{Hess}_{uu}\, L(\hat{y},\hat{u},\hat{p})\end{bmatrix}$$

is positive semidefinite on the kernel of the linearized constraints, in the sense that

$$\left(\text{Hess}\, L(\hat{y},\hat{u},\hat{p})\begin{pmatrix}h_1\\h_2\end{pmatrix},\begin{pmatrix}h_1\\h_2\end{pmatrix}\right)_{Y\times U}\geq 0$$

for all $(h_1,h_2)\in Y\times U$ *with*

$$\left[(\partial c/\partial y)\ (\partial c/\partial u)\right]\begin{pmatrix}h_1\\h_2\end{pmatrix}=0.$$

(b) *If* $(\hat{y},\hat{u},\hat{p})$ *satisfy the necessary conditions of first order as stated in Theorem* 2.13, *and* Hess L *is coercive on the kernel of the equality constraints, that is, there is a constant* $c<\infty$ *such that*

$$\left(\text{Hess}\, L(\hat{y},\hat{u},\hat{p})\begin{pmatrix}h_1\\h_2\end{pmatrix},\begin{pmatrix}h_1\\h_2\end{pmatrix}\right)_{Y\times U}\geq c\left(\begin{pmatrix}h_1\\h_2\end{pmatrix},\begin{pmatrix}h_1\\h_2\end{pmatrix}\right)_{Y\times U}$$

for all $(h_1,h_2)\in Y\times U$ *with*

$$\left[(\partial c/\partial y)\ (\partial c/\partial u)\right]\begin{pmatrix}h_1\\h_2\end{pmatrix}=0,$$

then $(\hat{y},\hat{u})$ *is a local minimum, provided* Hess L *satisfies a Lipschitz condition as in Theorem* 2.11.

Proof. The proposition is an immediate consequence of Theorem 2.12 and (2.7). □

The expression on the right-hand side of (2.7) is often called the reduced Hessian, because it is derived from a reduction (i.e., projection) of the full Hessian onto the nullspace of the constraints. We see immediately that the well-posedness of the optimization problem is equivalent to the coercivity of the reduced Hessian (if there are no further constraints on the control).

For illustration of the concepts above, we will derive specific necessary optimality conditions in a finite-dimensional case and for the basic elliptic optimal control problem of PDE constrained optimization.

Example 2.16. (a) We study the finite-dimensional optimization problem

$$\min J(y,u),$$
$$c(y,u)=0,$$

where $u \in \mathbb{R}^{n_u}$, $y \in \mathbb{R}^{n_y}$, $c(y,u) \in \mathbb{R}^{n_y}$, and the functions J, c are at least twice differentiable. We use in $\mathbb{R}^{n_u}$ and $\mathbb{R}^{n_y}$ the Euclidean scalar product $(x,y)_0 = x^\top y$. With that definition of the spaces, the necessary optimality conditions of first order can be written as

$$
\begin{aligned}
0 &= \nabla_y L(y,u,p) = \left(\frac{\partial J}{\partial y}\right)^\top + \left(\frac{\partial c}{\partial y}\right)^\top p, \\
0 &= \nabla_u L(y,u,p) = \left(\frac{\partial J}{\partial u}\right)^\top + \left(\frac{\partial c}{\partial y}\right)^\top p, \\
0 &= \nabla_p L(y,u,p) = c(y,u),
\end{aligned}
$$

where the Lagrangian is

$$L(y,u,p) := J(y,u) + p^\top c(y,u).$$

(b) In many cases, one studies PDE optimization problems in variational form, e.g., of the type

$$
\begin{aligned}
&\min J(y,u), \\
&(y,p)_Y = b(u,p) \quad \forall p \in Y,
\end{aligned}
$$

where $(.,.)_Y$ is a scalar product in Y and $b(.,.) : U \times Y \to \mathbb{R}$ is a bilinear form. We can use the engineering approach to arrive at a concise formulation of the necessary conditions. Since $b(u,p) - (y,p)_Y = 0$ for all $p \in Y$, we can add it to the objective

$$\hat{J}(u) = J(y(u),u) = J(y(u),u) - (y(u),p)_Y + b(u,p).$$

The necessary condition is

$$
\begin{aligned}
0 &= \left.\frac{d}{dt}\right|_{t=0} f(u+th) \\
&= \left.\frac{d}{dt}\right|_{t=0} \left(J(y(u+th),u+th) - (y(u+th),p)_Y + b((u+th),p)\right) \\
&= \left(\frac{\partial J}{\partial u}\right) h + \left(\frac{\partial J}{\partial y}\right)\left(\frac{\partial y}{\partial u}\right) h - \left(\left(\frac{\partial y}{\partial u}\right) h, p\right)_Y + b(h,p) \\
&= \left(\frac{\partial J}{\partial u}\right) h + \left[\left(\frac{\partial J}{\partial y}\right)\left(\frac{\partial y}{\partial u}\right) h - \left(\left(\frac{\partial y}{\partial u}\right) h, p\right)_Y\right] + b(h,p) \\
&= (\nabla_u J, h) + b(h,p) \quad \forall h \in U
\end{aligned}
$$

if we can find $p \in Y$ such that

$$\left(\frac{\partial J}{\partial y}\right) y - (y,p)_Y = 0 \quad \forall y \in Y$$

and therefore $p = \nabla_y J$ in the Y-scalar product. Now, by use of the Lagrangian

$$L(y,u,p) := J(y,u) - (y,p)_Y + b(u,p)$$

we state the necessary conditions

$$
\begin{array}{llllllll}
\text{(adjoint eq.)} & 0 & = & \nabla_y L & \Leftrightarrow & (h,p)_Y & = & (\partial J/\partial y)h & \forall h \in Y, \\
\text{(design eq.)} & 0 & = & \nabla_u L & \Leftrightarrow & (\nabla_u J, w)_U & = & -b(w,p) & \forall w \in U, \\
\text{(state eq.)} & 0 & = & \nabla_p L & \Leftrightarrow & (y,q)_Y & = & b(u,q) & \forall q \in Y.
\end{array}
$$

(c) In this example, we derive necessary optimality conditions for a basic elliptic optimal control given by

$$
\min J(y,u) = \frac{1}{2}\int_\Omega (y(x) - y_\Omega(x))^2 dx + \frac{\nu}{2}\int_\Omega u(x)^2 dx \tag{2.8a}
$$

$$
\begin{aligned}
-\Delta y(x) &= u(x) \quad \forall x \in \Omega, \\
y(x) &= 0 \quad \forall x \in \Gamma := \partial\Omega
\end{aligned} \tag{2.8b}
$$

with the computational domain $\Omega := [0,1]^2$ and $y_\Omega : \Omega \to \mathbb{R}$ a given function, and $\nu > 0$. It is well known that the solution y of the state equation is in $H^1(\Omega)$ if $u \in L^2(\Omega)$. The homogeneous Dirichlet boundary condition can be taken into account by reduction to the space $Y := H_0^1(\Omega) = \{h \in H^1(\Omega) \,|\, h|_\Gamma = 0\}$. We note that the bilinear forms

$$
(y,p)_1 := \int_\Omega \nabla y(x)^\top \nabla p(x) dx \quad \text{and} \quad (y,p)_0 := \int_\Omega y(x)\, p(x) dx
$$

are both scalar products on $H_0^1(\Omega)$, and $(u,w)_0$ is a scalar product in $U := L^2(\Omega)$. With these definitions, we can write the problem in the equivalent variational form

$$
\begin{gathered}
\min J(y,u) = \frac{1}{2}(y - y_\Omega, y - y_\Omega)_0 + \frac{\nu}{2}(u,u)_0, \\
(y,q)_1 = (u,q)_0 \quad \forall q \in H_0^1(\Omega).
\end{gathered}
$$

Here, we can apply the results of example (b) above and arrive at the necessary conditions in variational form: determine $(y,u,p) \in H_0^1(\Omega) \times L^2(\Omega) \times H_0^1(\Omega)$ which satisfy

$$
\begin{array}{rcll}
(h,p)_1 & = & (h, y - y_\Omega)_0 & \forall h \in Y, \\
\nu(u,w)_0 & = & -(p,w)_0 & \forall w \in U, \\
(y,q)_1 & = & (u,q)_0 & \forall q \in Y.
\end{array}
$$

We notice that the adjoint variational equation is equivalent to the boundary value problem

$$
\begin{gathered}
-\Delta p(x) = y(x) - y_\Omega(x) \quad \forall x \in \Omega, \\
p(x) = 0 \quad \forall x \in \Gamma
\end{gathered}
$$

so that we can rewrite the necessary conditions in differential form

$$
\begin{array}{rclllcll}
-\Delta p(x) & = & y(x) - y_\Omega(x) & \forall x \in \Omega, & p(x) & = & 0 & \forall x \in \Gamma, \\
\nu u(x) + p(x) & = & 0 & \forall x \in \Omega, & & & & \\
-\Delta y(x) & = & u(x) & \forall x \in \Omega, & y(x) & = & 0 & \forall x \in \Gamma.
\end{array}
$$

It might be illustrative to compute the reduced Hessian in this example. First, we have to compute

$$D^2L\left[\begin{pmatrix}\tilde{y}_1\\ \tilde{u}_1\end{pmatrix},\begin{pmatrix}\tilde{y}_2\\ \tilde{u}_2\end{pmatrix}\right]=\frac{d}{dt_1}\bigg|_{t_1=0}\frac{d}{dt_2}\bigg|_{t_2=0}L(y+t_1\tilde{y}_1+t_2\tilde{y}_2,u+t_1\tilde{u}_1+t_2\tilde{u}_2,p)$$

for perturbations which satisfy

$$(\tilde{y}_1,q)_1=(\tilde{u}_1,q)_0 \quad \forall q\in Y,$$
$$(\tilde{y}_2,q)_1=(\tilde{u}_2,q)_0 \quad \forall q\in Y.$$

Obviously, we have

$$D^2L\left[\begin{pmatrix}\tilde{y}_1\\ \tilde{u}_1\end{pmatrix},\begin{pmatrix}\tilde{y}_2\\ \tilde{u}_2\end{pmatrix}\right]=(\tilde{y}_1,\tilde{y}_2)_0+\nu(\tilde{u}_1,\tilde{u}_2)_0.$$

Since

$$(\tilde{y},q)_1=(-\Delta\tilde{y},q)_0 \quad \forall\tilde{y},q\in H_0^1(\Omega),$$

we observe that $\tilde{y}_1=(-\Delta)^{-1}\tilde{u}_1$ and $\tilde{y}_2=(-\Delta)^{-1}\tilde{u}_2$ in $H_0^1(\Omega)$, which means that

$$\begin{aligned}D^2L\left[\begin{pmatrix}\tilde{y}_1\\ \tilde{u}_1\end{pmatrix},\begin{pmatrix}\tilde{y}_2\\ \tilde{u}_2\end{pmatrix}\right]&=((-\Delta)^{-1}\tilde{u}_1,(-\Delta)^{-1}\tilde{u}_1)_0+\nu(\tilde{u}_1,\tilde{u}_2)_0\\ &=(((-\Delta)^{-2}+\nu I_U)\tilde{u}_1,\tilde{u}_2)_0\end{aligned}$$

and, therefore, the reduced Hessian in the $(.,.)_0$ scalar product is

$$Hess\, f(u)=(-\Delta)^{-2}+\nu I_U.$$

Note that $(-\Delta)^{-2}$ is a compact operator and, in particular, not coercive. System solution with these operators is ill-posed. The reduced Hessian, however, is a compact perturbation of the operator νI_U, and the complete reduced Hessian is obviously coercive for $\nu>0$. This structure of the reduced Hessian operator is typical in PDE-constrained optimization problems and shows that, in general, regularization (here $\nu>0$) is necessary for the sake of well-posedness. However, there are exceptions, like in shape optimization (see later). ∎

2.3 The Formal Lagrangian Approach

In [339] a formal and straightforward approach to the derivation of necessary optimality conditions is given. We briefly sketch this approach and apply it to determine necessary conditions for slightly more complex problems as in the section above.

The key essence in deriving necessary conditions is the proper definition of a Lagrangian function. Different forms of this Lagrangian function are possible, as we have seen above. In [339] a canonical but formal Lagrangian approach is given, where, after formally deriving the necessary conditions, one has to decide whether the expressions derived make any sense at all. In optimization problems which are similar to the basic elliptic optimal control example above, this formal Lagrangian approach means using extensions of the $(.,.)_0$ scalar product in the image space of the differential operator and then playing with the resulting formulas. We use and demonstrate this approach in the following two examples.

Example 2.17. (a) For the sake of getting acquainted with the formal Lagrangian approach, we derive again the necessary conditions for the basic elliptic optimal control problem

$$\min J(y,u) = \frac{1}{2}\int_\Omega (y(x) - y_\Omega(x))^2 dx + \frac{\nu}{2}\int_\Omega u(x)^2 dx$$

$$\begin{aligned} -\Delta y(x) &= u(x) \quad \forall x \in \Omega, \\ y(x) &= 0 \quad \forall x \in \Gamma := \partial\Omega. \end{aligned}$$

According to the formal Lagrangian approach, we define an adjoint variable p in the domain Ω and an adjoint variable p_Γ on the boundary $\Gamma = \partial\Omega$ such that we get the formal expression

$$L(y,u,p,p_\Gamma) = J(y,u) + \int_\Omega p(-\Delta y - u)dx + \int_\Gamma p_\Gamma\, y ds.$$

We will have to recognize expressions in the $(.,.)_0$ with y alone, which means without a differential operator. Therefore, we first swap the differential operator over to the adjoint variable by the use of Green's formula

$$L(y,u,p,p_\Gamma) = J(y,u) + \int_\Omega (-\Delta p)y - p u dx + \int_\Gamma p\frac{\partial y}{\partial \vec{n}} - y\frac{\partial p}{\partial \vec{n}} ds + \int_\Gamma p_\Gamma\, y ds.$$

The adjoint boundary value problem is derived from the expression for perturbations $\tilde{y}$

$$\begin{aligned} 0 &= \left.\frac{d}{dt}\right|_{t=0} L(y + t\,\tilde{y}, u, p, p_\Gamma) \\ &= \underbrace{(y - y_\Omega, \tilde{y})_0 + ((-\Delta p), \tilde{y})_0}_{(1)} + \underbrace{\int_\Gamma p_\Gamma\, \tilde{y} ds - \int_\Gamma \tilde{y}\frac{\partial p}{\partial \vec{n}} ds}_{(2)} + \underbrace{\int_\Gamma p\frac{\partial \tilde{y}}{\partial \vec{n}} ds}_{(3)}. \end{aligned}$$

This equation should hold, in particular, for all perturbations $\tilde{y} \in C^\infty(\Omega)$, that is, all infinitely often differential functions with compact support. Focusing on these perturbations eliminates the expressions (2) and (3) for the moment, which results in expression (1) in

$$y - y_\Omega + (-\Delta p) = 0.$$

This result eliminates expression (1) permanently. Next, we focus on perturbations $\tilde{y}$ which satisfy the Dirichlet boundary condition ($\tilde{y}|_\Gamma = 0$) but have arbitrary Neumann values $\partial\tilde{y}/\partial\vec{n}$. This eliminates expression (2) temporarily and gives us from expression (3)

$$p(x) = 0 \quad \forall x \in \Gamma.$$

This eliminates expression (3) permanently. Now, we have already derived the adjoint boundary value problem. For the sake of completeness, we also try to give an expression for p_Γ. We do this by leaving the perturbations $\tilde{y}$ completely free. Thus we conclude from (2) that

$$p_\Gamma(x) - \frac{\partial \tilde{y}}{\partial \vec{n}} = 0 \quad \forall x \in \Gamma.$$

The design equation is found out analogously by stating

$$0 = \left.\frac{d}{dt}\right|_{t=0} L(y, u + t\,\tilde{u}, p, p_\Gamma) = \nu(u,\tilde{u})_0 - (p,\tilde{u})_0$$

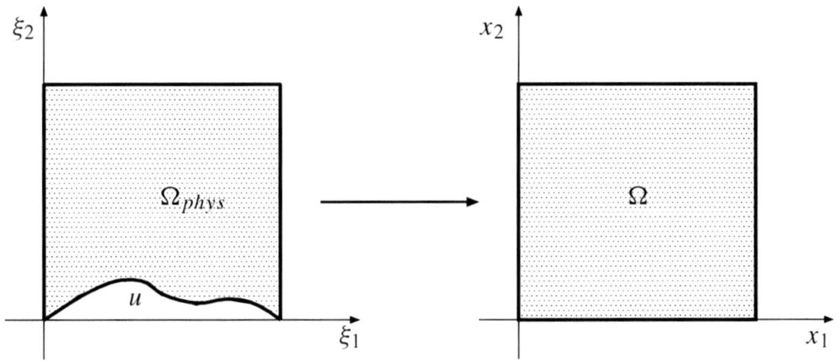

Figure 2.1. *Method of mapping in shape optimization.*

for arbitrary perturbations $\tilde{u} \in L^2(\Omega)$, which gives us the already known design equation

$$\nu u - p = 0.$$

(b) Now, we use the formal Lagrangian approach to derive necessary conditions for optimality of a typical model problem for aerodynamic shape optimization methods. The left-hand side of Figure 2.1 is imagined as the physical domain, which is thought as part of the free air stream along a geometric detail of an airplane which is aligned with the ξ_1-axis. The free stream is supposed to go from left to right and is modeled by the potential flow $-\Delta y = 0$ with a Neumann condition at the shape boundary. The shape boundary is supposed to be described by the function $u(\xi_1)$. In order to simplify the problem, we map the physical domain onto the computational domain on the right-hand side of Figure 2.1 by stretching it appropriately in the vertical direction, such that we can write

$$(x_1, x_2) = \left(\xi_1, \frac{\xi_2 - u(\xi_1)}{1 - u(\xi_1)} \right)$$

and $y(\xi)$ is thought of as $y(x(\xi))$. That means that the Neumann condition in the physical plane at the shape boundary $\{(\xi_1, \xi_2) \,|\, \xi_2 = u(\xi_1)\}$

$$0 = \frac{\partial y}{\partial \vec{n}} = \frac{\partial y}{\partial \xi_1} \dot{u} - \frac{\partial y}{\partial \xi_2}, \qquad \dot{u} := \frac{\partial u}{\partial \xi_1}$$

is mapped to the boundary condition

$$0 = \dot{u} \frac{\partial y}{\partial \xi_1} - \frac{1 + \dot{u}^2}{1 - u^2} \frac{\partial y}{\partial x_2}.$$

For the purpose of simplification, we assume that the shape u is almost a straight line $u(x_1) \approx 0$, such that squared expressions like u^2 and $\dot{u}^2$ can be neglected. Furthermore, the tangential velocity $\frac{\partial y}{\partial x_1}$ can be assumed to be constant (e.g., 1). This results in the boundary condition

$$\frac{\partial y}{\partial x_2} = \frac{\partial u}{\partial x_1}.$$

The goal of the problem is to match a certain pressure profile at the shape boundary. By the gas law, pressure is inversely correlated to the tangential velocity, which means we aim at matching a certain tangential velocity profile at the shape boundary. The complete problem formulation in the computational domain is then

$$\min J(y,u) = \int_{\Gamma_1} \left(\frac{\partial y(x_1,x_2)}{\partial x_1} - v(x_1) \right)^2 dx_1, \tag{2.9}$$

$$\begin{aligned} -\Delta y(x) &= 0 \quad \forall x \in \Omega, \\ \frac{\partial y}{\partial \vec{n}}(x) &= \frac{\partial u}{\partial x_1}(x) \quad \forall x \in \Gamma_1, \\ y(x) &= x_1 \quad \forall x \in \Gamma_2 := \partial\Omega \backslash \Gamma_1, \end{aligned} \tag{2.10}$$

where $\Omega = (0,1) \times (0,1)$, $\Gamma_1 := [0,1] \times \{0\}$, and $v : [0,1] \to \mathbb{R}$ is a given function. We build up the formal Lagrangian as follows

$$\begin{aligned} L(y,u,p,p_1,p_2) &= J(y,u) + \int_\Omega p(-\Delta y) dx + \int_{\Gamma_1} p_1 \left(\frac{\partial y}{\partial \vec{n}} - \frac{\partial u}{\partial \xi_1} \right) ds + \int_{\Gamma_2} p_2 (y - x_1) ds \\ &= \int_{\Gamma_1} \left(\frac{\partial y(x_1,x_2)}{\partial x_1} - v(x_1) \right)^2 dx_1 + \int_\Omega (-\Delta p) y dx \\ &\quad + \int_{\Gamma_1 \cup \Gamma_2} p \frac{\partial y}{\partial \vec{n}} - y \frac{\partial p}{\partial \vec{n}} ds + \int_{\Gamma_1} p_1 \left(\frac{\partial y}{\partial \vec{n}} - \frac{\partial u}{\partial \xi_1} \right) ds + \int_{\Gamma_2} p_2 (y - x_1) ds. \end{aligned}$$

The adjoint boundary value problem is again derived from the expression for perturbations $\tilde{y}$

$$\begin{aligned} 0 &= \left. \frac{d}{dt} \right|_{t=0} L(y + t\,\tilde{y}, u, p, p_1, p_2) \\ &= \int_{\Gamma_1} \left(\frac{\partial y}{\partial x_1} - v(x_1) \right) \frac{\partial \tilde{y}}{\partial x_1} dx_1 + \int_\Omega (-\Delta p) \tilde{y} dx \\ &\quad + \int_{\Gamma_1 \cup \Gamma_2} p \frac{\partial \tilde{y}}{\partial \vec{n}} - \tilde{y} \frac{\partial p}{\partial \vec{n}} ds + \int_{\Gamma_1} p_1 \frac{\partial \tilde{y}}{\partial \vec{n}} ds + \int_{\Gamma_2} p_2 \tilde{y} ds \\ &= \left[\left(\frac{\partial y}{\partial x_1} - v(x_1) \right) \tilde{y} \right]_0^1 - \int_{\Gamma_1} \tilde{y} \frac{\partial}{\partial x_1} \left(\frac{\partial y}{\partial x_1} - v(x_1) \right) dx_1 + \int_\Omega (-\Delta p) \tilde{y} dx \\ &\quad + \int_{\Gamma_1 \cup \Gamma_2} p \frac{\partial \tilde{y}}{\partial \vec{n}} - \tilde{y} \frac{\partial p}{\partial \vec{n}} ds + \int_{\Gamma_1} p_1 \frac{\partial \tilde{y}}{\partial \vec{n}} ds + \int_{\Gamma_2} p_2 \tilde{y} ds. \end{aligned}$$

(1) We focus on perturbations $\tilde{y} \in C^\infty(\Omega)$, which results in

$$-\Delta p = 0.$$

(2) Next we focus on perturbations with fixed Dirichlet value 0 and variable Neumann values, which gives

$$p|_{\Gamma_1} = -p_1|_{\Gamma_1}, \qquad p|_{\Gamma_2} = 0.$$

(3) Finally, we perturb in all respects (with the exception of the corner points), which gives

$$\left. \frac{\partial p}{\partial \vec{n}} \right|_{\Gamma_1} = \frac{\partial}{\partial x_1} \left(\frac{\partial y}{\partial x_1} - v \right), \qquad p_2 = \left. \frac{\partial p}{\partial \vec{n}} \right|_{\Gamma_2}.$$

The adjoint boundary value problem now reads completely as

$$-\Delta p = 0 \quad \text{in } \Omega,$$
$$\left.\frac{\partial p}{\partial \vec{n}}\right|_{\Gamma_1} = \frac{\partial}{\partial x_1}\left(\frac{\partial y}{\partial x_1} - v\right),$$
$$p|_{\Gamma_2} = 0.$$

For the design condition, we have to perturb the Lagrangian also in the control direction, which means

$$0 = \left.\frac{d}{dt}\right|_{t=0} L(y, u + t\,\tilde{u}, p, p_1, p_2) = -\int_{\Gamma_1} p_1 \frac{\partial \tilde{u}}{\partial x_1} ds$$
$$= -\left[p_1\,\tilde{u}\right]_0^1 + \int_{\Gamma_1} \frac{\partial p_1}{\partial x_1}\tilde{u}\,dx_1 = \int_{\Gamma_1} \frac{\partial p_1}{\partial x_1}\tilde{u}\,dx_1.$$

Free variation of $\tilde{u}$ gives the design equation on Γ_1,

$$0 = \frac{\partial p_1}{\partial x_1} = -\frac{\partial p}{\partial x_1},$$

which also gives the gradient of the objective with respect to u in the L^2 scalar product on Γ_1. In order to decide about well-posedness of the problem, we might be interested in the reduced Hessian of $\hat{J}(u) = J(y(u,u))$. Here it seems impossible to perform analytical manipulations of the optimality conditions in order to come up with an explicit expression. We note that

$$\nabla \hat{J}(u) = -\left.\frac{\partial p(y(u))}{\partial x_1}\right|_{\Gamma_1} = -\left.\frac{\partial}{\partial x_1}\right|_{\Gamma_1}\left(\left.\frac{\partial y(u)}{\partial x_1}\right|_{\Gamma_1} - v\right),$$

where the mapping $u \mapsto y$ is described by the forward problem and the mapping $y \mapsto p$ by the adjoint problem.

All these mappings are affine linear. Therefore, a Fourier analysis of the homogeneous parts gives us the highest differential order of the Hessian operator. If this order is zero or greater, we can conclude well-posedness of the shape optimization problem. For the Fourier analysis, we neglect the boundary Γ_2 and stretch the boundary Γ_1 infinitely wide. We start with an arbitrary Fourier mode for u, as follows

$$u = e^{i\omega_u x_1} \Rightarrow \frac{\partial u}{\partial x_1} = i\omega_u e^{i\omega_u x_1}.$$

The solution of the forward problem is assumed to be of the form

$$y(x_1, x_2) = r\,e^{i\omega_1 x_1 + i\omega_2 x_2} \Rightarrow \left.\frac{\partial y}{\partial \vec{n}}\right|_{\Gamma_1} = i\omega_2\, r\, e^{i\omega_1 x_1} \quad \text{and} \quad \left.\frac{\partial^2 y}{\partial x_1^2}\right|_{\Gamma_1} = -\omega_1^2\, r\, e^{i\omega_1 x_1}.$$

From the boundary condition of the flow problem, we obtain

$$i\omega_u e^{i\omega_u x_1} = i\omega_2\, r\, e^{i\omega_1 x_1}, \qquad x_1 \in \mathbb{R}.$$

For $x_1 = 0$ this implies $\omega_2 = \omega_u / r$ and thus also $\omega_1 = \omega_u$. The differential equation $-\Delta y = 0$ gives

$$0 = (\omega_1^2 + \omega_2^2)\, r\, e^{i\omega_1 x_1 + i\omega_2 x_2} \Leftrightarrow \frac{\omega_u^2}{r^2} + \omega_u^2 = 0,$$

which means $r = \pm i$. The adjoint solution is assumed to be of similar form

$$p(x_1, x_2) = s\, e^{i\phi_1 x_1 + i\phi_2 x_2} \Rightarrow \left.\frac{\partial p}{\partial \vec{n}}\right|_{\Gamma_1} = i\phi_2 s\, e^{i\phi_1 x_1}.$$

The Neumann boundary condition of the adjoint problem now gives

$$i\phi_2 s\, e^{i\phi_1 x_1} = -\omega_1^2 r\, e^{i\omega_1 x_1}, \qquad x_1 \in \mathbb{R};$$

with arguments similar to those above, we conclude that

$$\phi_2 = -\frac{\omega_1^2 r}{is} = \mp \frac{\omega_u^2}{s},$$
$$\phi_1 = \omega_1 = \omega_u.$$

From the adjoint differential equation, we obtain

$$0 = \phi_1^2 + \phi_2^2 \Rightarrow s = \pm i\omega_u.$$

Thus

$$\left.\frac{\partial p}{\partial x_1}\right|_{\Gamma_1}(x_1) = \mp\omega_u^2 u.$$

From the necessary conditions of optimality, the Hessian has to be positive (semi-) definite and therefore we have for large ω_u

$$\left.\frac{\partial p}{\partial x_1}\right|_{\Gamma_1}(x_1) = \omega_u^2 u.$$

From that, we conclude by Fourier analysis that the highest order of the Hessian operator is 2, such that the Hessian can be approximated by

$$Hess\, \hat{J} \approx -\frac{\partial^2}{\partial x_1^2},$$

which means that the problem is essentially well posed and needs no further regularization. A very similar structure of the Hessian arises in Section 7.2.1, where we will discuss realistic aerodynamic shape optimization problems. ∎

2.4 Control Constraints

In the previous sections, we have omitted inequality constraints. State constraints are a highly complicated issue theoretically. Therefore, we consider only the practical treatment of them in the applications section. Control constraints, however, are easily integrated in the context of this chapter. Let us again look at the generic problem (2.1)–(2.3). Now we assume that we allow only $u \in U_{ad}$, where $U_{ad} \subset U$ is a convex subset of the Hilbert space U. We consider the following

$$\min f(u), \tag{2.11}$$
$$u \in U_{ad}. \tag{2.12}$$

Theorem 2.18. *If $U_{ad} \subset U$ is a convex subset of the Hilbert space U and $f : U_{ad} \to \mathbb{R}$ is a differentiable function, then the solution $\hat{u}$ of the optimization problem* (2.11)–(2.12) *satisfies the variational inequality*

$$(\nabla f(\hat{u}), u - \hat{u}) \geq 0 \quad \forall u \in U_{ad}. \tag{2.13}$$

Proof. The proof follows the lines of the proof of Lemma 2.20 in [339]. We choose an arbitrary $u \in U_{ad}$ and consider the convex combination

$$u_s := \hat{u} + s(u - \hat{u}), \quad s \in [0,1].$$

We observe $u_s \in U_{ad}$, because of the convexity of U_{ad}. The optimality of $\hat{u}$ gives us $f(u_s) \geq f(\hat{u})$, from which we conclude that

$$\frac{1}{s}(f(u_s) - f(\hat{u})) = \frac{1}{s}(f(\hat{u} + s(u - \hat{u})) - f(\hat{u})) \geq 0$$

and therefore

$$0 \leq \lim_{s \to 0} \frac{1}{s}(f(\hat{u} + s(u - \hat{u})) - f(\hat{u})) = f'(\hat{u})(u - \hat{u}) = (\nabla f(\hat{u}), u - \hat{u}). \quad \square$$

Corollary 2.19. *If f in Theorem* 2.18 *is a convex function, then the condition* (2.13) *is sufficient for optimality of $\hat{u}$.*

Proof. f is assumed to be convex. Therefore

$$f(tu + (1-t)\hat{u}) \leq t f(u) + (1-t) f(\hat{u}) \quad \forall t \in [0,1].$$

Differentiating with respect to t, that is, applying $\frac{d}{dt}\big|_{t=0}$ on both sides, gives

$$0 \leq f'(\hat{u})(u - \hat{u}) \leq f(u) - f(\hat{u})$$

for arbitrary $u \in U_{ad}$, which means that $\hat{u}$ is optimal. $\square$

We study consequences of Theorem 2.18 in two frequently appearing instances: box constraints for control functions in L^2, and finite-dimensional controls and constraints. We first treat box constraints in L^2.

Theorem 2.20. *We consider the space $U = L^2(\Omega)$ for some bounded and open computational domain $\Omega \subset \mathbb{R}^n$ and the subset $U_{ad} \subset U$, defined in the following way*

$$U_{ad} := \left\{ u \in U \mid \underline{u}(x) \leq u(x) \leq \bar{u}(x) \text{ for almost all } x \in \Omega \right\},$$

where $\underline{u}, \bar{u} \in U$ with $\underline{u}(x) \leq \bar{u}(x)$ for almost all $x \in \Omega$ are functions defining lower and upper bounds for the control u. For $\lambda, \mu \in U$, we define the Lagrangian

$$L(u, \lambda, \mu) := f(u) + (\lambda, \underline{u} - u)_U + (\mu, u - \bar{u})_U.$$

If $\hat{u}$ is the solution to the problem

$$\min f(u),$$
$$u \in U_{ad},$$

then there exist $\lambda, \mu \in U$ *with* $\lambda(x), \mu(x) \geq 0$, *for almost all* $x \in \Omega$, *such that*

$$\nabla_u L(\hat{u}) = 0,$$
$$\lambda(x)(\hat{u}(x) - \underline{u}(x)) = 0,\ \mu(x)(\bar{u}(x) - \hat{u}(x)) = 0 \quad \textit{for almost all } x \in \Omega.$$

Proof. The convexity of the set U_{ad} is obvious. Therefore, we can apply Theorem 2.18. First, we define

$$\lambda(x) := (\nabla f(\hat{u}(x)))^+, \qquad \mu(x) := (\nabla f(\hat{u}(x)))^- \quad \forall x \in \Omega,$$

where we define for $r \in \mathbb{R}$

$$r^+ := (|r| + r)/2, \qquad r^- := (|r| - r)/2.$$

From that definition of λ, μ we obtain immediately $\lambda, \mu \in U$ and

$$\nabla_u L(\hat{u}) = \nabla_u f(\hat{u}) - \lambda + \mu = 0.$$

From Theorem 2.18, we now obtain

$$(\lambda - \mu, u - \hat{u})_U \geq 0 \quad \forall u \in U_{ad}.$$

We define

$$\Omega^+ := \{x \in \Omega \,|\, \nabla f(\hat{u}(x)) > 0\}.$$

Since $L^2(\Omega) \subset L_1(\Omega)$, this is a measurable subset of Ω, and we have $\mu|_{\Omega^+} = 0$. Now, we show that

$$\int_M \lambda(x)(\hat{u}(x) - \underline{u}(x))dx = 0$$

for all subsets $M \subset \Omega$ with nonzero measure: if we would find a measurable subset $M \subset \Omega$ with $\int_M \lambda(x)(\hat{u}(x) - \underline{u}(x))dx > 0$, then we can assume without loss of generality that $M \subset \Omega^+$, because the integrand is zero on $M \setminus (\Omega^+ \cap M)$. We define the function $\tilde{u}$ by

$$\tilde{u}(x) = \begin{cases} \underline{u}(x), & x \in M, \\ \hat{u}(x), & x \in \Omega \backslash M. \end{cases}$$

Then, $\tilde{u} \in U_{ad}$ and

$$0 \leq \int_M \lambda(x)(\hat{u}(x) - \underline{u}(x))dx = \int_M (\lambda(x) - \mu(x))(\hat{u}(x) - \underline{u}(x))dx = -(\lambda - \mu, \tilde{u} - \hat{u})_U \leq 0$$

and therefore $\lambda(x)(\hat{u}(x) - \underline{u}(x)) = 0$ for almost all $x \in \Omega$ and analogously $\mu(x)(\bar{u}(x) - \hat{u}(x)) = 0$. □

Remark. As we know from Section 2.2, the existence of a solution in U_{ad} can be guaranteed only if f is convex, which is the case in many model problems which are typically linear-quadratic. In practice, convexity can be assured only from the Taylor series locally in the vicinity of the optimal solution.

More general inequality constraints require the introduction of certain constraint qualifications, which are beyond the scope of this book. Instead, we consider general inequality conditions for $u \in \mathbb{R}^n$ together with the Euclidean scalar product.

Theorem 2.21. *For $u \in \mathbb{R}^n$, we consider the constrained optimization problem* (2.11), (2.12), *where U_{ad} is defined by*

$$U_{ad} = \left\{ u \in \mathbb{R}^n \mid h(u) \le 0 \right\},$$

where the inequality is meant componentwise. The mapping $h : \mathbb{R}^n \to \mathbb{R}^m$ is supposed to be twice differentiable, and its components h_i are supposed to be convex such that

$$U_{ad} = \bigcap_{i=1}^{m} \left\{ u \in \mathbb{R}^n \mid h_i(u) \le 0 \right\}$$

is a convex set. We assume that $\hat{u}$ is the solution of problem (2.11), (2.12). *At $\hat{u}$, we define the active set*

$$\mathcal{A} := \{ i \in \{1, \dots, m\} \mid h_i(\hat{u}) = 0 \}.$$

Furthermore, we assume that the functions h_i satisfy the linear independence constraint qualification (LICQ)

$$\{\nabla h_i(\hat{u}) \mid i \in \mathcal{A}\} \text{ is a linear independent set of vectors.}$$

Then, there is an adjoint vector $\lambda \in \mathbb{R}^m$ with components $\lambda_i \ge 0$, for all $i = 1, \dots, m$, such that the Lagrangian function

$$L(u, \lambda) := f(u) + \lambda^\top h(u)$$

satisfies

$$\nabla_u L(\hat{u}) = 0, \tag{2.14}$$

$$\lambda_i \, h_i(\hat{u}) = 0 \quad \forall i = 1, \dots, m. \tag{2.15}$$

Proof. The solution $\hat{u}$ is by definition also the solution to the problem

$$\min f(u),$$
$$h_i(u) = 0 \quad \forall i \in \mathcal{A}.$$

Because of the LICQ, we can separate the variables locally such that

$$u = \begin{pmatrix} u_d \\ u_f \end{pmatrix}, \quad u_d \in \mathbb{R}^{\#\mathcal{A}}, \quad \frac{\partial h_{\mathcal{A}}}{\partial u_d} \text{ nonsingular,}$$

where $h_{\mathcal{A}} = (h_i)_{i \in \mathcal{A}}$. We obtain the locally equivalent problem

$$\min f(u_d, u_f),$$
$$h_{\mathcal{A}}(u_d, u_f) = 0, \quad \frac{\partial h_{\mathcal{A}}}{\partial u_d} \text{ nonsingular.}$$

We know this problem structure already from Section 2.2 and conclude from Theorem 2.13 that there is $\mu \in \mathbb{R}^{\#\mathcal{A}}$ with

$$0 = \nabla_{u_d} f + \left(\frac{\partial h_{\mathcal{A}}}{\partial u_d} \right)^\top \mu,$$

$$0 = \nabla_{u_f} f + \left(\frac{\partial h_{\mathcal{A}}}{\partial u_f} \right)^\top \mu;$$

if we define $\lambda \in \mathbb{R}^m$ by

$$\lambda_i = \begin{cases} \mu_i, & i \in \mathcal{A}, \\ 0, & i \notin \mathcal{A}, \end{cases}$$

we can write this equivalently as

$$0 = \nabla_u f + \left(\frac{\partial h}{\partial u}\right)^\top \lambda = \nabla_u L.$$

The complementarity conditions (2.15) are now obvious: either $i \in \mathcal{A}$ and then $h_i(\hat{u}) = 0$, or $i \notin \mathcal{A}$ and then $\lambda_i = 0$. From a Taylor series expansion, we obtain now

$$h_i(u) = h_i(\hat{u}) + \nabla h_i(\hat{u})^\top (u - \hat{u}) + \mathcal{O}(\|u - \hat{u}\|^2) \quad \forall i = 1, \dots, m,$$

from which we conclude that

$$\nabla h_i(\hat{u})^\top (u - \hat{u}) \le \eta \quad \forall i \in \mathcal{A},$$

where $\eta = \mathcal{O}(\|u - \hat{u}\|^2)$. We choose $u^i \in U_{ad}$ close to $\hat{u}$ such that

$$\nabla h_i(\hat{u})^\top (u^i - \hat{u}) < 0, \text{ and } \nabla h_j(\hat{u})^\top (u^i - \hat{u}) = 0 \quad \forall i \in \mathcal{A}.$$

Now, we conclude from condition (2.13) that

$$0 \le \nabla f(\hat{u})^\top (u^i - \hat{u}) = -\sum_{j=1}^{m} \lambda_j \nabla h_j(\hat{u})^\top (u^i - \hat{u}) = -\lambda_i \nabla h_i(\hat{u})^\top (u^i - \hat{u})$$

and therefore $\lambda_i \ge 0$ for all $i = 1, \dots, m$. $\square$

We observe that the necessary conditions for inequalities are very similar to the necessary conditions for equality constraints: The differences lie in the predefined sign of the adjoint variables ($\lambda \ge 0$) and in the complementarity conditions. These complementarity conditions are the starting point for active set methods and for interior point methods for the treatment of inequalities.

Chapter 3

Discretization of Optimality Systems

3.1 Introduction

In this chapter, the analysis of discretization of PDE optimality systems is presented in the case of finite difference discretization. Elliptic and parabolic control problems are considered with second-order and higher-order discretization. The case of an optimal control problem governed by a Fredholm integral equation is also discussed. These classes of problems are especially of academic interest, and finite differences are preferred because they allow easy implementation, which makes this choice convenient for teaching purposes. On the other hand, the theoretical framework that we use is valid in a finite element or a finite volume context. In fact, other discretization schemes and other problem classes are discussed in later chapters. In the chapter dedicated to multigrid methods, in Section 5.7.3 we discuss the solution of an elliptic optimality system with FEM discretization. In Section 7.3.3, we discuss a modified Crank–Nicolson scheme for time-dependent bilinear control problems and address the importance of the choice of the optimization space for an efficient numerical optimization process. Further, in Section 7.4, we illustrate the finite difference time domain discretization of the Maxwell equations and of a related optimization problem.

It is clear that a PDE-based optimization problem is naturally defined with infinite-dimensional functional spaces that are Hilbert spaces that concern the PDE model and the formulation of the objective to be optimized. However, the solution of these problems becomes possible after an approximation procedure which involves both the PDE equation and the objective. Furthermore, we know that the solution of a PDE optimization problem is characterized as the solution of the first-order optimality conditions that consist of the PDE model, the corresponding adjoint PDE equation, and the condition of optimality. Now, the discretization step can be applied at the level of the PDE model and of the objective and then followed by the derivation of the optimality condition of the resulting discrete constrained problem, or one applies the discretization step later after the formulation of the optimality system. These two possible discretization procedures are named as follows. The former is called the discretize-before-optimize (DBO) approach, while the latter is called the optimize-before-discretize (OBD) approach. We remark that there are advantages and disadvantages with both approaches which we mention in the following.

On the one hand, the OBD approach has the advantage of allowing a substantially easier formulation of the optimality system which is considered at the continuous level and then each equation of the system is discretized with the preferred method, allowing any different choice of the discretization schemes for the forward and adjoint equations, and the optimality conditions. In particular, one could use approximations of different accuracy [48] and even different grids for the different PDE equations involved.

On the other hand, the DBO approach guarantees a consistent evaluation of the gradient and of the Hessian independently of the discretization used. In fact, the drawback of the OBD approach is the possible inconsistency between the discretized objective and the reduced gradient given by the discrete optimality condition; see [16, 99, 157]. This means discrepancy between the directional derivative $(\nabla \tilde{J}(u), \phi)_{\mathcal{H}}$ and its approximation $\frac{\tilde{J}(u+\epsilon\phi)-\tilde{J}(u-\epsilon\phi)}{2\epsilon}$. The other disadvantage of the OBD approach is that the Hessian may not be symmetric; see, e.g., [29]. This problem cannot arise in the DBO approach. However, it is reasonably clear but less investigated that the differences between the DBO and OBD frameworks can be controlled at the cost of increasing accuracy of discretization by using, e.g., finer meshes and higher-order schemes; see [99]. Another possibility to have the advantages of the OBD scheme with the consistency of the DBO scheme is to discretize the state and adjoint equations with appropriate schemes; see, e.g., [16, 29]. Later in the applications chapter, a dipole quantum control problem is discussed, where the DBO approach becomes advantageous.

For detailed discussions and further results and references on discretization schemes for PDE optimization problems, see [9, 29, 57, 58, 61, 99, 245, 254, 289, 356] and, in particular, [339].

3.2 Discretization of Elliptic Optimization Problems

Consider the following elliptic optimization problem

$$
\left\{
\begin{array}{rcll}
\min J(y,u) & := & \frac{1}{2}\|y-z\|^2_{L^2(\Omega)} + \frac{\nu}{2}\|u\|^2_{L^2(\Omega)}, & \\
-\Delta y & = & u+g & \text{in } \Omega, \\
y & = & 0 & \text{on } \partial\Omega.
\end{array}
\right.
\tag{3.1}
$$

Existence of a unique solution to (3.1) and its characterization are well known [235]. This solution is characterized by the following optimality system

$$
\begin{array}{rcll}
-\Delta y & = & \frac{1}{\nu}p+g & \text{in } \Omega, \\
y & = & 0 & \text{on } \partial\Omega, \\
-\Delta p & = & -(y-z) & \text{in } \Omega, \\
p & = & 0 & \text{on } \partial\Omega, \\
\nu u - p & = & 0 & \text{in } \Omega.
\end{array}
\tag{3.2}
$$

We have that if $z,\, g \in L^2(\Omega)$, then $(y,u,p) \in (H^1_0(\Omega) \cap H^2(\Omega))^3$. The fact that u attains the same regularity as the Lagrange multiplier p is due to the special form of the optimality condition. Next, we consider the finite difference discretization of (3.2) with $\nu > 0$.

Consider a sequence of grids $\{\Omega_h\}_{h>0}$ given by

$$
\Omega_h = \{\mathbf{x} \in \mathbb{R}^2 : x_i = s_i\, h,\; s_i \in \mathbb{Z}\} \cap \Omega.
$$

We assume that Ω is a rectangular domain and that the values of the mesh size h are chosen such that the boundaries of Ω coincide with grid lines. For grid functions v_h and w_h defined on Ω_h we introduce the discrete L^2-scalar product

$$(v_h, w_h)_{L_h^2} = h^2 \sum_{\mathbf{x}\in\Omega_h} v_h(\mathbf{x})\, w_h(\mathbf{x}),$$

with associated norm $|v_h|_0 = (v_h, v_h)_{L_h^2}^{1/2}$. We also need $|v_h|_\infty = \max_{\mathbf{x}\in\Omega_h} |v_h(\mathbf{x})|$.

We introduce first-order backward and forward partial space derivatives of v_h in the x_i direction that are denoted by ∂_i^- and ∂_i^+, respectively, and are given by

$$\partial_i^- v_h(\mathbf{x}) = \frac{v_h(\mathbf{x}) - v_h(\mathbf{x} - \hat{\imath}\, h)}{h} \qquad \text{and} \qquad \partial_i^+ v_h(\mathbf{x}) = \frac{v_h(\mathbf{x} + \hat{\imath}\, h) - v_h(\mathbf{x})}{h},$$

where $\hat{\imath}$ denotes the i coordinate direction vector and v_h is extended by 0 on grid points outside of Ω; see [174, 165, 327]. In this framework, the discrete H^1-norm is given by

$$|v_h|_1 = \left(|v_h|_0^2 + \sum_{i=1}^{2} |\partial_i^- v_h|_0^2 \right)^{1/2}.$$

The spaces L_h^2 and H_h^1 consist of the sets of grid functions v_h endowed with $|v_h|_0$, respectively, $|v_h|_1$, as norm. For the definition of H_h^2 we refer the reader to [174] as well. We have the inverse property $|v_h|_2 \le c h^{-1} |v_h|_1$.

Functions in $L^2(\Omega)$ and $H^2(\Omega)$ are approximated by grid functions defined through their mean values with respect to elementary cells $[x_1 - \frac{h}{2}, x_1 + \frac{h}{2}] \times [x_2 - \frac{h}{2}, x_2 + \frac{h}{2}]$. This gives rise to the restriction operators $\tilde{R}_h : L^2(\Omega) \to L_h^2$ and $R_h : H_0^1(\Omega) \cap H^2(\Omega) \to L_h^2$ defined in [174]. For the definition of H_h^2 we refer the reader to [174] as well. Further, we define $\tilde{R}_h^2 : L^2(\Omega) \times L^2(\Omega) \to L_h^2 \times L_h^2$ by $\tilde{R}_h^2 = (\tilde{R}_h, \tilde{R}_h)$ and analogously $R_h^2 = (R_h, R_h)$. For continuous functions $v \in C^k(\bar{\Omega})$, $k = 0, 1, \ldots$, we denote with $(R_h v)(x) = v(x)$ the restriction operator on $\bar{\Omega}_h$; that is, continuous functions are represented by their grid values.

The following property can be proved

$$|\tilde{R}_h v - R_h v|_0 \le c h^2 |v|_{H^2(\Omega)} \quad \forall v \in H^2(\Omega). \tag{3.3}$$

Here and below, c denotes a positive constant which does not depend on the discretization parameters. We denote with V_h the vector space of nodal functions v_h defined on Ω_h which are zero on the boundary. The system of nodal functions (v_h, w_h) is denoted by $\mathcal{V}_h = V_h \times V_h$.

We need the following lemma [327, 328].

Lemma 3.1 (Poincaré–Friedrichs inequality for finite differences). *For any grid function v_h, there exists a constant c_*, independent of $v_h \in V_h$ and h, such that*

$$|v_h|_0^2 \le c_* \sum_{i=1}^{2} |\partial_i^- v_h|_0^2. \tag{3.4}$$

In particular, for $\Omega = (a,b) \times (c,d)$ we have that

$$c_* = \left(\frac{2}{(b-a)^2} + \frac{2}{(d-c)^2} \right)^{-1};$$

see, e.g., [165, 166, 327].

The second-order five-point approximation to the Laplacian with homogeneous Dirichlet boundary conditions is defined by

$$\Delta_h = \partial_1^+ \partial_1^- + \partial_2^+ \partial_2^- .$$

We have the following consistency result

$$|\Delta_h R_h v - \tilde{R}_h \Delta v|_\infty \le c\, h^2 \|v\|_{C^4(\bar{\Omega})}; \tag{3.5}$$

see, e.g., [174].

Next, we consider an elliptic control problem where the control u has been eliminated using the optimality condition. After discretization, we obtain

$$\left\{ \begin{array}{rcl} -\nu\, \Delta_h y_h - p_h &=& \nu\, \tilde{R}_h g, \\ -\Delta_h p_h + y_h &=& \tilde{R}_h z. \end{array} \right. \tag{3.6}$$

To investigate the convergence of the solution of (3.6) to the solution of the continuous optimality system as $h \to 0^+$, we introduce the family of operators [61]

$$\mathcal{A}_h = \begin{pmatrix} -\nu\, \Delta_h & -I_h \\ I_h & -\Delta_h \end{pmatrix}, \tag{3.7}$$

where I_h is the identity operator on grid functions v_h. The operators $\mathcal{A}_h$ are defined between product spaces of grid functions. Here, the cases $\mathcal{A}_h : H_h^1 \times H_h^1 \to H_h^{-1} \times H_h^{-1}$ and $\mathcal{A}_h : H_h^2 \times H_h^2 \to L_h^2 \times L_h^2$ are important. Here H_h^{-1} denotes the dual space of H_h^1 with L_h^2 as pivot space.

The family $\{\mathcal{A}_h\}_{h>0}$ is called H_h^1-regular if $\mathcal{A}_h$ is invertible and there exists a constant C_1 independent of h such that

$$||\mathcal{A}_h^{-1}||_{\mathcal{L}(H_h^{-1} \times H_h^{-1}, H_h^1 \times H_h^1)} \le C_1,$$

and analogously it is called H_h^2-regular if

$$||\mathcal{A}_h^{-1}||_{\mathcal{L}(L_h^2 \times L_h^2, H_h^2 \times H_h^2)} \le C_2$$

for C_2 independent of h.

Lemma 3.2. *The family of operators $\{\mathcal{A}_h\}_{h>0}$, with h such that the boundaries of Ω are grid lines, is H_h^1-regular.*

Proof. Let $(v_h, w_h) \in \mathcal{V}_h$ be a pair of grid functions. Then

$$\begin{aligned} (\mathcal{A}_h(v_h, w_h), (v_h, w_h))_{L_h^2 \times L_h^2} &= \nu(-\Delta_h v_h, v_h)_{L_h^2} + (-\Delta_h w_h, w_h)_{L_h^2} \\ &\ge \min(\nu, 1)\, C \sum_{i=1}^{2} (|\partial_i^- v_h|_0^2 + |\partial_i^- w_h|_0^2), \end{aligned} \tag{3.8}$$

where C is independent of h and arises from the coercivity estimate for $-\Delta_h$, i.e.,

$$(-\Delta_h v_h, v_h)_{L_h^2} \geq C \sum_{i=1}^{2} |\partial_i^- v_h|_0^2 \quad \forall v_h; \tag{3.9}$$

see, e.g., [174]. Using the Poincaré inequality in (3.8) results in

$$(\mathcal{A}_h(v_h, w_h), (v_h, w_h))_{L_h^2 \times L_h^2} \geq C_1^{-2} \, |(v_h, w_h)|^2_{H_h^1 \times H_h^1} \quad \forall (v_h, w_h) \in L_h^2 \times L_h^2,$$

with $C_1^{-2} = \min(\nu, 1) C c_0$. Due to the Lax–Milgram lemma, $\mathcal{A}_h$ is invertible. Moreover

$$\|\mathcal{A}_h^{-1}\|_{\mathcal{L}(H_h^{-1} \times H_h^{-1}, H_h^1 \times H_h^1)} \leq C_1 \quad \forall h. \quad \square$$

The infinite-dimensional analogue of $\mathcal{A}_h$ is the operator

$$\mathcal{A} = \begin{pmatrix} -\nu \Delta & -I \\ I & -\Delta \end{pmatrix}, \tag{3.10}$$

where Δ is understood with homogeneous Dirichlet boundary conditions. It is well defined from $H_0^1(\Omega) \times H_0^1(\Omega)$ to $H^{-1}(\Omega) \times H^{-1}(\Omega)$ as well as from $(H^2(\Omega) \cap H_0^1(\Omega)) \times (H^2(\Omega) \cap H_0^1(\Omega))$ to $L^2(\Omega) \times L^2(\Omega)$. We have the following consistency result.

Lemma 3.3. *There exists a constant C_K independent of h such that*

$$\|\mathcal{A}_h R_h^2 - \tilde{R}_h^2 \mathcal{A}\|_{\mathcal{L}((H^2 \cap H_0^1)^2, (H_h^{-1} \times H_h^{-1}))} \leq C_K \, h.$$

Proof. Let $(v, w) \in (H^2(\Omega) \cap H_0^1(\Omega))^2$ and note that due to the consistency property of $-\Delta_h$ as discretization of $-\Delta$ we have

$$\begin{aligned}
&|\mathcal{A}_h R_h^2(v, w) - \tilde{R}_h^2 \mathcal{A}(v, w)|^2_{H_h^{-1} \times H_h^{-1}} \\
&\quad \leq \nu \, |(-\Delta_h) R_h v - \tilde{R}_h(-\Delta) v|^2_{H_h^{-1}} + |(-\Delta_h) R_h w - \tilde{R}_h(-\Delta) w|^2_{H_h^{-1}} \\
&\qquad + |R_h v - \tilde{R}_h v|^2_{H_h^{-1}} + |R_h w - \tilde{R}_h w|^2_{H_h^{-1}} \\
&\qquad \leq C_K^2 \, h^2 \, |(v, w)|_{H^2(\Omega)^2};
\end{aligned}$$

see [174, p. 232]. $\square$

We have the following result [61].

Theorem 3.4. *There exists a constant K_1, depending on Ω, g, z, and independent of h, such that*

$$|y_h - R_h y|_1 + |u_h - R_h u|_1 + |p_h - R_h p|_1 \leq K_1 \, h.$$

In case of a general convex domain and finite differences, attention must be paid to the discretization of $-\Delta$ along the boundary. The literature offers several options. For the Shortley–Weller discretization [174], $-\Delta_h$ is H_h^1-regular and consistent with $-\Delta$ from $H^2(\Omega)$ to H_h^{-1}. Using these facts the generalization of Theorem 3.4 to convex domains is straightforward.

In the following result [61] the assumption that the boundaries of Ω coincide with grid lines is used.

Theorem 3.5. *There exists a constant K_2, depending on Ω, g, z, and independent of h, such that*

$$|y_h - R_h y|_0 + |u_h - R_h u|_0 + |p_h - R_h p|_0 \le K_2 h^2.$$

Notice that this result remains valid in the case of a linear FEM discretization on regular triangulation as illustrated in Section 5.7.3; see [230, 339] for further details and references.

Next, we discuss higher-order approximations of the elliptic optimality system. As mentioned in [174], the five-point formula above is optimal in the sense that there is no compact nine-point formula which provides an order of accuracy higher than two. Indeed, Collatz's compact nine-point scheme provides fourth-order accuracy when used in combination with a five-point representation of the right-hand side. In the case of optimality systems the use of the compact nine-point scheme requires a five-point representation of the control $u \in C^2(\Omega)$ in the state equation and of y in the adjoint equation. However, in the cases where constraints on the control are active, the control u may result not sufficiently smooth to allow Collatz's approach, and in these cases the accuracy of the compact nine-point optimal control solution collapses to second order. For this reason, it is preferable to use the extended nine-point schema considered in [77, 299] which is suitable for less smooth controls. We use the following fourth-order nine-point approximation to the Laplacian

$$\Delta_h = \left(1 - \frac{h^2}{12}\partial_1^+\partial_1^-\right)\partial_1^+\partial_1^- + \left(1 - \frac{h^2}{12}\partial_2^+\partial_2^-\right)\partial_2^+\partial_2^-. \tag{3.11}$$

For more insight, the one-dimensional expanded form of this operator is given by

$$\left(1 - \frac{h^2}{12}\partial_1^+\partial_1^-\right)\partial_1^+\partial_1^- v(x) = \frac{1}{12h^2}(-v(x-2h) + 16\,v(x-h) - 30\,v(x) + 16\,v(x+h) - v(x+2h)).$$

In the following we denote with $\tilde{\Delta}_h$ the second-order five-point Laplacian.

At grid points with distance h from the boundary, the operator Δ_h must be modified. In [77] it is shown that the approximation of the Laplacian near the boundary needs to be only $\mathcal{O}(h^2)$ without destroying the overall fourth-order accuracy of the scheme. Thus the five-point Laplace operator could be used (in particular to implement boundary controls as in [58]). However, on coarse grids it turns out [48] that the following asymmetric fourth-order approximation of the second partial derivative is more accurate. For $\mathbf{x} \in \Omega_h$ next to the left-hand side boundary we have [299]

$$\frac{\partial^2 v}{\partial x_1^2} \approx \frac{1}{12h^2}(10\,v(x_1-h) - 15\,v(x_1) - 4\,v(x_1+h) + 14\,v(x_1+2h) - 6\,v(x_1+3h) + v(x_1+4h)). \tag{3.12}$$

The scheme (3.11)–(3.12) results in a matrix of coefficients which is neither diagonally dominant nor of nonnegative type [77]. Nevertheless, in the case that $\tilde{\Delta}_h$ is used close to the boundary, it is proved in [77] that the resulting problem satisfies a maximum principle. The same proof applies with few modifications to the case when (3.12) is used instead. Further we have that

$$|\Delta_h R_h v - \tilde{R}_h \Delta v|_\infty \le c\, h^4 \|v\|_{C^6(\bar{\Omega})}, \tag{3.13}$$

where c is independent of v and h.

Next, an a priori estimate of the accuracy of solutions to the optimality system (3.2) is illustrated. After discretization, we have the following discrete optimality system

$$-\Delta_h y_h - p_h/\nu = \tilde{g}_h, \tag{3.14}$$

$$-\Delta_h p_h + y_h = \tilde{z}_h, \tag{3.15}$$

where $\tilde{g}_h = \tilde{R}_h g$ and $\tilde{z}_h = \tilde{R}_h z$.

Now consider the inner product of (3.14) by $\nu\, y_h$ and of (3.15) by p_h and take the sum of the two resulting equations. We obtain

$$\nu(-\Delta_h y_h, y_h)_{L^2_h} + (-\Delta_h p_h, p_h)_{L^2_h} = \nu(\tilde{g}_h, y_h)_{L^2_h} + (\tilde{z}_h, p_h)_{L^2_h},$$

which implies that

$$\nu(-\Delta_h y_h, y_h)_{L^2_h} + (-\Delta_h p_h, p_h)_{L^2_h} \le \nu\, |(\tilde{g}_h, y_h)_{L^2_h}| + |(\tilde{z}_h, p_h)_{L^2_h}|.$$

By construction of (3.11), we have that $(-\Delta_h v_h, v_h)_{L^2_h} \ge (-\tilde{\Delta}_h v_h, v_h)_{L^2_h}$ for all functions v_h. Because $(-\tilde{\Delta}_h v_h, v_h)_{L^2_h} = \sum_{i=1}^{2} |\partial_i^- v_h|_0^2$ [327] and using Lemma 3.1, we obtain

$$\nu\, |y_h|_0^2 + |p_h|_0^2 \le c_*\, \nu\, |(\tilde{g}_h, y_h)_{L^2_h}| + c_*\, |(\tilde{z}_h, p_h)_{L^2_h}|.$$

Applying the Cauchy–Schwarz and Cauchy inequalities on the right-hand side of this expression results in

$$\nu\, |y_h|_0^2 + |p_h|_0^2 \le c(\nu\, |\tilde{g}_h|_0^2 + |\tilde{z}_h|_0^2), \tag{3.16}$$

where $c = c_*/(2 - c_*)$. We remark that the same inequality is obtained if we use the five-point Laplacian in place of Δ_h in (3.14) and/or (3.15).

Using (3.16), we are now able to determine the degree of accuracy of the optimal solution. For this purpose, notice that (3.14)–(3.15) hold true with y_h and p_h replaced by their respective error functions, and with $\tilde{g}_h$ and $\tilde{z}_h$ replaced by the truncation error for Δ_h estimated by (3.13) (resp., by (3.5)). Further notice that dividing (3.16) by ν and recalling that $u_h = p_h/\nu$, we obtain the estimate for the control from $|y_h|_0^2 + \nu\, |u_h|_0^2 \le c(|\tilde{f}_h|_0^2 + |\tilde{z}_h|_0^2/\nu)$. These statements are summarized in the following theorem.

Theorem 3.6. *Let* $y \in C^{k+2}(\bar{\Omega})$, $k = 2,4$, *and* $p \in C^{l+2}(\bar{\Omega})$, $l = 2,4$, *be solutions to* (3.2) *and let* y_h *and* p_h *be solutions to* (3.14)–(3.15). *Then there exists a constant* c, *depending on* Ω, *and independent of* h, *such that*

$$|y_h - R_h y|_0^2 + \frac{1}{\nu}\, |p_h - R_h p|_0^2 \le c \left(h^{2k}\, \|y\|^2_{C^{k+2}(\bar{\Omega})} + h^{2l}\, \frac{1}{\nu}\, \|p\|^2_{C^{l+2}(\bar{\Omega})} \right).$$

This estimate holds for optimality systems with a linear control mechanism. For results concerning a priori error estimates for elliptic optimal control problems with a bilinear control structure see [225] and the references given therein.

3.3 Discretization of Parabolic Optimization Problems

In this section, we discuss discretization issues concerning the following parabolic optimal control problem

$$\left\{\begin{array}{rcll} \min J(y,u) & := & \frac{1}{2}\|y-z\|^2_{L^2(Q)} + \frac{\nu}{2}\|u\|^2_{L^2(Q)}, & \\ -\partial_t y + \Delta y & = & u & \text{in } Q = \Omega\times(0,T), \\ y(\mathbf{x},0) & = & y_0(\mathbf{x}) & \text{in } \Omega \text{ at } t=0, \\ y(\mathbf{x},t) & = & 0 & \text{on } \Sigma = \partial\Omega\times(0,T), \end{array}\right. \tag{3.17}$$

where we take $y_0(\mathbf{x}) \in H^1_0(\Omega)$. Here, $\nu > 0$ is the weight of the cost of the control and $z \in L^2(Q)$ denotes the desired state. Then there exists a unique solution to the optimal control problem above; see [235, 339]. Corresponding to our setting we have $y^*(u^*) \in H^{2,1}(Q)$, where $H^{2,1}(Q) = L^2(0,T;H^2(\Omega)\cap H^1_0(\Omega))\cap H^1(0,T;L^2(\Omega))$.

The solution to (3.17) is characterized by the following optimality system

$$\begin{aligned} -\partial_t y + \Delta y &= u, \\ \partial_t p + \Delta p + (y-z) &= 0, \\ \nu u - p &= 0, \end{aligned} \tag{3.18}$$

with initial condition $y(\mathbf{x},0) = y_0(\mathbf{x})$ for the state equation (evolving forward in time) and terminal condition $p(\mathbf{x},T) = 0$ for the adjoint equation (evolving backward in time). From (3.18) we have $p, u \in H^{2,1}(Q)$.

In the following, the numerical solution of the optimality system (3.18) in the framework of finite differences and backward Euler schemes is considered. We call $\overline{\Omega}_h$ the mesh, Ω_h is the set of interior mesh-points, and Γ_h is the set of boundary mesh-points. We consider the negative Laplacian with homogeneous Dirichlet boundary conditions approximated by the five-point stencil.

Let $\delta t = T/N_t$ be the time step size. Define

$$Q_{h,\delta t} = \{(\mathbf{x},t_m) : \mathbf{x}\in\Omega_h,\ t_m = (m-1)\delta t,\ 1\le m\le N_t+1\}.$$

On this grid, y_h^m and p_h^m denote grid functions at time level m. The action of the one-step backward and forward time-discretization operator on these functions is defined as follows

$$\partial^+ y_h^m := \frac{y_h^m - y_h^{m-1}}{\delta t} \quad \text{and} \quad \partial^- p_h^m := -\frac{p_h^m - p_h^{m+1}}{\delta t}.$$

For grid functions defined on $Q_{h,\delta t}$ we use the discrete $L^2(Q)$ scalar product with norm $\|v_{h,\delta t}\| = (v_{h,\delta t}, v_{h,\delta t})^{1/2}_{L^2_{h,\delta t}(Q_{h,\delta t})}$.

For convenience, it is assumed that there exist positive constants $c_1 \le c_2$ such that $c_1 h^2 \le \delta t \le c_2 h^2$. Hence h can be considered as the only discretization parameter. Therefore, in the following, the subscript δt is omitted.

On the cylinder Q_h define the family of functions piecewise constant on intervals $[t_m, t_{m+1})$ as follows

$$V_h = \{v_h \mid v_h(t) = v_h(t_m) \text{ for } t\in[t_m,t_{m+1}),\ v_h(t_m)\in L^2_h(\Omega_h)\}.$$

The space-time extension of the operators $\tilde{R}_h$ and R_h is denoted by

$$\tilde{R}_{h,Q} : L^2(Q) \to V_h \quad \text{and} \quad R_{h,Q} : H^{2,1}(Q) \to V_h.$$

Condition (3.3) implies

$$||\tilde{R}_{h,Q}\, v - R_{h,Q}\, v|| \le c\, h^2\, |v|_{H^{2,1}(Q)}. \tag{3.19}$$

With this preparation, we can formulate the discrete version of the parabolic optimality system (3.18). We have

$$\begin{aligned} -\partial_t^+ y_h + \Delta_h y_h &= u_h, \\ \partial_t^- p_h + \Delta_h p_h &= -(y_h - \tilde{R}_{h,Q} z), \\ \nu u_h - p_h &= 0. \end{aligned} \tag{3.20}$$

Next we eliminate u_h from this system. In expanded form, we obtain

$$\begin{aligned} -\left[1+4\gamma\right] y_{i\,j\,m} + \gamma\left[y_{i+1\,j\,m} + y_{i-1\,j\,m} + y_{i\,j+1\,m} + y_{i\,j-1\,m}\right] + y_{i\,j\,m-1} \\ = \frac{\delta t}{\nu}\, p_{i\,j\,m}, \quad 2 \le m \le N_t + 1, \end{aligned} \tag{3.21}$$

$$\begin{aligned} -\left[1+4\gamma\right] p_{i\,j\,m} + \gamma\left[p_{i+1\,j\,m} + p_{i-1\,j\,m} + p_{i\,j+1\,m} + p_{i\,j-1\,m}\right] + p_{i\,j\,m+1} \\ + \delta t\,(y_{i\,j\,m} - \tilde{z}_{i\,j\,m}) = 0, \quad 1 \le m \le N_t, \end{aligned} \tag{3.22}$$

where $\gamma = \frac{\delta t}{h^2}$, $2 \le i, j \le N_{\mathbf{x}}$ index the internal grid points and $\tilde{z} = \tilde{R}_{h,Q} z$. The implementation of the boundary conditions on Σ, of the initial condition at $t = 0$, and of the terminal condition at $t = T$ should be clear.

In [50], the theory of [245] is elaborated on to prove that the solution of (3.20) is second-order accurate. For this purpose, in [50] the approach of [245] is extended to the present finite difference framework. The following estimates are obtained

$$\|u_h - R_{h,Q}\, u\| \le c\, h^2 \tag{3.23}$$

and

$$\|y_h - R_{h,Q}\, y\| \le c\, h^2 \quad \text{and} \quad \|p_h - R_{h,Q}\, p\| \le c\, h^2. \tag{3.24}$$

In [50], results of numerical experiments are reported to validate this accuracy estimate.

Next, we discuss the case of second-order time discretization. For this purpose, we follow [150] and consider the second-order backward differentiation formula (BDF2) together with the Crank–Nicolson (CN) method in order to obtain a second-order time-discretization scheme. Notice that the techniques discussed here can be generalized to higher-order BDF schemes. We remark that CN schemes are strictly nondissipative but easily oscillatory in contrast to BDF schemes that introduce numerical dissipation and thus are more appropriate in a multigrid framework. Therefore, we use the CN scheme only as a second-order one-step method for the purpose of initialization. For a detailed discussion of the BDF and CN schemes, see [13, 121].

To illustrate the BDF2 approach, we use the framework in [170, 174, 337] and assume that Ω is a square domain. The action of the BDF2 time-difference operators on these functions is as follows

$$\partial_{BD}^{+} y_h^m := \frac{3y_h^m - 4y_h^{m-1} + y_h^{m-2}}{2\delta t} \quad \text{and} \quad \partial_{BD}^{-} p_h^m := -\frac{3p_h^m - 4p_h^{m+1} + p_h^{m+2}}{2\delta t}.$$

The coefficients in the last two expressions above are given by the classical BDF2 formula (see, e.g., [13]), while the minus sign in the second operator allows us to discretize the adjoint variable taking into account its backward evolution in time.

With this setting, the following discrete optimality system is obtained

$$\begin{aligned} -\partial_{BD}^{+} y_h^m + \sigma \Delta_h y_h^m &= f_h^m + u_h^m, \\ \partial_{BD}^{-} p_h^m + \sigma \Delta_h p_h^m + \alpha(y_h^m - y_{dh}^m) &= 0, \\ \nu u_h^m - p_h^m &= 0, \end{aligned} \tag{3.25}$$

where we assume sufficient regularity of the data, y_d, y_T, and f, such that these functions are properly approximated by their values at grid points.

Notice that the BDF2 scheme is a multistep method, and therefore we need to combine it with a second-order one-step scheme for initialization of the state and adjoint equations at their initial and terminal time steps, respectively. For this purpose, we use the CN method.

Therefore, at $t = \delta t$ $(m = 2)$, the optimality system results in the following

$$\begin{aligned} -\partial^{+} y_h^m = \tfrac{1}{2}\Big[&\left(-\sigma \Delta_h y_h^m + u_h^m + f_h^m\right) \\ &+ \left(-\sigma \Delta_h y_h^{m-1} + u_h^{m-1} + f_h^{m-1}\right)\Big], \\ \partial_{BD}^{-} p_h^m &+ \sigma \Delta_h p_h^m + \alpha(y_h^m - y_{dh}^m) = 0, \\ \nu u_h^m &- p_h^m = 0. \end{aligned} \tag{3.26}$$

On the other hand, at $t = T - \delta t$ $(m = N_t)$, the optimality system results in the following

$$\begin{aligned} -\partial_{BD}^{+} y_h^m &+ \sigma \Delta_h y_h^m = f_h^m + u_h^m, \\ \partial^{-} p_h^m = \tfrac{1}{2}\Big[&\left(-\sigma \Delta_h p_h^{m+1} - \alpha(y_h^{m+1} - y_{dh}^{m+1})\right) \\ &+ \left(-\sigma \Delta_h p_h^m - \alpha(y_h^m - y_{dh}^m)\right)\Big], \\ \nu u_h^m &- p_h^m = 0. \end{aligned} \tag{3.27}$$

Now, using the theory in [245] and the BDF2 estimates theory in [121] one can prove that the scheme above guarantees a second-order accurate approximation [150]. We have

$$\max_{2 \le m \le N_{t+1}} |y(t_m) - y_h^m|^2 \le c\left(|y(t_0) - y_0|^2 + |y(\delta t) - y_h^1|^2 + O(\delta t^6)\right). \tag{3.28}$$

Similarly, for the adjoint variable we have

$$\max_{N_{t+1} \le m \le 2} |p(t_m) - p_h^m|^2 \le c\left(|p(T) - p_h^{N t+1}|^2 + |p(t_{Nt}) - p_h^{Nt}|^2 + O(\delta t^6)\right). \tag{3.29}$$

The two estimates (3.28) and (3.29) allow us to state that if the initial and terminal conditions, the first-step initial approximations of the state, and the adjoint variables are second-order accurate, then the proposed approach guarantees an optimal-order accuracy.

3.4 Discretization of Optimization Problems with Integral Equations

In this section, we discuss optimization with a model given by integral equations. A motivating reference for this particular class of optimization problems is [338], where optimal control problems are considered with PDE models in the integral formulation. Here, we follow [6] to discuss an optimal control problem with a Fredholm integral equation of the second kind.

Consider a Fredholm integral equation of the second kind with a linear distributed control mechanism. With a tracking objective, the purpose of the control is to determine a control function such that the resulting state $y \in L^2(\Omega)$ tracks as closely as possible a desired target configuration $z \in L^2(\Omega)$, where Ω is the domain. The corresponding optimal control problem is formulated as the minimization of a cost functional J subject to the constraint given by the integral equation. We have

$$\min J(y,u) := \frac{1}{2}\|y - z\|^2_{L^2(\Omega)} + \frac{\nu}{2}\|u\|^2_{L^2(\Omega)}, \tag{3.30}$$

$$y = f(y) + u + g \qquad \text{in } \Omega. \tag{3.31}$$

Here $g \in L^2(\Omega)$ is given, $u \in L^2(\Omega)$ is the control function, and the optimization parameter $\nu > 0$ is the weight of the cost of the control. The term $f(y)$ represents the integral operator, and it is given by

$$f(y)(x) = \int_\Omega K(x,t)\, y(t)\, dt. \tag{3.32}$$

Regarding the governing model, we assume an integral equation of the second kind, where the kernel K satisfies the conditions of the Fredholm alternative theorem [224] such that existence and uniqueness of solution for a given u is guaranteed. In particular, we consider a symmetric integral operator $f(\cdot) = f^T(\cdot)$, where $f^T(y)(x) := \int_\Omega K(t,x)\, y(t)\, dt$, and we require that

$$\|K\|_{L^2(\Omega\times\Omega)} = \int\!\!\int_{\Omega\times\Omega} |K(x,t)|^2 dx\, dt < 1. \tag{3.33}$$

This condition itself is sufficient to prove existence and uniqueness of solution to the integral equation [224], and it can be easily verified in application. We have [6] the following.

Theorem 3.7. *Let K be such that existence and uniqueness of solution for (3.31) are guaranteed and define the gradient $\nabla \hat{J}(u) := \nu u - p$, where p is the solution to the integral equation $p = f^T(p) - y + z$ in Ω. Then the control problem (3.30)–(3.31) has a unique solution in $L^2(\Omega)$ if and only if $\nabla \hat{J}(u) = 0$. Therefore, the optimal solution is characterized as the solution of the following first-order optimality system*

$$\left\{ \begin{array}{rcl} y - f(y) - u & = & g, \\ p - f^T(p) + y & = & z, \\ \nu u - p & = & 0. \end{array} \right. \tag{3.34}$$

Using the scalar equation $\nu u - p = 0$, we can replace $u = p/\nu$ in the state equation and obtain the following equivalent system:

$$\left\{ \begin{array}{lcl} y - f(y) - p/\nu & = & g, \\ p - f(p) + y & = & z. \end{array} \right. \tag{3.35}$$

Notice that system (3.35) corresponds to two coupled integral equations that can be recast as a unique integral equation system, as follows

$$\begin{pmatrix} 1 & -1/\nu \\ 1 & 1 \end{pmatrix} \begin{pmatrix} y \\ p \end{pmatrix} = \begin{pmatrix} f(y)+g \\ f^T(p)+z \end{pmatrix},$$

that is,

$$\begin{pmatrix} y \\ p \end{pmatrix} = \frac{1}{1+\nu} \begin{pmatrix} \nu & 1 \\ -\nu & \nu \end{pmatrix} \begin{pmatrix} f(y)+g \\ f^T(p)+z \end{pmatrix}. \tag{3.36}$$

The advantage of this formulation is the possibility to prove existence and uniqueness of solutions to (3.35) using condition (3.33). Notice that the coefficient matrix is never singular for $\nu > 0$. Following [224], we have that, in general, the integral system

$$\phi_i(x) = \lambda \sum_{j=1}^{n} \int_\Omega \tilde{K}_{ij}(x,t)\phi_j(t)\,dt + g_i(x), \qquad i = 1, \dots, n,$$

where $\Omega = (a,b)$, $g_i \in L^2(\Omega)$, and $\tilde{K} \in L^2(\Omega \times \Omega)$, has a unique solution $\phi_i \in L^2(\Omega)$ provided that $|\lambda| < 1/C$, where $C^2 = \sum_{i,j=1}^{n} \int\int_{\Omega\times\Omega} |\tilde{K}_{ij}(x,t)|^2 dx\,dt$. In our case, we have

$$C^2 = \frac{1}{(1+\nu)^2} \int\int_{\Omega\times\Omega} (2\nu^2 |K(x,t)|^2 + (1+\nu^2)|K(x,t)|^2)\,dx\,dt.$$

Therefore, applying (3.33) to the optimality system, we find that this system admits a unique solution provided that

$$\int\int_{\Omega\times\Omega} |K(x,t)|^2 dx\,dt < \frac{(1+\nu)^2}{1+3\nu^2}, \tag{3.37}$$

which is less restrictive than (3.33) when ν is sufficiently small. This result shows that controlled solutions may exist under weaker conditions than those required for the uncontrolled problem. A similar result is obtained in [57], in the case of singular elliptic control problems.

Now, we discuss the discretization of the Fredholm optimality system by the Nyström method. In [6], the optimality system (3.35) is discretized on a finite difference grid using direct quadrature (DQ) with the Nyström method [14, 176]. Take $x \in \Omega = (-D, D)$ and set the grid points

$$\Omega_h := \{x_i = i\,h,\, i = -N, -N+1, \dots, 0, 1, \dots, N-1, N\},$$

where $h = D/N$. On this grid, the following semidiscrete version of (3.35) is considered

$$\left\{ \begin{array}{lcl} y_N(x) - f_N(y_N)(x) - p_N(x)/\nu & = & g(x), \\ p_N(x) - f_N(p_N)(x) + y_N(x) & = & z(x), \end{array} \right. \tag{3.38}$$

where $x \in \Omega$. Here $y_N(x), p_N(x)$ are approximations to the solutions $y(x), p(x)$, for $x \in \Omega$, and $f_N(y_N)$ is the approximation to the integral with a direct quadrature: $f(y)(x) \approx f_N(y_N)(x) = h\sum_{j=-N}^{N} v_j\, K(x,t_j)\, y_N(t_j)$, where the v_j are the weights of a DQ of order q.

Following the Nyström method, the full discretization of (3.38) is obtained setting $y_i = y_N(x_i)$, $p_i = p_N(x_i)$, having suppressed the evidence of N for an easier notation. Thus, one obtains the following discrete optimality system

$$\begin{aligned} y_i - h\sum_{j=-N}^{N} w_{ij}\, y_j - p_i/\nu &= g_i, \\ p_i - h\sum_{j=-N}^{N} w_{ij}\, p_j + y_i &= z_i, \end{aligned} \tag{3.39}$$

where $i = -N,\dots,N$ and $w_{ij} = v_j\, K(x_i,t_j)$, with v_j given by the quadrature rule. In the following, we denote $y_h = \{y_{-N},\dots,y_N\}$ and $p_h = \{p_{-N},\dots,p_N\}$. The solution of (3.39) gives the approximate solution (y_h, p_h) of (3.35) at the mesh-points. Assuming that $K(x,t)\,y(t)$ is q-time continuously differentiable in t, and uniformly differentiable in x [14, 176], and assuming g is continuous, the following solution error estimate is obtained:

$$\|y - y_h\|_\infty + \|p - p_h\|_\infty \le O(h^q).$$

In a semidiscrete setting, approximation formulas for $y_N(x)$ and $p_N(x)$, $x \in \Omega$, can be found from (3.38), i.e., by the Nyström interpolation formula

$$\begin{cases} y_N = \frac{1}{1+\nu}\left[z + f_N(p_N) + \nu(g + f_N(y_N))\right], \\ p_N = \frac{\nu}{1+\nu}\left[z + f_N(p_N) - (g + f_N(y_N))\right]. \end{cases} \tag{3.40}$$

Under the same regularity condition on the kernel as given above, uniqueness of the solution of (3.39) results from the uniqueness of solution to (3.40). For these functions, one obtains a q-order convergence $\|y - y_N\|_\infty + \|p - p_N\|_\infty \le O(h^q)$.

Chapter 4

Single-Grid Optimization

4.1 Introduction

In this section, we recall some well-known unconstrained optimization methods. Our purpose is to emphasize some aspects of the implementation and application of these methods in the case of PDE optimization. In fact, apart from being large-sized, PDE optimization problems are naturally formulated on functional spaces where the objectives and the solution of the PDE model are defined. Thus, depending on the inner product and norms used, the formulation and implementation of any optimization method will change. For an excellent introduction to optimization methods in Euclidean spaces see [265].

Here, black-box methods represent the class of optimization schemes that apply to any optimization problem independently of the nature and structure of the problem, provided that some essential operators are available. Specifically, we assume that corresponding to a given value of the optimization variable, we can evaluate the objective and its gradient with respect to this variable. In addition, we may require computing the product of the Hessian with a given vector. In PDE optimization, these operations require a particular effort.

4.2 Black-Box Methods

Consider the following class of PDE optimization problems

$$\left\{ \begin{array}{rcl} \min J(y,u) & := & h(y) + \nu g(u), \\ c(y,u) & = & 0, \end{array} \right. \tag{4.1}$$

where $c : Y \times U \to Z$ represents the PDE operator and the objective $J : Y \times U \to \mathbb{R}$, with appropriate Hilbert spaces Y, U, and Z. We denote with $y \in Y$ the state variable, and $u \in U$ denotes the optimization variable. The solution of the PDE equation provides the differentiable mapping $u \to y(u)$. Here, g and h are required to be continuously differentiable, bounded from below, and such that $g(u) \to \infty$ as $\|u\| \to \infty$. Further, denote by $\hat{J}(u) = J(y(u),u)$ the reduced objective and assume that $u \to \hat{J}(u)$ is uniformly convex. This implies existence of a unique optimizer. The corresponding optimal solution is characterized as the solution to the first-order necessary conditions given by the following optimality system

$$\begin{aligned} c(y,u) &= 0, \\ c_y(y,u)^* p &= -h'(y), \\ \nu g'(u) + c_u^* p &= 0. \end{aligned} \tag{4.2}$$

Notice that the third equation of this system represents the optimality condition, in the sense that

$$\hat{J}'(u^*)\delta u = (\nabla \hat{J}(u), \delta u)_U := (\nu g'(u) + c_u^* p, \delta u)_U = 0$$

for all variations δu.

We see that the solution of the PDE problem given by the mapping $u \to y(u)$ allows us to transform the PDE-constrained optimization problem in an unconstrained optimization problem as follows

$$\min_{u \in U} \hat{J}(u). \tag{4.3}$$

Thus the PDE equation has formally disappeared, although we need to solve the PDE problem to evaluate the objective at u. Moreover, to compute $\nabla \hat{J}(u)$ for a given u, we have to first solve the state equation and then the adjoint equation to obtain the adjoint variable to construct the gradient. This procedure is summarized in the following.

ALGORITHM 4.1. Evaluation of the gradient at *u*.

1. Solve (exactly) the state equation $c(y,u) = 0$.
2. Solve (exactly) the adjoint equation $c_y(y,u)^* p = -h'(y)$;
3. Compute the gradient $\nabla \hat{J}(u) = \nu g'(u) + c_u^* p$;
4. End.

Now, we focus on black-box numerical methodologies that apply to (4.3) assuming that we solve exactly all related PDE equations approximated by a discretization scheme. Clearly, under these working conditions the optimization properties of the black-box schemes do not depend on the size and structure of the PDE equations except that this structure determines the properties of the underlying Hessian. On the other hand, it is not reasonable to require the exact solution of PDE problems, and this fact represents the true limitation in the use of black-box methods.

Most of these methods implement as the first iteration step a step in the direction of the negative gradient. Recalling the discussion in Section 2.2 about the dependency of the gradient on the metric chosen, one has to decide about a proper scalar product. Of course, the Hessian would be an ideal choice for the scalar product, and in a function space setting, one often has good idea about it, as we see, e.g., in Sections 2.2 and 7.2.1. In finite dimensions, often one does not have the faintest idea about the Hessian, which is then an appropriate positive definite matrix A, if we, e.g., define the scalar product in $\mathbb{R}^2$ as $(u,v)_A = u^\top A v$. This scalar product is equivalent to a change of variables (i.e., scaling) in the form

$$\bar{u} = A^{1/2} u$$

such that the standard scalar product in $\bar{u}$ is equal to the scalar product in u defined by A. It has proved successful in many applications to heuristically choose the scaling in such a manner that the first gradient does not change the control by more than 10%. That means $A^{1/2}u = (s_1 u_1, \dots, s_n u_n)^\top, s_i \in \mathbb{R}_+$, so that $\frac{1}{s_i^2}|\frac{\partial f}{\partial u_i}| \leq 0.1 \cdot |u_i|$ and therefore $s_i = \sqrt{10 \cdot |\frac{\partial f}{\partial u_i}/u_i|}$ if the expression under the square root is nonzero (otherwise $s_i = |u_i|$) and $|u_i| \neq 0$ (otherwise $s_i = 1$). This scaling strategy has proved invaluable in practical applications like [143, 312].

In the following, we discuss the steepest descent, nonlinear conjugate gradient (NCG), quasi-Newton, and Newton methods and should keep in mind that these methods possess increasingly local convergence properties from linear over superlinear to quadratic [265].

4.2.1 Steepest Descent and NCG Methods

The steepest descent method represents the simplest gradient-based optimization scheme and surely the most popular scheme that is used to validate an optimization framework. The steepest descent algorithm is given in Algorithm 4.2.

ALGORITHM 4.2. Steepest descent scheme.

- Input: initial approx. u_0, $d_0 = -\nabla \hat{J}(u_0)$, index $k = 0$, maximum k_{max}, tolerance tol.
  1. While ($k < k_{max}$ && $\|d_k\|_U > tol$) do
  2. Evaluate steepest descent $d_k = -\nabla \hat{J}(u_k)$;
  3. Compute steplength α_k along d_k by a given rule;
  4. Set $u_{k+1} = u_k + \alpha_k d_k$;
  5. Set $k = k + 1$;
  6. End while

Convergence of the steepest descent scheme is established with the largest steplength $\alpha_k \in \{\alpha^\ell, \ell = 0, 1, \dots\}$, $\alpha \in (0, 1)$, that satisfies the Armijo condition of sufficient decrease of $\hat{J}$'s value given by

$$\hat{J}(u_k + \alpha_k d_k) \leq \hat{J}(u_k) + \delta \alpha_k (\nabla \hat{J}(u_k), d_k)_U \tag{4.4}$$

together with the Wolfe condition

$$(\nabla \hat{J}(u_k + \alpha_k d_k), d_k)_U > \sigma (\nabla \hat{J}(u_k), d_k)_U, \tag{4.5}$$

where $0 < \delta < \sigma < 1/2$; see [265]. The last condition means that the graph of $\hat{J}$ should not increase too fast beyond the minimum. Notice that we use the inner product of the U space.

NCG schemes represent extensions of linear conjugate gradient methods to non-quadratic problems; see, e.g., [319, 145]. In the common variants, the basic idea is to avoid matrix operations and express the search directions recursively as

$$d_{k+1} = -g_{k+1} + \beta_k d_k, \tag{4.6}$$

where $g_k = \nabla \hat{J}(u_k)$, $k = 0, 1, 2, \ldots$, with $d_0 = -g_0$. The iterates for a minimum point are given by

$$u_{k+1} = u_k + \alpha_k d_k, \tag{4.7}$$

where $\alpha_k > 0$ is a steplength. The parameter β_k is chosen so that (4.6)–(4.7) reduces to the linear CG scheme if $\hat{J}$ is a strictly convex quadratic function and α_k is the exact one-dimensional minimizer of $\hat{J}$ along d_k. In this case the NCG scheme terminates in at most n steps in exact arithmetic. This case provides a lower bound to the computational complexity of NCG schemes.

There are many different formulas for β_k which result in different performances depending on the (nonlinear) problem. Well-known formulas are the following:

Fletcher–Reeves, $\beta^{FR} = \frac{(g_{k+1}, g_{k+1})_U}{(g_k, g_k)_U}$;

Polak–Ribière, $\beta^{PR} = \frac{(g_{k+1}, g_{k+1} - g_k)_U}{(g_k, g_k)_U}$;

Hestenes–Stiefel, $\beta^{HS} = \frac{(g_{k+1}, g_{k+1} - g_k)_U}{(d_k, g_{k+1} - g_k)_U}$.

A more recent formulation is due to Dai and Yuan [107] as follows

$$\beta_k^{DY} = \frac{(g_{k+1}, g_{k+1})_U}{(d_k, y_k)_U},$$

where $y_k = g_{k+1} - g_k$, and another one is due to Hager and Zhang [181] based on the formula

$$\beta_k^{HZ} = \frac{(\sigma_k, g_{k+1})_U}{(d_k, y_k)_U}, \qquad \sigma_k = y_k - 2d_k \frac{(y_k, y_k)_U}{(y_k, d_k)_U}. \tag{4.8}$$

Results in [63, 354] suggest that the last two formulations are advantageous in PDE optimization. We note the definition of β_k based on U-space inner product. For optimization in complex Hilbert spaces, the inner products above are replaced with $\Re e(\cdot,\cdot)_U$; see [63].

The NCG scheme is implemented as follows.

ALGORITHM 4.3. NCG scheme.

- Input: initial approx. u_0, $d_0 = -\nabla \hat{J}(u_0)$, index $k = 0$, maximum k_{max}, tolerance tol.

  1. While ($k < k_{max}$ && $\|g_k\|_U > tol$) do
  2. Evaluate steplength $\alpha_k > 0$ along d_k satisfying (4.13)–(4.14);
  3. Set $u_{k+1} = u_k + \alpha_k d_k$;
  4. Compute $g_{k+1} = \nabla \hat{J}(u_{k+1})$;
  5. Compute β_k by, e.g., (4.8);
  6. Let $d_{k+1} = -g_{k+1} + \beta_k d_k$;
  7. Set $k = k + 1$;
  8. End while

4.2.2 Quasi-Newton Methods

Quasi-Newton methods can be seen as extensions of the conjugate gradient method, in which additional storage is used to accelerate convergence. In these methods, approximations of the Hessian matrix are constructed using low-rank updates based on gradient evaluations.

The BFGS method is a quasi-Newton method which makes successive rank-two updates to a matrix B such that it serves as an approximation to the true Hessian. Typically, the BFGS scheme exhibits convergence rates superior to those of NCG schemes at the expense of additional computational effort.

Denote with B_k the kth BFGS approximation to the Hessian. Then the BFGS search direction at the kth step is given by $p_k = -B_k^{-1} g_k$. Further, denote the difference between two successive updates of u as $s_k = \alpha_k p_k$, where α_k is the steplength. The matrix B can be formed explicitly via the well-known recurrence formula

$$B_{k+1} = B_k - \frac{(B_k s_k)(B_k s_k)^\top}{s_k^\top B_k s_k} + \frac{y_k y_k^\top}{y_k^\top s_k}. \tag{4.9}$$

To compute the search direction, it is necessary to invert the matrix B. We denote its inverse as $H = B^{-1}$. Using the Sherman–Morrison–Woodbury formula, we can also establish a recurrence for H

$$H_{k+1} = H_k + \frac{s_k^\top y_k + y_k^\top H_k y_k}{(s_k^\top y_k)^2}(s_k s_k^\top) - \frac{H_k y_k s_k^\top + s_k y_k^\top H_k}{s_k^\top y_k}. \tag{4.10}$$

In the case where the control u and the gradient of the objective function $\nabla \hat{J}$ are elements in a function space, it is not immediately obvious how to directly use this formula since it requires forming outer products. Moreover, to compute the search direction, we only need the action of H on a vector g and it is not necessary to construct any matrix. These facts are discussed in [354], where a matrix-free BFGS is formulated; see also [248, 264, 252].

Suppose U is either $L^2(0,T;\mathbb{R})$ or $H^1(0,T;\mathbb{R})$ and that $x, y \in U$. Then we can denote the function space analogue of the outer product as a dyadic operator $x \otimes y : U \to U$. The action of this operator on a third element $z \in U$ can be expressed in terms of the inner product $(x \otimes y)z = (y,z)_U\, x$.

Now, we illustrate the computational steps to compute the BFGS solution. We start with the initial approximation u_0 and correspondingly determine $g_0 = \nabla \hat{J}(u_0)$. We set $p_0 = -g_0$ and minimize along p_0 with line search with steplength α_0. We obtain u_1 and g_1, and therefore we can compute $y_0 = g_1 - g_0$ and $s_0 = \alpha_0\, p_0$. The first step requires the initialization $H_0 = I$, and we have $z_0 = H_0\, y_0$ and $p_1 = -H_0\, g_1$. Correspondingly, minimizing along p_1 with line search with steplength α_1, we obtain $y_1 = g_2 - g_1$ and $s_1 = \alpha_1\, p_1$.

From the recursion relation for H, we obtain the following summation formula for the search direction for $k > 0$. The term $z_k = H_k\, y_k$ is computed as follows

$$z_k = H_0\, y_k + \sum_{j=1}^{k-1} \left[c_j\, (s_j, y_k)_U\, r_j - (z_j, y_k)_U\, s_j \right], \tag{4.11}$$

where $c_j = (s_j, y_j)_U^{-1}$, $d_j = 1 + c_j (y_j, z_j)_U$, and $r_j = d_j s_j - z_j$. The BFGS search directions are computed as follows

$$p_{k+1} = -H_0 g_{k+1} - \sum_{j=1}^{k} \left[c_j (s_j, g_{k+1})_U r_j - (z_j, g_{k+1})_U s_j \right]. \tag{4.12}$$

In the numerical implementation, these functions are approximated on a grid and the L^2- and H^1-inner products are in either case approximated by a vector inner product with a weighting matrix. This can be written as $(u,v)_{L^2} \approx u^\top M v$ and $(u,v)_{H^1} \approx u^\top K v$, where M is the mass matrix and K is the stiffness matrix.

Notice that with both the NCG and BFGS schemes, the new control u_{k+1} is composed of a linear combination of the original control and the gradients at every step. Further, in the BFGS approach the current approximation to the inverse of the Hessian is stored, whereas in the matrix-free BFGS method the vectors $\{s_j, y_j, z_j\}$ are stored. These are the s_j vectors which are the search steps themselves, the y_j which are the differences between successive gradients, and the z_j vectors which are elements in the space spanned by $\{s_0, \ldots, s_{j-1}\}$. As a counterpart, the matrix-free BFGS formula requires progressively more computation for each optimization step, so it is important that the improved convergence properties at least compensate for the increased computational effort. The matrix-free BFGS algorithm is given below.

ALGORITHM 4.4. BFGS scheme.

- Input: choose $H_0 = I$, initial approx. u_0, $g_0 = \nabla \hat{J}(u_0)$, $p_0 = -g_0$, index $k = 0$, maximum k_{max}, tolerance tol.
- Compute $u_1 = u_0 + \alpha_0 p_0$ with α_0 satisfying (4.13)–(4.14); compute $g_1 = \nabla \hat{J}(u_1)$, $y_0 = g_1 - g_0$, $s_0 = \alpha_0 p_0$, and $p_1 = -H_0 g_1$;
- Set $k = 1$;

1. While ($k < k_{max}$ && $\|g_{k-1}\|_U > tol$) do
2. Compute $u_{k+1} = u_k + \alpha_k p_k$ with α_k satisfying (4.13)–(4.14);
3. Compute $g_{k+1} = \nabla \hat{J}(u_{k+1})$, $y_k = g_{k+1} - g_k$, $s_k = \alpha_k p_k$;
4. Compute z_k with (4.11);
5. Compute and save $c_k = (s_k, y_k)_U^{-1}$, $d_k = 1 + c_k (y_k, z_k)_U$, and $r_k = d_k s_k - z_k$;
6. Compute new search direction p_{k+1} with (4.12).
7. Set $k = k + 1$;
8. End while

Note that the typically superlinear convergence properties of quasi-Newton methods are in Hilbert spaces obtained only if the initial Hessian approximation is chosen as a compact perturbation of the Hessian at the optimal solution (cf. [228, 297, 153]).

4.2.3 Krylov–Newton Methods

Newton methods aim at solving the necessary first-order optimality condition that the gradient $\nabla \hat{J}(u)$ vanish at a local minimum. The Newton method consists of the following iterative procedure

$$\nabla^2 \hat{J}(u_k)\delta u = -\nabla \hat{J}(u_k),$$
$$u_{k+1} = u_k + \delta u,$$

where $\nabla^2 \hat{J}(u_k)$ denotes the reduced Hessian. Because of the typical large size of PDE optimization problems, the Newton equation is solved iteratively using a Krylov method. This solution process involves solving PDE problems that can be done reasonably only to a given tolerance. This means that Newton methods applied to PDE optimization problems are always inexact [111]. Moreover, due to discretization and other approximation introduced in the computation of the Hessian and of the gradient, we should refer to inexact perturbed Newton methods. See [102] for a related discussion and the investigation of the natural connection between quasi-Newton, inexact, and inexact perturbed Newton methods.

In the following, we discuss the Krylov–Newton scheme that is a particular instance of the inexact Newton method. In order to anticipate the possible lack of positive definiteness and still exploit the symmetry of the Hessian, we use the Krylov-type symmetric LQ (SYMMLQ) method [130]. In our experience, this method robustly computes search directions in less time than other Krylov methods, such as GMRES or BiCG, if no preconditioner or a symmetric preconditioner is used. If the Hessian has negative eigenvalues, the SYMMLQ scheme may compute an ascent direction. Whether the direction δu is an ascent or descent can be determined from the sign of its projection onto the gradient. In the cases where δu is an ascent direction, we use $-\delta u$ as a descent direction.

Notice that we refer to δu as the descent direction and not as increment of the optimization variable. In fact, the objective may be nonconvex and the constraint nonlinear, and therefore the approximate solution of the Newton equation cannot guarantee a correct update. Robustness of the Newton approach is obtained using a globalization strategy with a robust linesearch procedure.

In this section, we illustrate a globalized Krylov–Newton scheme; see [355, 356] for theoretical and implementation details. The workflow of the optimization scheme is given by Algorithm 4.5, which requires the computation of the Newton descent direction given by Algorithm 4.7 and a linesearch scheme. A MATLAB implementation of this solution procedure for solving optimal quantum control problems is given in [355].

ALGORITHM 4.5. Optimization scheme.

- Input: initial approx. u_0, index $k = 0$, maximum k_{max}, tolerance tol.

  1. While ($k < k_{max}$ && $\|\nabla \hat{J}(u_k)\|_U > tol$) do
  2. Compute search direction δu with Algorithm 4.7;
  3. Compute α with a linesearch algorithm;
  4. Set $u_{k+1} = u_k + \alpha\, \delta u$;
  5. Set $k = k + 1$;
  6. End while

ALGORITHM 4.6. Apply the reduced Hessian to δu.

- At (y, u, p) and δu given
  1. Solve $c_y \, \delta y = -c_u \, \delta u$;
  2. Solve $c_y^* \, \delta p = -[L_{yy} \, \delta y + L_{yu} \, \delta u]$;
  3. Assemble $\nabla^2 \hat{J}(u) \delta u := L_{uu} \, \delta u + c_u^* \, \delta p + L_{uy} \, \delta y$;
  4. End

ALGORITHM 4.7. Solve the Hessian problem.

- Given u and an initial guess to δu, e.g., $\delta u = -L_{uu}^{-1} \nabla \hat{J}(u)$
  1. Iteratively solve $\nabla^2 \hat{J}(u) \delta u = -\nabla \hat{J}(u)$ with SYMMLQ (Use Algorithm 4.6 to apply the Hessian);
  2. If ascent direction $(\delta u, \nabla \hat{J}(u))_U > 0$ set $\delta u := -\delta u$;
  3. End

With a Newton method, it is recommended to begin the line search with $\alpha = 1$; see, e.g., [265]. This is a reasonable choice when the functional is locally quadratic and knowing that the Newton update is automatically scaled in the functional space where the Hessian and the gradient are defined. However, in nonconvex optimization problems the desired steplength can be orders of magnitude smaller. Therefore, it is advantageous to have an upper bound α_{max} on the maximum feasible steplength. See [356] for details on this issue and the use of α_{max} to design a robust bisection linesearch scheme. The usage of left-sided block-preconditioners for GMRES in constrained problems is illustrated in Section 4.4.

4.2.4 Cascadic Black-Box Schemes

The cascadic approach results from combining nested iteration techniques with single-grid solvers. To illustrate this approach, consider a hierarchy of grids with index $lev = lev_0, \ldots, lev_f$. The idea is to start from a coarse grid with index lev_0, where the problem is small-sized and still well represented. On this grid, the PDE optimization problem can be solved by an iterative black-box scheme with a small computational effort. Let us denote with x_{lev_0} the solution obtained by this process with initialization given by $x^*_{lev_0}$. The step that follows is to interpolate the solution x_{lev_0} to the next finer grid, using an interpolation operator I_{lev}^{lev+1}. Therefore we obtain an initialization of the black-box iterative process on the finer grid that is given by

$$x^*_{lev+1} = I_{lev}^{lev+1} x_{lev},$$

where $lev = lev_0$. With this initialization and using the black-box scheme we obtain the solution x_{lev+1}. This process is repeated until the finest grid is reached and the desired solution is obtained. This method is summarized in Algorithm 4.8, where we denote with $x_{lev} = BB_{lev}(x^*_{lev})$ the result of the black-box iteration, with x^*_{lev} as initialization.

ALGORITHM 4.8. Cascadic method.

- Given $lev = lev_0$ and coarse initialization $x^*_{lev_0}$.
  1. Compute $x_{lev} = BB_{lev}(x^*_{lev})$;
  2. If $lev = lev_f$ then stop;
  3. Else if $lev < lev_f$ then interpolate $x^*_{lev+1} = I^{lev+1}_{lev} x_{lev}$;
  4. Set $lev = lev + 1$, goto 1.;
  5. End

The original motivation for using the cascadic approach comes from computational experience and the results given in [43, 318], where a cascadic conjugate gradient method is discussed and optimal computational complexity for elliptic problems is proved. See, e.g., [63, 354] for successful implementation of the cascadic scheme with NCG and BFGS schemes.

Notice that we have discussed black-box methods for PDE optimization problems without constraints on the optimization variable. In the case where the admissible set $U_{ad} \subset U$ is a closed convex subset of the optimization space, a projection operator $\mathcal{P}_{U_{ad}}(u)$ on U_{ad} of the updates of u should be applied. In particular, an update step becomes

$$u_{k+1} = \mathcal{P}_{U_{ad}}(u_k + \alpha_k d_k),$$

where d_k is the kth optimization direction and α_k is obtained with a linesearch procedure which includes the projection. Specifically, the Armijo condition of sufficient decrease of $\hat{J}$'s value is given by

$$\hat{J}(\mathcal{P}_{U_{ad}}(u_k + \alpha_k d_k)) \leq \hat{J}(u_k) + \delta\, \alpha_k (\nabla \hat{J}(u_k), \tilde{d}_k)_U \tag{4.13}$$

and the Wolfe condition becomes

$$(\nabla \hat{J}(\mathcal{P}_{U_{ad}}(u_k + \alpha_k d_k)), d_k)_U > \sigma\, (\nabla \hat{J}(u_k), \tilde{d}_k)_U, \tag{4.14}$$

where $0 < \delta < \sigma < 1/2$, and $\tilde{d}_k$ is as follows

$$\tilde{d}_k = \begin{cases} d_k, & u_k \in int(U_{ad}), \\ 0, & u_k \in \partial U_{ad} \text{ and } d_k \text{ points outwards } U_{ad}. \end{cases}$$

4.3 Semismooth Newton Methods

In this section, we briefly discuss an extension of the Newton method to the case of PDE optimization problems where additional constraints to the optimization variables result in variational inequalities in the optimality system. In particular, we consider constraints such that the KKT operator equation is Lipschitz continuous but not C^1 regular. For a detailed discussion on this topic and for references see [207].

To illustrate the case of Lipschitz continuous but not C^1 regular function, we consider the following complementarity problem

$$g(u) \leq 0, \qquad u \leq \overline{u}, \qquad g(u)^\top (u - \overline{u}) = 0, \tag{4.15}$$

where $g:\mathbb{R}^n \to \mathbb{R}^n$ and $\overline{u} \in \mathbb{R}^n$ and the inequalities must be interpreted componentwise. The solution to this problem can be formulated as the solution to the following equation

$$F(u) := g(u) + \max(0, -g(u) + u - \overline{u}) = \max(g(u), u - \overline{u}) = 0.$$

If g is locally Lipschitz and C^1 regular, it results that F is a locally Lipschitz continuous function, but it is not C^1. A function is called locally Lipschitz continuous if it is Lipschitz continuous on every bounded subset of its domain.

Next, consider the following elliptic distributed optimal control problem

$$\left\{\begin{array}{rcll} \min_{u\in U_{ad}} J(y,u) & := & \frac{1}{2}\|y-z\|^2_{L^2(\Omega)} + \frac{\nu}{2}\|u\|^2_{L^2(\Omega)}, & \\ -\Delta y & = & u+g & \text{in } \Omega, \\ y & = & 0 & \text{on } \partial\Omega, \end{array}\right. \tag{4.16}$$

where we require that the set of admissible controls be the closed convex subset of $L^2(\Omega)$ given by

$$U_{ad} = \{u \in L^2(\Omega) \,|\, \underline{u}(\mathbf{x}) \le u(\mathbf{x}) \le \overline{u}(\mathbf{x}) \text{ a.e. in } \Omega\}, \tag{4.17}$$

where $\underline{u}$ and $\overline{u}$ are elements of $L^\infty(\Omega)$.

We have that the solution is characterized by the following optimality system

$$\begin{array}{rll} -\Delta y = & u+g & \text{in } \Omega, \\ y = & 0 & \text{on } \partial\Omega, \\ -\Delta p = & -(y-z) & \text{in } \Omega, \\ p = & 0 & \text{on } \partial\Omega, \\ (\nu u - p, v-u) \ge & 0 & \forall v \in U_{ad}. \end{array} \tag{4.18}$$

We see that in the presence of control constraints $\underline{u} \le u \le \overline{u}$, the optimality condition $\nabla \hat{J}(u) = 0$ is replaced by the variational inequality $(\nabla \hat{J}(u), v-u) \ge 0$ for all admissible $v \in U_{ad}$. This condition can be reformulated introducing Lagrange multipliers as follows

$$\nabla \hat{J}(u) + \lambda_H - \lambda_L = 0, \tag{4.19}$$

$$\lambda_L \ge 0, \qquad \underline{u} - u \le 0, \qquad (\lambda_L, \underline{u} - u) = 0, \tag{4.20}$$

$$\lambda_H \ge 0, \qquad u - \overline{u} \le 0, \qquad (\lambda_H, u - \overline{u}) = 0. \tag{4.21}$$

Also in this case, we can reformulate (4.19) and the pointwise (a.e. in Ω) complementarity problems (4.20) and (4.21) using max and min functions as follows

$$\nabla \hat{J}(u) + \lambda = 0, \tag{4.22}$$

$$\max\{0, \lambda + c(u-\overline{u})\} + \min\{0, \lambda + c(u - \underline{u})\} - \lambda = 0, \tag{4.23}$$

where $\lambda = \lambda_H - \lambda_L$ and any $c > 0$.

Notice that the operator equation (4.23) is Lipschitz continuous but not differentiable in the classical sense. However, in the case $c = \nu$ it enjoys a property called Newton differentiability that generalizes the concept of differentiability to a.e. differentiable Lipschitz continuous functions; see, e.g., [207].

To illustrate this concept in the finite-dimensional case, consider a locally Lipschitz continuous function $F:\mathbb{R}^n \to \mathbb{R}^n$ and let D_F denote the set of points at which F is differentiable. For $u \in \mathbb{R}^n$ we define $\partial_B F(x)$ as

$$\partial_B F(u) = \left\{ d : \lim_{u_i \to u,\, u_i \in D_F} \nabla F(u_i) \right\}.$$

Here, B stands for Bouligand differentiable. We can state that a locally Lipschitz function F is B-differentiable at u if and only if it is directionally differentiable at this point. Further, we denote with $\partial F(u)$ the generalized derivative at u defined as the convex hull $\partial F(u) = co\, \partial_B F(u)$; see [104]. Then the directional derivative $F'(u;\delta u)$ is Lipschitz continuous, and there exists a $\tilde{V}(u) \in \partial F(u)$ such that

$$F'(u;\delta u) = \tilde{V}(u)\delta u.$$

In this case, F is said to be semismooth at u. Furthermore, for every $V(u) \in \partial F(u)$, we have

$$V(u)\delta u - F'(u;\delta u) = o(|\delta u|) \qquad \text{as } \delta u \to 0.$$

Now, we can discuss a generalized Newton iteration for solving $F(u) = 0$. We have

$$u_{k+1} = u_k - V_k^{-1} F(u_k), \qquad V_k = V(u_k) \in \partial_B F(u_k). \tag{4.24}$$

For example, consider the complementarity problem discussed at the beginning of this section. We have $F(u) = g(u) + \max(0, -g(u) + u - \overline{u})$, and the corresponding semismooth derivative results as follows

$$V(u) = \begin{cases} 1 & \text{if} \quad -g(u) + u - \overline{u} > 0, \\ g'(u) & \text{if} \quad -g(u) + u - \overline{u} \le 0, \end{cases} \tag{4.25}$$

and this definition must be interpreted componentwise.

We remark that the result (4.25) is representative of an important fact. We have that a semismooth Newton iteration can be equivalently formulated as a primal-dual active set method [191]. In fact, consider the semismooth Newton iteration (4.24) and define the following active and inactive sets (componentwise)

$$\mathcal{A} = \{i : -g(u)(i) + u(i) - \overline{u}(i) > 0\} \qquad \text{and} \qquad \mathcal{I} = \{i : -g(u)(i) + u(i) - \overline{u}(i) \le 0\},$$

where $i = 1, \ldots, n$. On these sets, we obtain the δu increment as the solution to the following

$$\delta u + u - \overline{u} = 0 \text{ on } \mathcal{A} \qquad \text{and} \qquad g'(u)\delta u + g(u) = 0 \text{ on } \mathcal{I}.$$

We see that the u update obtained with these increments corresponds to the semismooth Newton update (4.24)–(4.25).

Next, we discuss the Newton method with a primal-dual active set strategy. For this purpose, we rewrite the complementarity problem (4.15), in terms of primal and dual variables, as $P(u,\lambda) = 0$, where

$$P(u,\lambda) = \begin{pmatrix} g(u) + \lambda \\ \max\{0, \lambda + c(u - \overline{u})\} - \lambda \end{pmatrix}, \tag{4.26}$$

where $c > 0$.

One step of the primal-dual active set strategy applied to this problem proceeds as follows

$$\begin{aligned} &\text{Solve} \quad g(u_{k+1}) + \lambda_{k+1} = 0 \qquad \text{with} \\ &u_{k+1} = \overline{u} \qquad \text{in } \mathcal{A} = \{i : \lambda_k(i) + c(u_k(i) - \overline{u}(i)) > 0\}, \\ &\lambda_{k+1} = 0 \qquad \text{in } \mathcal{I} = \{i : \lambda_k(i) + c(u_k(i) - \overline{u}(i)) \le 0\}. \end{aligned}$$

On the other hand, a semismooth Newton step on the same problem is given by

$$\begin{array}{lll} \text{Solve} & g'(u_k)(u_{k+1}-u_k)+g(u_k)+\lambda_{k+1}=0 & \text{with} \\ u_{k+1}=\overline{u} & \text{in } \mathcal{A}=\{i: \lambda_k(i)+c(u_k(i)-\overline{u}(i))>0\}, & \\ \lambda_{k+1}=0 & \text{in } \mathcal{I}=\{i: \lambda_k(i)+c(u_k(i)-\overline{u}(i))\leq 0\}. & \end{array}$$

Now, we focus on (4.22)–(4.23) representing the elliptic distributed optimal control problem with bilateral control constraints. The two active sets and one inactive set at the kth Newton iteration are given by

$$\mathcal{A}_k^L=\{x\in\Omega: \lambda_k(x)+c(u_k(x)-\underline{u}(x))<0\}$$

and

$$\mathcal{A}_k^H=\{x\in\Omega: \lambda_k(x)+c(u_k(x)-\overline{u}(x))>0\},$$

and $\mathcal{A}_k=\mathcal{A}_k^L\cup\mathcal{A}_k^H$, whereas the inactive set $\mathcal{I}_k=\Omega\setminus\mathcal{A}_k$. With this setting, a semismooth Newton step is formulated as follows

$$\begin{bmatrix} \nabla^2\hat{J}(u_k) & \chi_{\mathcal{A}_k} \\ \chi_{\mathcal{A}_k} & 0 \end{bmatrix}\begin{pmatrix} \delta u \\ \delta\lambda_{\mathcal{A}_k} \end{pmatrix}=-\begin{pmatrix} \nabla\hat{J}(u_k)+\lambda_k \\ \chi_{\mathcal{A}_k^L}(u_k-\underline{u})+\chi_{\mathcal{A}_k^H}(u_k-\overline{u}) \end{pmatrix}, \tag{4.27}$$

where $\chi_{\mathcal{U}}$ denotes the characteristic function of a set $\mathcal{U}$, and $\delta\lambda_{\mathcal{A}_k}$ denotes the restriction of $\delta\lambda$ to the active set $\mathcal{A}_k$.

We summarize the semismooth Newton iteration in the following algorithm.

ALGORITHM 4.9. Newton method with primal-dual active set strategy.

- Input: initial approx. u_0 and λ_0, index $k=0$, maximum k_{max}.
  1. While ($k<k_{max}$ && $\mathcal{A}_k\neq\mathcal{A}_{k+1}$) do
  2. Evaluate $\nabla\hat{J}(u_k)$;
  3. Determine the active/inactive sets, $\mathcal{A}_k^L$, $\mathcal{A}_k^H$, $\mathcal{A}_k$, and $\mathcal{I}_k$;
  4. Assemble the right-hand side of the Newton equation;
  5. Solve the Newton equation (4.27) iteratively;
  6. Set $u_{k+1}=u_k+\delta u$, $\lambda_{k+1}=u_k+\delta u$ on $\mathcal{A}_k$, and $\lambda_{k+1}=0$ on $\mathcal{I}_k$ (globalization may be required);
  7. Set $k=k+1$;
  8. End while

Next, we reformulate the optimality system (4.18) in view of the preceding discussion and formulate a primal-dual iteration on this system which represents a realization of the Newton iteration described above. We have

$$\begin{aligned} -\Delta y &= u+g, \\ y &= 0, \\ -\Delta p &= -(y-z), \\ p &= 0, \\ \nu u-p+\lambda &= 0, \\ \max\{0,\lambda+c(u-\overline{u})\}+\min\{0,\lambda+c(u-\underline{u})\}-\lambda &= 0. \end{aligned}$$

ALGORITHM 4.10. Primal-dual active set strategy for bilateral control constraints.

- Input: initial approx. u_0 and λ_0, index $k = 0$, maximum k_{max}, $c > 0$.

  1. While ($k < k_{max}$ && $\mathcal{A}_k \neq \mathcal{A}_{k+1}$) do
  2. Determine the active/inactive sets

$$\mathcal{A}_k^L = \{x \in \Omega : \lambda_k(x) + c(u_k(x) - \underline{u}(x)) < 0\}$$

and

$$\mathcal{A}_k^H = \{x \in \Omega : \lambda_k(x) + c(u_k(x) - \overline{u}(x)) > 0\},$$

and $\mathcal{A}_k = \mathcal{A}_k^L \cup \mathcal{A}_k^H$, $\mathcal{I}_k = \Omega \setminus \mathcal{A}_k$.

  3. Solve the following system

$$\begin{aligned} -\Delta y_k &= f_k + g, \\ y_k &= 0, \\ -\Delta p_k &= -(y_k - z), \\ p_k &= 0, \end{aligned}$$

where

$$f_k(x) = \begin{cases} \underline{u}(x) & \text{if } x \in \mathcal{A}_k^L, \\ p_k(x)/\nu & \text{if } x \in \mathcal{I}_k, \\ \overline{u}(x) & \text{if } x \in \mathcal{A}_k^H. \end{cases}$$

  4. Set $\lambda_{k+1} = p_k - \nu u_k$;
  5. Set $k = k + 1$;
  6. End while

For further discussion on semismooth Newton schemes and applications, we refer the reader to [207]. A case of polygonal constraints is discussed in [226]. Convex control constraints are discussed in [357]. A case of a problem with control constraints with L^1 cost solved with a semismooth Newton scheme is discussed in [323].

4.4 Preconditioning

A generic problem class in PDE constrained optimization is defined by linear-quadratic problems. Either they arise as such or as a subproblem within SQP methods. In any case, the solution of this type of problem is important, and if it is not performed by multigrid methods, most often Krylov subspace methods are employed, like GMRES, or, using the specific symmetric problem structure, SYMMLQ and MINRES. Because of the spread-out eigenvalue structure of quadratic programming (QP) problems in PDE constrained optimization, we expect rather slow convergence of those Krylov subspace methods. Therefore, preconditioning techniques have to be employed in order to speed up convergence. The general guideline for preconditioning is to try to squeeze the eigenvalues of the preconditioned systems into a small number of tight clusters, whereas the action of the preconditioner itself should be as cheap as possible. In particular, we focus on the following generic quadratic optimization problem with linear equality constraints

$$\begin{aligned}&\min_{y,u} \tfrac{1}{2}\left(y^\top H_{yy}\, y + y^\top H_{yu}\, u + u^\top H_{uy}\, y + u^\top H_{uu}\, u\right) + f_y^\top x + f_u^\top u,\\ &C_y\, y + C_u\, u + c = 0,\end{aligned} \tag{QP}$$

where $y \in \mathbb{R}^{n_y}$, $u \in \mathbb{R}^{n_u}$ are the variable vectors of the optimization problem, the vectors $c, f_y \in \mathbb{R}^{n_y}$, $f_u \in \mathbb{R}^{n_u}$, and the matrices arising are of consistent dimensions:

$$\begin{aligned}&H_{yy} \in \mathbb{R}^{n_y \times n_y}, \quad H_{uu} \in \mathbb{R}^{n_u \times n_u}, \quad H_{yu} = H_{uy}^\top \in \mathbb{R}^{n_y \times n_u},\\ &C_y \in \mathbb{R}^{n_y \times n_y}, \quad C_u \in \mathbb{R}^{n_y \times n_u}.\end{aligned}$$

We assume as an important structural property that C_y is nonsingular and utilize a nullspace basis Z such that $[C_y\ C_u]Z = 0$:

$$Z = \begin{bmatrix} -C_y^{-1} C_u \\ I \end{bmatrix}.$$

Considering the Hessian of (QP), we have

$$H := \begin{bmatrix} H_{yy} & H_{yu} \\ H_{uy} & H_{uu} \end{bmatrix},$$

and we assume that the respective reduced Hessian

$$S = Z^\top H Z = H_{uu} - H_{uy} C_y^{-1} C_u - C_u^\top C_y^{-\top} H_{yu} + C_u^\top C_u^{-\top} H_{yy} C_y^{-1} C_u \tag{4.28}$$

is positive definite, which guarantees the unique solvability of (QP). These linear-quadratic problems typically arise as subproblems within large-scale model-based optimization tasks, where the constraint consists of a discretized PDE. In the following, we focus on the linear-quadratic problem itself.

Problem (QP) is known to satisfy a KKT system of the form

$$\begin{bmatrix} H_{yy} & H_{yu} & C_x^\top \\ H_{uy} & H_{uu} & C_p^\top \\ C_y & C_u & 0 \end{bmatrix} \begin{pmatrix} y \\ u \\ p \end{pmatrix} = - \begin{pmatrix} f_y \\ f_u \\ c \end{pmatrix}, \tag{4.29}$$

and we should note that the reduced Hessian S can be interpreted as the Schur complement of the KKT matrix in (4.29) with respect to the variables (y, p). The block structure in the system matrix has motivated research in block-structured preconditioners for Krylov subspace methods. In [26] three different preconditioners for the solution of (QP) with MINRES and SYMMLQ [271] are analyzed and tested. In [37] several preconditioners based on reduced space techniques are investigated and applied to accelerate GMRES [295] iterations for the linear system. In [162] again a block-structured preconditioner is used to accelerate QMR [130] for model-based output least-squares problems. The article [114] again studies approximate block factorizations in the context of Krylov iterations.

Equation (4.29) is a saddle-point problem with specific features. The only natural assumptions for this system are the assumption that C_y is invertible and that the reduced Hessian in (4.28) is positive definite. These assumptions result from the fact that (4.29) is derived from a PDE-constrained optimization problem. Often, (4.29) is compared with variational optimization problems, like the Stokes problem. The key feature of variational optimizations problem is, however, the positivity of the $\begin{bmatrix} H_{yy} & H_{yu} \\ H_{uy} & H_{uu} \end{bmatrix}$-block. This is essential also for preconditioning methods. Although the typical test problem also possesses this property, we will not discuss any further methods which rely on the positivity of this block, since there are many meaningful PDE-constrained optimization problems with rank deficient $\begin{bmatrix} H_{yy} & H_{yu} \\ H_{uy} & H_{uu} \end{bmatrix}$-block. In [285] so-called optimal preconditioners are presented which rely on multigrid iterations for subblocks and also assume the positivity of the Hessian block.

Readers interested in an overview of various methods for preconditioning of optimization problems are referred to [32, 64]. In this section, we focus on preconditioning methods, which are considered as highly successful by many authors [37, 26, 211, 122] for the particular problem class under investigation.

If one aims at preserving the symmetry structure in (4.29), which we denote with $Kz = r$, one is searching for inexpensively invertible linear transformations P such that the system matrix in

$$P^{-1} K P^{-\top} \tilde{z} = P^{-1} r \qquad (\tilde{z} = P^{\top} z) \tag{4.30}$$

is still symmetric and possesses clustered eigenvalues. Thus, SYMMLQ or MINRES can be applied to the preconditioned equation (4.30). Obviously, symmetric preconditioners have to involve roots of arising operators and therefore mostly rely on positive definiteness of the Hessian part. Thus, typically they are not focused on the most reliable part of the KKT system, which is the PDE system matrix, which has to be invertible and of which in most cases good approximations are known. Also, in [188] a promising indefinite preconditioner for a projected preconditioned conjugate gradient method is introduced which is still independent of the PDE system operator.

Those preconditioners, which take into account available knowledge of the PDE system operator, are of the form of a left-sided preconditioner

$$P_L^{-1} K z = P_L^{-1} r. \tag{4.31}$$

The resulting preconditioned system is no longer symmetric. Therefore, general Krylov subspace methods like typically GMRES are applied. Most authors (see, e.g., [37, 26, 211, 122]) agree on the following form of this left-sided preconditioner which can be derived from reduced SQP methods or nullspace factorizations

$$P_L = \begin{bmatrix} 0 & 0 & A^{\top} \\ 0 & B & C_p^{\top} \\ A & C_p & 0 \end{bmatrix}, \tag{4.32}$$

where A is an approximation to the forward system matrix and B is an approximation to the reduced Hessian. In [26, 211], it is observed that the choice $A = C_y$ and $B = S$ results in a nilpotent iteration matrix $I - P_L^{-1} K$ of nilpotency degree 3, and thus a Krylov subspace method converges within three steps in this case. The results in [211, 122] indicate that the matrix B should be an approximation to the approximate reduced Hessian S_A,

$$S_A = H_{uu} - H_{uy} A^{-1} C_{uu} - C_{uu}^{\top} A^{-\top} H_{yu} + C_{uu}^{\top} A^{-\top} H_{yy} A^{-1} C_{uu}, \tag{4.33}$$

rather than to the exact reduced Hessian (4.28). This should be kept in mind if the reduced Hessian approximation is generated by quasi-Newton-like update strategies. On the other hand, analytic information can often be obtained for the exact reduced Hessian. And this information can be profitably used as an approximation to S_A.

Next, we illustrate the application of the preconditioning technique discussed above to the basic elliptic control problem. Consider

$$\begin{aligned} \min \frac{1}{2}\int_\Omega (y(\xi)-\bar{y}(\xi))^2 d\xi + \frac{\nu}{2}\int_\Omega u(\xi)^2 d\xi \\ -\triangle y(\xi) = u(\xi) \quad \forall \xi \in \Omega, \\ y(\xi) = 0 \quad \forall \xi \in \partial\Omega. \end{aligned}$$

The variables y and u are functions defined on the domain Ω, and $\triangle$ denotes the Laplacian operator. The aim of the problem is to track the given target function $\bar{y}$. We choose $\Omega = [0,1]$. Then, this model problem simplifies to

$$\begin{aligned} \min \frac{1}{2}\int_0^1 (y(\xi)-\bar{y}(\xi))^2 d\xi + \frac{\nu}{2}\int_0^1 u(\xi)^2 d\xi \\ -y''(\xi) = u(\xi) \quad \forall \xi \in [0,1], \\ y(0) = 0,\ y(1) = 0. \end{aligned}$$

The function $\bar{y}$ is chosen as (cf. Figure 4.1)

$$\bar{y}(\xi) = \begin{cases} 0.8-\xi, & 0 \le \xi \le 0.4, \\ -2.6+2\xi, & 0.4 < \xi \le 1. \end{cases}$$

This problem is discretized by finite differences on a regular mesh with mesh size $h = 1/(N-1)$, where $N = n_y$:

$$\begin{aligned} y_\ell &:= y(\ell h), \quad \ell = 0,\dots,N, \\ u_\ell &:= u(\ell h), \quad \ell = 0,\dots,N, \\ -y''(\ell h) &\approx \frac{1}{h^2}(-y_{\ell-1}+2y_\ell - y_{\ell+1}), \quad \ell = 1,\dots,N-1, \\ \int_0^1 (y(\xi)-\bar{y}(\xi))^2 d\xi &\approx h\sum_{\ell=1}^{N-1}(y_\ell - \bar{y}(\ell h))^2, \\ \int_0^1 u(\xi)^2 d\xi &\approx h\sum_{\ell=1}^{N-1} u(\xi)^2. \end{aligned}$$

For the sake of simplicity, we omit values at 0 and 1 so that our vectors of unknowns are

$$y = (y_1,\dots,y_{N-1})^\top, \qquad u = (u_1,\dots,u_{N-1})^\top.$$

The discretized problem is now of the form (QP) with

$$H_{yy} = hI, \quad H_{yu} = H_{uy}^\top = 0, \quad H_{uu} = \nu h I, \qquad C_u = I, \qquad f_y = -\bar{y}, \quad f_u = 0, \quad c = 0,$$

where I is the identity in $\mathbb{R}^{N-1}$ and

$$C_y = \frac{1}{h^2}\begin{bmatrix} -2 & 1 & & & \\ 1 & -2 & 1 & & \\ & \ddots & \ddots & \ddots & \\ & & 1 & -2 & 1 \\ & & & 1 & -2 \end{bmatrix}.$$

In order to perform numerical convergence tests with varying approximations A to C_y, we construct these approximations by Jacobi steps; i.e., for $D := \mathrm{diag}(C_y)$ we define

$$A_0^{-1} := D^{-1}, \tag{4.34}$$
$$A_1^{-1} := D^{-1}(I + (I - C_y D^{-1})), \tag{4.35}$$
$$A_2^{-1} := D^{-1}(I + (I + (I - C_y D^{-1}))(I - C_y D^{-1})), \tag{4.36}$$
$$\vdots \quad \vdots$$

and therefore

$$A_i^{-1} = A_0^{-1}((A_0 - C_y)A_{i-1}^{-1} + I).$$

Thus

$$1 > \rho(I - A_1^{-1}C_y) > \rho(I - A_2^{-1}C_y) = \rho(I - A_1^{-1}C_y)^2 > \rho(I - A_3^{-1}C_y)$$
$$= \rho(I - A_1^{-1}C_y)^3 > \cdots .$$

In the case of A_0, we can give the Schur complement analytically:

$$S_{A_0} = H_{uu} + C_u^\top A_0^{-\top} H_{yy} A_0^{-1} C_u = \left(\nu h + \frac{h^5}{4}\right) I. \tag{4.37}$$

In all other cases, the formulas become more complicated. We treat B analogously: we choose $B_0 := H_{uu}$ and B_j^{-1} as the approximation to S_A^{-1} after j Richardson iterations with H_{uu}. That means

$$B_0 := H_{uu} = \mu h I, \tag{4.38}$$
$$B_j^{-1} := B_0^{-1}(I - C_u^\top A_i^{-\top} H_{yy} A_i^{-1} C_u B_{j-1}^{-1}). \tag{4.39}$$

Additionally, we investigate the cases $B = S_A$ and $B = S$ from (4.28).

We investigate for the setting $N = 101$ ($h = \frac{1}{100}$) and $\nu = 0.001$ the convergence performance. We perform GMRES iterations until the residual is below 10^{-6}, where we start at an iteration vector constant to zero. The problem solution is plotted in Figure 4.1.

Table 4.1 summarizes the results. In column i, we denote how many Jacobi iterations are performed for obtaining A_i^{-1} in (4.34)–(4.36). Column j gives analogous information for B_j^{-1}. The notation $j = S_A$ means that we choose $B = S_A$, and $j = S$ means that we choose $B = S$. Furthermore, ρ_A denotes the spectral radius of the matrix $I - A_i^{-1}C_y$. Analogously, ρ_S denotes the spectral radius of the iteration matrix of the Schur complement; cf. (4.39). In the last column "# it" refers to the number of GMRES iterations either without limitations on the history (GMRES(∞)) or with a history of only 30 vectors

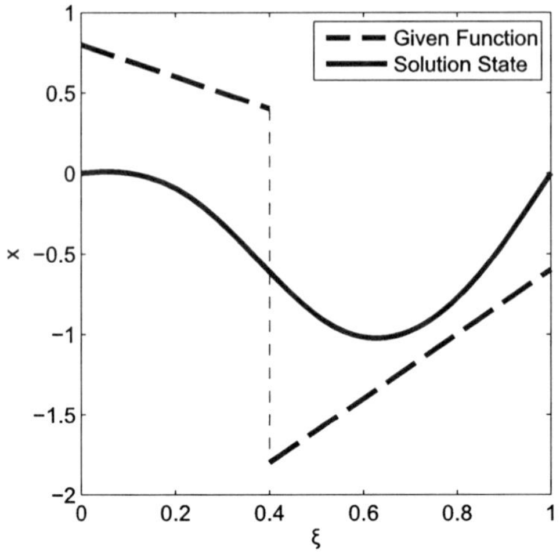

Figure 4.1. *Solution (state, solid) and desired function $\bar{y}$ (dashed) to be tracked. This figure first appeared in K. Ito, K. Kunisch, V. Schulz, and I. Gherman, Approximate nullspace iterations for KKT systems, SIAM J. Matrix Anal. Appl.,* 31(4) (2010), 1835–1847.

Table 4.1. *Convergence results for* $N = 101$ *and* $\nu = 0.001$.

i	j	ρ_A	ρ_S	GMRES(∞) # inner It.	GMRES(30) # outer It.
—	—	—	—	201	—
0	0	0.9995	0.0000	181	—
0	1	0.9995	0.0000	181	—
0	S_A	0.9995	0.0000	181	—
0	S	0.9995	0.9113	182	—
10	0	0.9946	0.0003	56	7
10	1	0.9946	0.0000	56	7
10	S_A	0.9946	0.0000	56	7
10	S	0.9946	0.9112	57	8
20	0	0.9897	0.0011	41	3
20	1	0.9897	0.0000	41	3
5	S_A	0.9897	0.0000	41	3
5	S	0.9897	0.9112	42	4
100	0	0.9514	0.0243	19	1
100	1	0.9514	0.0006	19	1
100	S_A	0.9514	0.0000	19	1
100	S	0.9514	0.9091	19	1
1000	0	0.6101	0.1560	6	1
1000	S_A	0.6101	0.0000	7	1
1000	S	0.6101	0.7728	7	1

(GMRES(30)). A line in the last column means that convergence could not be achieved within 100 outer iterations of GMRES. The first line gives the result of the GMRES method without any preconditioning.

We observe that the GMRES iterations profits very much from a better approximation of the system matrix and also a slight advantage of the choice $B = S_A$.

4.5 SQP Methods and Variants

The idea of SQP for optimization problems of the type

$$\min f(y,u), \tag{4.40}$$
$$c(y,u) = 0, \tag{4.41}$$
$$h(y,u) \geq 0 \tag{4.42}$$

is to compute the solution in an iterative fashion, where a linear-quadratic subproblem has to be solved in each iteration. Since linear-quadratic problems are much better understood than general nonlinear problems, this approach is very attractive. SQP methods can be applied in very general settings concerning the vector spaces involved—in the most general setup they are derived from set-valued Newton methods. The embedding of SQP methods within appropriate function spaces is of major importance if one wants to guarantee a discretization independent good performance of the SQP iterations. We do not want to go into those details and refer the reader interested in an up-to-date overview to the book [194]. Here, we focus on the finite-dimensional setting and will add further aspects later, if appropriate.

For ease of presentation, we drop the inequalities in problem (4.40)–(4.42) and lump together the variables as $z := (y,u)$ in order to investigate the generic problem

$$\min f(z), \tag{4.43}$$
$$c(z) = 0. \tag{4.44}$$

We nevertheless keep in mind that all or some of the equality constraints might later represent so-called active inequality constraints. Optimization theory provides the following necessary conditions in terms of the Lagrangian

$$\partial L(z,p)/\partial z = 0, \qquad \text{where } L(z,p) := f(z) + p^\top c(z). \tag{4.45}$$

The principle of SQP methods starts at Newton's method for the necessary conditions (4.45) together with the constraints (4.44). This method iterates over z and the adjoint variables p in the form $(z^{k+1}, p^{k+1}) = (z^k, p^k) + (\Delta z^{k+1}, \Delta p^{k+1})$, where the increments solve the linear system

$$\begin{bmatrix} H & c_z^\top \\ c_z & 0 \end{bmatrix} \begin{pmatrix} \Delta z \\ \Delta p \end{pmatrix} = \begin{pmatrix} -\nabla f^k - c_z(z^k)^\top p^k \\ -c(z^k) \end{pmatrix}. \tag{4.46}$$

Here H denotes the Hessian of the Lagrangian L with respect to z, i.e., $H = L_{zz}(z^k, p^k)$, and subscripts denote respective derivatives. This equation is equivalent to a linear-quadratic optimization problem—essentially in two different formulations. We see immediately that the necessary conditions for QP

$$\min \frac{1}{2}\Delta z^\top H \Delta z + \left(\nabla f^k + c_z(z^k)^\top p^k\right)^\top \Delta z, \tag{4.47}$$

$$c_z \Delta z + c(z^k) = 0 \tag{4.48}$$

are equivalent to (4.46), where Δp are the adjoint variables in QP (4.47), (4.48). The solution of a QP is a challenging task by itself, and, in particular, in the presence of inequality constraints, the adjoint variables are involved in the solution process. For instance, in active-set strategies, the sign of the adjoint variables indicates whether an inequality is active or not. However, in formulation (4.47), (4.48), the adjoint variables Δp have no meaning, since they converge to zero anyway.

If we write Δp in (4.46) in the form $\Delta p = p^{k+1} - p^k$ and subtract the term $c_z(z^k)^\top p^k$ from both sides of the first row in (4.46), we end up with equivalent formulation

$$\begin{bmatrix} H & c_z^\top \\ c_z & 0 \end{bmatrix} \begin{pmatrix} \Delta z \\ p^{k+1} \end{pmatrix} = \begin{pmatrix} -\nabla f^k \\ -c(z^k) \end{pmatrix}. \tag{4.49}$$

The equivalent linear-quadratic program is now

$$\min 1/2 \Delta z^\top H \Delta z + \nabla {f^k}^\top \Delta z, \tag{4.50}$$

$$c_z \Delta z + c(z^k) = 0, \tag{4.51}$$

with adjoint variable p^{k+1}. If we assume that the algorithm converges to the solution of the optimization problem, then p^{k+1} converges to the adjoint variables of the nonlinear problem and can also be used in order to decide about the activity of inequality constraints.

Because the derivative generation needed to establish the matrices H and c_z often turns out to be prohibitively expensive, one often uses approximations instead, i.e., $G :\approx H$ and $A :\approx c_z$ (of course, these approximations may also change from iteration to iteration). This substitution will deteriorate the convergence behavior of the resulting SQP method. However, in view of (4.46) the fixed points of the iteration are not changed as long as the resulting matrix $\begin{bmatrix} G & A^\top \\ A & 0 \end{bmatrix}$ on the left-hand side is nonsingular. If the approximations G and A are sufficiently accurate, one may expect at least linear local convergence of the iteration method. In the implementation, one will terminate the iteration as soon as $(\Delta z^{k+1}, \Delta p^{k+1})$ is sufficiently close to zero in a suitable norm. If an estimate of the convergence rate is known, an a priori estimate for the distance to the limit point can be given.

In order to arrive at a generalization of this approach again to inequality constrained problems, we first reformulate the system of equations

$$\begin{bmatrix} G & A^\top \\ A & 0 \end{bmatrix} \begin{pmatrix} \Delta z \\ \Delta p \end{pmatrix} = \begin{pmatrix} -\nabla f^k - c_z(z^k)^\top p^k \\ -c(z^k) \end{pmatrix} \tag{4.52}$$

in the form of an equivalent linear-quadratic problem. Rewriting again $\Delta p = p^{k+1} - p^k$ we obtain the formulation

$$\begin{bmatrix} G & A^\top \\ A & 0 \end{bmatrix} \begin{pmatrix} \Delta z \\ p^{k+1} \end{pmatrix} = \begin{pmatrix} -\nabla f^k + (A - c_z(z^k))^\top p^k \\ -c(z^k) \end{pmatrix}.$$

If G is symmetric and positive definite on the nullspace of A, this is equivalent to the linear-quadratic problem

$$\min 1/2\Delta z^\top G\Delta z + \left(\nabla f^k - (A - c_z(z^k))^\top p^k\right)^\top \Delta z, \tag{4.53}$$

$$A\Delta z + c(z^k) = 0, \tag{4.54}$$

where the adjoint variable for (4.54) is p^{k+1}. In this way, e.g., accurate evaluations of the matrices c_z or H can be avoided. Nevertheless, the matrix c_z appears in the objective of the QP, however, only in the form of a matrix-vector product with p^k, which can be efficiently realized, e.g., in the adjoint mode of automatic differentiation [154].

This formulation is generalizable to inequality constraints (cf. [40, 41]), e.g., of the form $c(z) \geq 0$ in (4.44), which leads to linear-quadratic subproblems of the form

$$\min 1/2\Delta z^\top G\Delta z + \left(\nabla f^k - (A - c_z(z^k))^\top p^k\right)^\top \Delta z, \tag{4.55}$$

$$A\Delta z + c(z^k) \geq 0. \tag{4.56}$$

The adjoint variables of (4.56) converge to the adjoint variables of the inequality constraints at the solution. Therefore, they provide a proper decision criterion within an active-set strategy. This formulation of SQP methods allowing for approximations of derivatives also forms the basis for the real-time investigations in [113].

By usage of the rather general formulation (4.55), (4.56), we can discuss several variants of SQP methods. We identify essentially two different cases.

Case 1: $A = c_z$ and H is approximated by G

This is the standard case around which the vast majority of research on SQP methods revolves. The interested reader should consult [265] for an in-depth discussion. Here, we highlight three major representatives for numerical methods:

Projected gradient method: $G = I$

In the unconstrained case, this method amounts to $\Delta z = -\tau \nabla f^k$ with a steplength parameter τ. In the constrained case, the step direction is chosen as the gradient projected to the feasible directions. Linear convergence speed towards the solution can be expected from this method.

Quasi-Newton method: G is constructed from updates

The change in the gradients from iteration to iteration gives an approximation to second order information in the step direction. This idea is used in update techniques. The most prominent is the BFGS-update constructed from the differences $q = \nabla_z L(z^{k+1}, p^{k+1}) - \nabla_z L(z^k, p^{k+1})$ and $s = z^{k+1} - z^k$ as [265]

$$G^{k+1} = G^k + \frac{qq^\top}{q^\top s} - \frac{G^k s (G^k s)^\top}{s^\top G^k s}.$$

It is essential for this method that the approximation G^k stay positive definite all over the iterations, although this can be expected only for H projected to the nullspace of the constraints. One can cope with this problem, e.g., by a modified update rule (Powell). Alternatively, trust region approaches are often employed for guaranteeing global convergence, which can tolerate lack of positive definiteness of G^{k+1}. The typical convergence speed for this SQP variant is superlinear.

Gauss–Newton methods for nonlinear least squares

If the objective $f(z)$ has a special structure like $f(z) = F(z)^\top F(z)$ for vector-valued mapping F, it is convenient to use the approximation $G = J^\top J$, where J is the Jacobian of F. This approximation ignores second-derivative terms of the Hessian of the Lagrangian, which is the reason for the convenience of this method. However, in particular in parameter estimation problems with a small final residual, the second-order terms present in $H - G$ are rather small such that the resulting method possesses a rather small linear convergence rate. This method is called the Gauss–Newton method in the unconstrained case and the generalized Gauss–Newton method [39] in the constrained case.

Reduced SQP

Reduced SQP methods focus on the Hessian of the Lagrangian projected to the nullspace of the linearized constraints. If this nullspace is the range space of a linear mapping T, then reduced SQP methods use the approximation $G = TBT^\top$, where B is an approximation to the reduced Hessian such that $B \approx T^\top HT$. If T spans the nullspace of all active constraints, we know that $T^\top HT$ is positive definite in the neighborhood of the solution if the optimization guarantees the existence of a locally unique solution. Thus, quasi-Newton update techniques as discussed above can be profitably applied for the generation of B. Furthermore, $T^\top HT$ inherits nonlinear least squares structures if they are present. If T spans a larger space than the kernel of all active constraints, the resulting method is called a partially reduced SQP method [310] and convergence results for quasi-Newton approximations depend on similarly unrealistic positive definiteness assumptions as in the case of full-space SQP methods. Partially reduced SQP methods are of particular advantage in the presence of inequality constraints. Reduced SQP methods are the basis for one-shot methods as discussed in the next section.

Case 2: Both c_z and H are approximated by A (resp., G)

If also c_z is approximated in (4.55), (4.56), the resulting methods can be called "approximate" methods, e.g., approximate reduced SQP methods. The convergence properties are not known in detail so far. The literature cited in the section on preconditioning ensures that the resulting approximate SQP variants are algorithms, which converge to the solution of the nonlinear optimization problem, provided the distance $c_z - A$ is small enough in some norm. However, the assumptions posed for convergence proofs typically are much too restrictive and those methods often do a good job also if these restrictive assumptions are not satisfied. Thus, there is still significant room for research.

The theoretical situation is better if the approximation quality of c_z is improved in each step of the SQP method in the fashion of an inexact Newton method. Then superlinear or even quadratic convergence properties can be regained, but with a cost significantly higher than in the approximate case. Most inexact Newton approaches are based on the investigations in [111]. Inexact (reduced) SQP methods based on trust region globalization have been extensively investigated in [187].

4.6 Reduced SQP and One-Shot Methods

The term "one-shot method" is used for solution methods for optimization problems, which solve the optimization problem during the solution of the state equation. It was coined by Ta'asan, Kuruvila, and Salas in [332] with a focus on multigrid methods. Various other names for the same algorithmic paradigm are in usage: boundary value problem approach, simultaneous design and analysis, piggy-back iterations. Thus the state equation and all other restrictions are never feasible during the optimization iterations besides at the optimal solution. From this point of view, already SQP methods are one-shot methods. However, usually the term one-shot methods is used for methods where not even linear subproblems are feasible during the iterations

We discuss again the separability framework, where problems of the type (2.1)–(2.3) are considered. First, assume the case without inequality constraints

$$\min f(y,u), \tag{4.57}$$

$$c(y,u) = 0. \tag{4.58}$$

The essential feature is the nonsingularity of c_y, which means that we can look at the problem as an unconstrained problem of the form

$$\min_u f(y(u),u). \tag{4.59}$$

When applied to the necessary optimality condition $\nabla_u f(y(u),u) = 0$, Newton's method, or its variants, yields good local convergence properties. Every iteration consists of two steps:

(1) Solve $B\Delta u = -\nabla_u f(y(u^k),u^k) = 0$, where $B \approx \nabla_u^2 f(y(u^k),u^k)$.

(2) Update $u^{k+1} = u^k + \tau\,\Delta u$, where τ is an appropriate step length.

The approximation B of the Hessian of f is often performed by quasi-Newton update formulas as discussed, e.g., in [265]. Step (1) of this algorithm involves two costly operations which, however, can be performed in a highly modular way. First $y(u^k)$ has to be computed, which means basically a full nonlinear solution of the flow problem abbreviated by (4.58). Furthermore, the corresponding gradient, $\nabla_u f(y(u^k),u^k)$, has to be determined. One of the most efficient methods for this purpose is the adjoint method, which means the solution of the linear adjoint problem, since for $y^k = y(u^k)$ one obtains

$$\nabla_u f(y(u^k),u^k) = f_u(y^k,u^k)^\top + c_u(y^k,u^k)^\top p,$$

where p solves the adjoint problem

$$c_y(y^k,u^k)^\top p = -f_y(y^k,u^k)^\top.$$

Assuming that (4.58) is solved for y by Newton's method, one might wonder whether it may be enough to perform only one Newton step per optimization iteration in the algorithm above. This results in the following algorithmic steps:

(1) Solve $\begin{bmatrix} 0 & 0 & c_y^\top \\ 0 & B & c_u^\top \\ c_y & c_u & 0 \end{bmatrix} \begin{pmatrix} \Delta y \\ \Delta u \\ p \end{pmatrix} = \begin{pmatrix} -f_y^\top \\ -f_u^\top \\ -c \end{pmatrix}.$

(2) Update $(y^{k+1},u^{k+1}) = (y^k,u^k) + \tau(\Delta y^{k+1},\Delta u^{k+1})$.

This algorithm is called a reduced SQP algorithm. The local convergence can be again of quadratic, superlinear, or linear type [228, 310], depending on how well B approximates the so-called reduced Hessian of the Lagrangian

$$B \approx L_{uu} - L_{uy}c_y^{-1}c_u - (L_{uy}c_y^{-1}c_u)^\top + c_u^\top c_y^{-\top} L_{yy} c_y^{-1} c_u.$$

The vector p produced in each step of the reduced SQP algorithm converges to the adjoint variable vector of problem (4.57), (4.58) at the solution. This iteration again can be written in the form of a Newton-type method as

$$\begin{bmatrix} 0 & 0 & c_y^\top \\ 0 & B & c_u^\top \\ c_y & c_u & 0 \end{bmatrix} \begin{pmatrix} \Delta y \\ \Delta u \\ \Delta p \end{pmatrix} = \begin{pmatrix} -L_y^\top \\ -L_u^\top \\ -c \end{pmatrix}, \quad \begin{pmatrix} y^{k+1} \\ u^{k+1} \\ p^{k+1} \end{pmatrix} = \begin{pmatrix} y^k \\ u^k \\ p^k \end{pmatrix} + \tau \begin{pmatrix} \Delta y \\ \Delta u \\ \Delta p \end{pmatrix}. \tag{4.60}$$

This iteration can be generalized to inexact linear solves with an approximate matrix $A \approx c_y$ such that the approximate reduced SQP iteration reads as

$$\begin{bmatrix} 0 & 0 & A^\top \\ 0 & B & c_u^\top \\ A & c_u & 0 \end{bmatrix} \begin{pmatrix} \Delta y \\ \Delta u \\ \Delta p \end{pmatrix} = \begin{pmatrix} -L_y^\top \\ -L_u^\top \\ -c \end{pmatrix}, \quad \begin{pmatrix} y^{k+1} \\ u^{k+1} \\ p^{k+1} \end{pmatrix} = \begin{pmatrix} y^k \\ u^k \\ p^k \end{pmatrix} + \tau \begin{pmatrix} \Delta y \\ \Delta u \\ \Delta p \end{pmatrix}. \tag{4.61}$$

It is shown in [211] that in this case the use of an approximation of the consistent reduced Hessian, i.e.,

$$B \approx L_{uu} - L_{uy}A^{-1}c_u - (L_{uy}A^{-1}c_u)^\top + c_u^\top A^{-\top} L_{yy} A^{-1} c_u,$$

is recommended. Let us compare this with the SQP formulation of (4.46) in the separability framework

$$\begin{bmatrix} L_{yy} & L_{yu} & c_y^\top \\ L_{uy} & L_{uu} & c_u^\top \\ c_y & c_u & 0 \end{bmatrix} \begin{pmatrix} \Delta y \\ \Delta u \\ \Delta p \end{pmatrix} = \begin{pmatrix} -L_y^\top \\ -L_u^\top \\ -c \end{pmatrix}, \quad \begin{pmatrix} y^{k+1} \\ u^{k+1} \\ p^{k+1} \end{pmatrix} = \begin{pmatrix} y^k \\ u^k \\ p^k \end{pmatrix} + \tau \begin{pmatrix} \Delta y \\ \Delta u \\ \Delta p \end{pmatrix}. \tag{4.62}$$

Because of the triangular structure of the system matrix in (4.60) or (4.61), the reduced SQP formulation is much more modular than the full SQP formulation (4.62). This is the reason, why the matrices in (4.60) or (4.61) are used as preconditioners in large-scale linear-quadratic optimal control problems [26] or as preconditioners in Lagrange–Newton–Krylov methods as discussed in [37, 38]. Indeed, one observes for the (iteration) matrix

$$M = I - \begin{bmatrix} 0 & 0 & c_y^\top \\ 0 & B & c_u^\top \\ c_y & c_u & 0 \end{bmatrix}^{-1} \begin{bmatrix} L_{yy} & L_{yu} & c_y^\top \\ L_{uy} & L_{uu} & c_u^\top \\ c_y & c_u & 0 \end{bmatrix}$$

the fact that $M \neq 0$ in general, but $M^3 = 0$, which is the basis of the convergence considerations in [211].

Now, let us discuss one-shot methods in the context of optimization with PDEs as in, e.g., [183, 182]. We have

$$\min J(y,u) \quad \text{(objective)}, \tag{4.63}$$

$$c(y,u) = 0 \quad \text{(PDE model)}, \tag{4.64}$$

$$h(y,u) \geq 0 \quad \text{(scalar constraints)}, \tag{4.65}$$

where y collects, e.g., the state variables of the model equation model and u is a finite-dimensional vector parameterizing the degrees of freedom for optimization. Often engineering knowledge tells us that the scalar constraint will be active at the solution. Therefore, we can formulate the constraint right from the beginning in the form of an equality constraint. In this context, a full SQP approach as in (4.62) reads as

$$\begin{bmatrix} L_{yy} & L_{yu} & h_y^\top & c_y^\top \\ L_{uy} & L_{uu} & h_u^\top & c_u^\top \\ h_y & h_u & 0 & 0 \\ c_y & c_u & 0 & 0 \end{bmatrix} \begin{pmatrix} \Delta y \\ \Delta u \\ \Delta \mu \\ \Delta p \end{pmatrix} = \begin{pmatrix} -L_y^\top \\ -L_u^\top \\ -h \\ -c \end{pmatrix}, \quad \begin{pmatrix} y^{k+1} \\ u^{k+1} \\ \mu^{k+1} \\ p^{k+1} \end{pmatrix} = \begin{pmatrix} y^k \\ u^k \\ \mu^k \\ p^k \end{pmatrix} + \tau \begin{pmatrix} \Delta y \\ \Delta u \\ \Delta \mu \\ \Delta p \end{pmatrix}. \tag{4.66}$$

This approach is not implementable in general, because one usually starts out with a PDE solver for $c(y,u) = 0$ and seeks a modular coupling with an optimization approach, which does not necessitate a change to the whole code structure, as would be the case with formulation (4.62). A modular but nevertheless efficient alternative is an approximate reduced SQP approach as in (4.61), which is adapted to the case of the additional lift (or pitching) constraint, as established in [143],

$$\begin{bmatrix} 0 & 0 & 0 & A^\top \\ 0 & B & \gamma & c_u^\top \\ 0 & \gamma^\top & 0 & 0 \\ A & c_u & 0 & 0 \end{bmatrix} \begin{pmatrix} \Delta y \\ \Delta u \\ \Delta \mu \\ \Delta p \end{pmatrix} = \begin{pmatrix} -L_y^\top \\ -L_u^\top \\ -h \\ -c \end{pmatrix}, \quad \begin{pmatrix} y^{k+1} \\ u^{k+1} \\ \mu^{k+1} \\ p^{k+1} \end{pmatrix} = \begin{pmatrix} y^k \\ u^k \\ \mu^k \\ p^k \end{pmatrix} + \tau \begin{pmatrix} \Delta y \\ \Delta u \\ \Delta \mu \\ \Delta p \end{pmatrix}, \tag{4.67}$$

where

$$\gamma = h_u^\top + c_u^\top \alpha \text{ such that } A^\top \alpha = -h_y^\top.$$

An algorithmic version of this modular formulation of a generic one-shot scheme is given by the following algorithm.

ALGORITHM 4.11. One-shot scheme.

(1) generate p^k by performing N iterations of an adjoint solver with right-hand side $J_y^\top(y^k, u^k)$ starting in p^k

(2) generate α^k by performing N iterations of an adjoint solver with right-hand side $h_y^\top(y^k, u^k)$ starting in α^k

(3) compute approximate reduced gradients

$$g = J_u^\top + c_u^\top p^{k+1}, \quad \gamma = h_u^\top + c_u^\top \alpha^{k+1}$$

(4) generate B_{k+1} as an approximation of the consistent reduced Hessian

(5) solve the QP

$$\begin{bmatrix} B & \gamma \\ \gamma^\top & 0 \end{bmatrix} \begin{pmatrix} \Delta u \\ \mu^{k+1} \end{pmatrix} = \begin{pmatrix} -g \\ -h \end{pmatrix}$$

(6) update $u^{k+1} = u^k + \Delta u$

(7) adjust the computational mesh, if necessary

(8) generate y^{k+1} by performing N iterations of the forward state solver starting from an interpolation of y^k at the new mesh.

This highly modular algorithmic approach is not an exact transcription of (4.67), but is shown in [143] to be asymptotically equivalent and to converge to the same solution. The overall algorithmic effort for this algorithm is typically in the range of factor 7 to 10 compared to a forward stationary simulation. Usually, step (6) has to be enhanced by a step relaxation which aims at keeping the forward residual below a desirable level. In Section 7.2.5, we discuss an application of this technique with a particular choice for the Hessian approximation B.

Another one-shot approach favored in [137] is strongly correlated with automatic differentiation. In contrast to the block-Gauss–Seidel-type iteration (4.61), this one-shot method applied to problem (4.57), (4.58) takes the form of a block Jacobi iteration

$$\begin{bmatrix} 0 & 0 & A^\top \\ 0 & B & 0 \\ A & 0 & 0 \end{bmatrix} \begin{pmatrix} \Delta y \\ \Delta u \\ \Delta p \end{pmatrix} = \begin{pmatrix} -L_y^\top \\ -L_u^\top \\ -c \end{pmatrix}, \quad \begin{pmatrix} y^{k+1} \\ u^{k+1} \\ p^{k+1} \end{pmatrix} = \begin{pmatrix} y^k \\ u^k \\ p^k \end{pmatrix} + \tau \begin{pmatrix} \Delta y \\ \Delta u \\ \Delta p \end{pmatrix}. \tag{4.68}$$

A step damping is performed on the basis of a primal and dual augmented Lagrangian. Furthermore, it is generalized to additional scalar restrictions by the usage of a penalty multiplier method as in [138].

Chapter 5

Multigrid Methods

5.1 Introduction

From the point of view of the multigrid methodology, we can roughly distinguish between the direct multigrid approach where the optimization problem is implemented within the hierarchy of grid levels and the use of multigrid schemes as inner solvers within an outer optimization loop. The focus of this chapter is on multigrid techniques for optimization, and therefore we do not attempt to discuss other multilevel approaches to optimization such as domain decomposition methods [186], space-mapping and surrogate model techniques [119, 192], multigrid schemes designed to cope with the ill-posedness of inverse problems [42, 214, 221, 251], and other multigrid optimization schemes [256].

Before we address multigrid strategies for optimization problems, we provide an introduction to general multigrid techniques and related theoretical tools, with a focus on the optimization properties of the multigrid components.

5.2 Multigrid Methods for Linear Problems

We start introducing the basic components of a multigrid algorithm and discuss two standard iterative techniques: the Jacobi and the Gauss–Seidel schemes. These two classical iterative methods are characterized by poor convergence rates. However, for errors whose length scales are comparable to the mesh size, they provide rapid damping, leaving behind smooth, longer wavelength errors. These smooth components are responsible for the slow global convergence. A multigrid algorithm, employing grids of different mesh sizes, allows us to solve all wavelength components and provides rapid convergence rates. The multigrid strategy combines two complementary schemes. The high-frequency components of the error are reduced applying iterative methods like Jacobi or Gauss–Seidel schemes. For this reason these methods are called smoothers. On the other hand, low-frequency error components are effectively reduced by a coarse-grid correction scheme. Because the action of a smoothing iteration leaves only smooth error components, it is possible to represent these components as the solution of an appropriate coarser system. Once this coarser problem is solved, its solution is interpolated back to the fine grid to correct the fine-grid approximation for the low-frequency errors.

5.2.1 Iterative Methods and the Smoothing Property

Consider a large sparse linear system of equations $Au = f$, where A is a symmetric positive definite $n \times n$ matrix and u and f are n-dimensional vectors. Iterative methods for solving this problem are formulated as follows

$$u^{(\nu+1)} = M u^{(\nu)} + N f, \tag{5.1}$$

where M and N have to be constructed in such a way that given an arbitrary initial vector $u^{(0)}$, the sequence $u^{(\nu)}$, $\nu = 0, 1, \ldots,$ converges to the solution $u = A^{-1} f$. Define the solution error at the sweep ν as $e^{(\nu)} = u - u^{(\nu)}$; then the iteration (5.1) is equivalent to $e^{(\nu+1)} = M e^{(\nu)}$. M is called the iteration matrix. We have the following convergence criterion based on the spectral radius $r(M)$ of the matrix; see, e.g., [367].

Theorem 5.1. *The method* (5.1) *converges for any initial iterate* $u^{(0)}$ *if and only if* $r(M) < 1$.

A general framework to define iterative schemes of type (5.1) is based on the concept of splitting [353]. Assume the splitting $A = B - C$, where B is nonsingular. By setting $Bu^{(\nu+1)} - Cu^{(\nu)} = f$ and solving with respect to $u^{(\nu+1)}$, one obtains

$$u^{(\nu+1)} = B^{-1} C u^{(\nu)} + B^{-1} f.$$

Thus $M = B^{-1} C$ and $N = B^{-1}$. Typically, one considers the regular splitting $A = D - L - U$, where $D = \text{diag}(a_{11}, a_{22}, \ldots, a_{nn})$ denotes the diagonal part of the matrix A, and $-L$ and $-U$ are the strictly lower and upper parts of A, respectively. Based on this splitting many choices for B and C are possible, leading to different iterative schemes. For example, the choice $B = \frac{1}{\omega} D$ and $C = \frac{1}{\omega}[(1-\omega)D + \omega(L+U)]$, with $0 < \omega \leq 1$, leads to the damped Jacobi iteration

$$u^{(\nu+1)} = (I - \omega D^{-1} A) u^{(\nu)} + \omega D^{-1} f. \tag{5.2}$$

Choosing $B = D - L$ and $C = U$, one obtains the Gauss–Seidel iteration

$$u^{(\nu+1)} = (D - L)^{-1} U u^{(\nu)} + (D - L)^{-1} f. \tag{5.3}$$

Later on we denote the iteration matrices corresponding to (5.2) and (5.3) with $M_J(\omega)$ and M_{GS}, respectively.

It is useful to define and analyze the smoothing property of these iterations by introducing a simple model problem. Consider the finite difference approximation of a one-dimensional Dirichlet boundary value problem. We have

$$\begin{cases} -u''(x) = f(x) & \text{in} \quad \Omega = (0,1), \\ \quad u(x) = g(x) & \text{on} \quad \{0,1\}. \end{cases} \tag{5.4}$$

Let Ω be represented by a grid Ω_h with grid size $h = \frac{1}{n+1}$ and grid points $x_j = jh$, $j = 0, 1, \ldots, n+1$. A discretization scheme for the second derivative at x_j is $h^{-2}[u(x_{j-1}) - 2u(x_j) + u(x_{j+1})] = u''(x_j) + O(h^2)$. Set $f_j^h = f(jh)$, and $u_j^h = u(jh)$. We obtain the following tridiagonal system of n equations

$$\begin{aligned} 2u_1^h - u_2^h &= h^2 f_1^h + g(0), \\ -u_{j-1}^h + 2u_j^h - u_{j+1}^h &= h^2 f_j^h, \qquad j = 2, \ldots, n-1, \\ -u_{n-1}^h + 2u_n^h &= h^2 f_n^h + g(1). \end{aligned} \tag{5.5}$$

Let us denote (5.5) (with all terms divided by h^2) by $A_h u_h = f_h$.

We discuss the solution to this problem by means of the damped Jacobi iteration with iteration matrix $M_J(\omega) = I - \omega D_h^{-1} A_h$. Consider the eigenvalue problem $M_J(\omega)v^k = \mu_k v^k$. The eigenvectors of $M_J(\omega)$ (and equivalently of A_h) are given by

$$v^k = \sqrt{2h}\,(\sin(k\pi hj))_{j=1,n}, \qquad k = 1,\ldots,n. \tag{5.6}$$

The eigenvalues of A_h are $\lambda_k = 4\sin^2(k\pi h/2)/h^2$ and the corresponding eigenvalues of $M_J(\omega)$ are

$$\mu_k(\omega) = 1 - \omega(1 - \cos(k\pi h)), \qquad k = 1,\ldots,n. \tag{5.7}$$

We have that $r(M_J(\omega)) < 1$ for $0 < \omega \le 1$, guaranteeing convergence. In particular, for the Jacobi iteration with $\omega = 1$ we have $r(M_J(1)) = 1 - \frac{1}{2}\pi h^2 + O(h^4)$, showing how the convergence of the Jacobi iteration deteriorates (i.e., r tends to 1) as $h \to 0$.

The purpose of an iteration in a multigrid algorithm is primarily to be a smoothing operator. In order to characterize this property, we need to distinguish between low- and high-frequency eigenvectors. We define

- *low-frequency* (LF) components: v^k with $1 \le k < \frac{n}{2}$;
- *high-frequency* (HF) components: v^k with $\frac{n}{2} \le k \le n$.

We now define the smoothing factor μ as the worst factor by which the amplitudes of HF components are damped per iteration. In the case of the Jacobi iteration we have

$$\begin{aligned}\mu &= \max\left\{|\mu_k|, \frac{n}{2} \le k \le n\right\} = \max\{1-\omega, \quad |1-\omega(1-\cos(\pi)|\} \\ &\le \max\{1-\omega, |1-2\omega|\}.\end{aligned}$$

Using this result we find that the optimal (smallest) smoothing factor $\mu = 1/3$ is obtained by choosing $\omega^* = 2/3$. This means that using $M_J(\omega^*)$ the HF error components are reduced by at least a factor of one-third after any sweep and this factor does not depend on the mesh size. Therefore if we use the expansion $e^{(\nu)} = \sum_k e_k^{(\nu)} v^k$, we have that a few sweeps of (5.2) give $|e_k^{(\nu)}| \ll |e_k^{(0)}|$ for HF error components. For this reason, although the global error decreases slowly by iteration, it is smoothed very quickly.

Most often, instead of a Jacobi method other iterations are used that suppress the HF components of the error more efficiently. This is the case of the Gauss–Seidel iteration (5.3). The smoothing property of this scheme is conveniently analyzed by using local Fourier analysis introduced by Brandt [83, 85]. This is an effective tool for analyzing the multigrid process even though it is based on certain idealized assumptions and simplifications: Boundary conditions are neglected, and the problem is considered on infinite grids $G^h = \{jh,\ j \in \mathbb{Z}\}$ and represented in terms of the (continuous) Fourier functions $\varphi(\theta, x) = e^{i\theta x/h}$ with $\theta \in (-\pi, \pi]$. Equivalently, we could refer to the Fourier functions on a unit interval with periodic boundary conditions. Notice that on G^h only the components $e^{i\theta x/h}$ with $\theta \in (-\pi, \pi]$ are visible; i.e., there is no other component with frequency $\theta_0 \in (-\pi, \pi]$ with $|\theta_0| < \theta$ such that $e^{i\theta_0 x/h} = e^{i\theta x/h}$, $x \in G^h$. The notion of LF and HF components on the grid G^h is related to a coarser grid denoted by G^H. In this way $e^{i\theta x/h}$ on G^h is said to be an HF component, with respect to the coarse grid G^H, if its restriction (projection) to G^H is not visible there. If $H = 2h$, then the high frequencies are those with $\frac{\pi}{2} \le |\theta| \le \pi$ and we have $e^{i\theta x/h} = e^{i(2\theta)x/H}$.

In this framework, in order to analyze a given iteration we represent solution errors in terms of their θ components $e^{(\nu)} = \sum_\theta \mathbb{E}_\theta^{(\nu)} e^{i\theta x/h}$ and $e^{(\nu+1)} = \sum_\theta \mathbb{E}_\theta^{(\nu+1)} e^{i\theta x/h}$ (with formal summation on θ), where $\mathbb{E}_\theta^{(\nu)}$ and $\mathbb{E}_\theta^{(\nu+1)}$ denote the error amplitudes of the θ component before and after smoothing, respectively. The action of the iteration matrix M is $e^{(\nu+1)} = M\, e^{(\nu)}$. In the Fourier space this action is represented by $\mathbb{E}_\theta^{(\nu+1)} = \hat{M}(\theta)\mathbb{E}_\theta^{(\nu)}$, and $\hat{M}(\theta)$ is the so-called Fourier symbol of M; see [340].

In the local Fourier analysis framework, the smoothing factor is then defined by

$$\mu = \max\left\{\left|\frac{\mathbb{E}_\theta^{(\nu+1)}}{\mathbb{E}_\theta^{(\nu)}}\right|, \ \frac{\pi}{2} \le |\theta| \le \pi\right\} = \max\left\{|\hat{M}(\theta)|, \ \frac{\pi}{2} \le |\theta| \le \pi\right\}. \tag{5.8}$$

Later, we consider the entire frequency domain spanned by the two sets of frequencies $\theta \in [-\pi/2, \pi/2)$ and $\overline{\theta} = \theta - sign(\theta)\pi$. Here θ represents LF components while $\overline{\theta}$ represents the HF components. This choice results in a representation with respect to two harmonics $e^{i\theta x/h}$ and $e^{i\overline{\theta} x/h}$. In this framework, a way to characterize the smoothing property of the smoothing operator M is to consider the action of M on both sets of frequencies,

$$\widehat{M}(\theta) = \begin{bmatrix} \hat{M}(\theta) & 0 \\ 0 & \hat{M}(\overline{\theta}) \end{bmatrix},$$

and to assume an ideal coarse-grid correction which annihilates the LF error components and leaves the HF error components unchanged. That is, one defines the projection operator $\widehat{Q}$ as follows

$$\widehat{Q}(\theta) = \begin{bmatrix} 0 & 0 \\ 0 & 1 \end{bmatrix}.$$

In this framework the smoothing property of M is defined as follows

$$\mu = \max\{r(\widehat{Q}(\theta)\widehat{M}(\theta)) : \theta \in [-\pi/2, \pi/2)\}, \tag{5.9}$$

where r is the spectral radius.

For illustration, consider the Gauss–Seidel scheme applied to our discretized model problem. A smoothing sweep starting with an initial approximation $u^{(\nu)}$ produces a new approximation $u^{(\nu+1)}$ such that the corresponding error satisfies

$$B_h e^{(\nu+1)}(x) - C_h e^{(\nu)}(x) = 0, \qquad x \in G^h, \tag{5.10}$$

where $B_h = D_h - L_h$ and $C_h = U_h$. For a given θ, (5.10) at $x = jh$ becomes

$$\sum_\theta \left[(2 - e^{-i\theta})\mathbb{E}_\theta^{(\nu+1)} - e^{i\theta}\mathbb{E}_\theta^{(\nu)}\right] e^{i\theta j} = 0,$$

which must hold for all j; therefore we obtain $\hat{M}(\theta) = e^{i\theta}/(2 - e^{-i\theta})$. Hence we have

$$\mu = \max\left\{\left|\frac{\mathbb{E}_\theta^{(\nu+1)}}{\mathbb{E}_\theta^{(\nu)}}\right|, \ \frac{\pi}{2} \le |\theta| \le \pi\right\} = \max\left\{\left|\frac{e^{i\theta}}{2 - e^{-i\theta}}\right|, \ \frac{\pi}{2} \le |\theta| \le \pi\right\} = 0.45.$$

Similar values are obtained for the Gauss–Seidel iteration applied to the two- and three-dimensional version of our model problem. For a two-dimensional model Poisson problem the effect of smoothing can be seen in Figure 5.1.

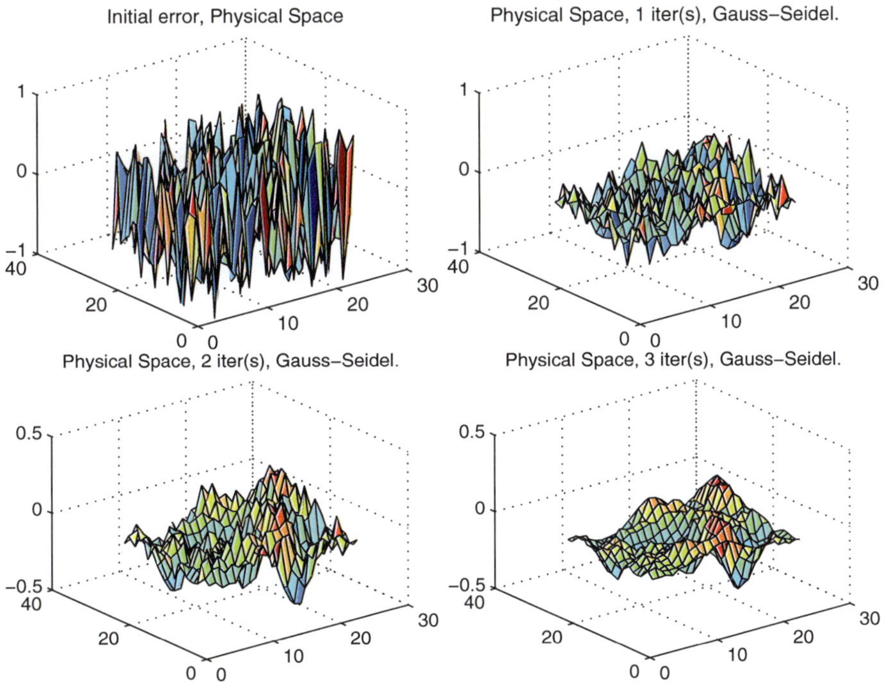

Figure 5.1. *Smoothing by Gauss–Seidel iteration.*

Another definition of smoothing property of an iterative scheme is due to Hackbusch [173]. Let M be the iteration matrix of a smoothing procedure and recall the relation $e^{(\nu)} = M^\nu e^{(0)}$. One can measure the smoothness of $e^{(\nu)}$ by a norm involving differences of the value of this error on different grid points. A natural choice is to take the second-order difference matrix A_h above. Then the following smoothing factor is defined

$$\mu(\nu) = \|A_h M^\nu\| / \|A_h\|.$$

The iteration defined by M is said to possess the smoothing property if there exists a function $\eta(\nu)$ such that, for sufficiently large ν, we have

$$\|A_h M^\nu\| \leq \eta(\nu) h^{-\alpha}, \tag{5.11}$$

where $\alpha > 0$ and $\eta(\nu) \to 0$ as $\nu \to \infty$. This is the case for our model problem where A_h is the discretization of the minus Laplacian and using the damped Jacobi iteration, $M = I_h - \omega h^2 A_h$, $\omega \in (0, 1/2)$. To show this fact, recall the following lemma [173].

Lemma 5.2. *Let B be real symmetric semipositive definite matrix such that $0 \leq B \leq I$, $0 < \gamma \leq 1$, and ν is a positive integer. Then*

$$\|B(I - \gamma B)^\nu\| \leq \hat{\eta}(\nu), \qquad \hat{\eta}(\nu) = \frac{\nu^\nu}{\gamma(\nu+1)^{\nu+1}}.$$

Proof. Notice that the spectrum $\sigma(B) \in (0,1)$ and that $\|f(B)\| = \max\{|f(\lambda)|, \lambda \in \sigma(B)\}$. Find the maximum of the function $f(x) = x(1-\gamma x)^{\nu}$. $\square$

Now, we have that there exists a constant C such that $\|h^2 A_h\| \leq C$; therefore the matrix $B = \frac{h^2}{C} A_h$ satisfies the conditions of Lemma 5.2. Hence, the smoothing property is given for $\omega < 1/C$ and with $\alpha = 2$ and $\eta(\nu) = (\frac{1}{\omega})\nu^{\nu}/(\nu+1)^{(\nu+1)}$. For the Gauss–Seidel iteration one can prove that the smoothing property holds with $\alpha = 2$ and $\eta(\nu) \approx 1/\nu$.

5.2.2 Iterative Methods as Minimization Schemes

The discussion on iterative schemes given above is typical within the classical multigrid framework where multigrid operators are characterized by their properties on the Fourier space. In our context, however, we are particularly interested in the optimization properties of the various multigrid components. In this section, we consider iterative schemes from this point of view. We use the equivalence between solving the problem $Au = f$, where A is symmetric and positive definite, and minimizing the functional

$$J(u) = \frac{1}{2} u^\top A u - u^\top f. \tag{5.12}$$

Recall that many iterative methods like Jacobi and Gauss–Seidel schemes can be written in terms of a nonsingular matrix R as follows

$$u^{(\nu+1)} = (I - RA)u^{(\nu)} + Rf = u^{(\nu)} + R(f - Au^{(\nu)}) = u^{(\nu)} + R\,r^{(\nu)}, \tag{5.13}$$

where $r^{(\nu)} = f - Au^{(\nu)} = -\nabla J(u^{(\nu)})$ is the residual for the $u^{(\nu)}$ approximation. With $R^{-1} = D/\omega$ we have the damped Jacobi iteration; choosing $R^{-1} = D - L$ (resp., $R^{-1} = D + U$), the forward (resp., backward) Gauss–Seidel scheme is obtained.

Using (5.13) in (5.12), we have

$$\begin{aligned}
J(u^{(\nu+1)}) &= \frac{1}{2}(u^{(\nu+1)})^\top A u^{(\nu+1)} - (u^{(\nu+1)})^\top f \\
&= \frac{1}{2}(u^{(\nu)} + R\,r^{(\nu)})^\top A(u^{(\nu)} + R\,r^{(\nu)}) - (u^{(\nu)} + R\,r^{(\nu)})^\top f \\
&= J(u^{(\nu)}) + \frac{1}{2}(R\,r^{(\nu)})^\top A\,R\,r^{(\nu)} + (R\,r^{(\nu)})^\top (Au^{(\nu)} - f) \\
&= J(u^{(\nu)}) + \frac{1}{2}(R\,r^{(\nu)})^\top A\,R\,r^{(\nu)} - (R\,r^{(\nu)})^\top r^{(\nu)}.
\end{aligned}$$

We obtain the following

$$J(u^{(\nu+1)}) = J(u^{(\nu)}) - (R\,r^{(\nu)})^\top \left[\left(R^{-1} - \frac{1}{2}A\right) R\,r^{(\nu)}\right].$$

In the case of the Gauss–Seidel iteration we have

$$R^{-1} - \frac{1}{2}A = \frac{1}{2}D - \frac{1}{2}(L - U)$$

and therefore

$$(R\,r^{(\nu)})^\top \left[\left(R^{-1} - \frac{1}{2}A\right) R\,r^{(\nu)}\right] = \frac{1}{2}(R\,r^{(\nu)})^\top [D\,R\,r^{(\nu)}] \geq c\,\|\nabla J(u^{(\nu)})\|^2 > 0,$$

where we used the fact that $(Rr^{(\nu)})^\top[(L-U)Rr^{(\nu)}]=0$ because $L-U$ is antisymmetric and we set $c=\lambda_{min}(R^\top D R)/2$. Hence, we find that the Gauss–Seidel scheme is a minimizer in the sense that

$$J(u^{(\nu)})-J(u^{(\nu+1)})\geq c\,\|\nabla J(u^{(\nu)})\|^2.$$

Summing this inequality over ν results in the following

$$J(u^{(0)})-J(u^{(k+1)})\geq c\sum_{\nu=0}^{k}\|\nabla J(u^{(\nu)})\|^2.$$

Then taking the limit as ν goes to $+\infty$, we obtain

$$\lim_{\nu\to+\infty}\nabla J(u^{(\nu)})=0,$$

as $J(u)$ is bounded below.

Next, consider the case of the damped Jacobi iteration where

$$R^{-1}-\frac{1}{2}A=\frac{1}{\omega}\left(D-\frac{\omega}{2}A\right).$$

We have the following lemma [367].

Lemma 5.3. *Let A be real symmetric with $a_{ii}>0$, and let $\omega>0$. The matrix $2\omega^{-1}D-A$, where $D=$ diag A, is positive definite if and only if ω satisfies*

$$0<\omega\leq\frac{2}{1-\mu_{min}},$$

where $\mu_{min}\leq 0$ is the minimum eigenvalue of $I-D^{-1}A$.

Proof. Let $B=I-D^{-1}A$. The matrix $2\omega^{-1}D-A$ is positive definite if and only if

$$2\omega^{-1}I-D^{-1/2}AD^{-1/2}=(2\omega^{-1}-1)I+D^{1/2}B\,D^{-1/2}=H$$

is positive definite. The eigenvalues of H are $2\omega^{-1}-1+\mu_i$, where μ_i are the eigenvalues of B. Since $Tr\,B=0$ and the μ_i are real, it follows that $\mu_{min}\leq 0$. Therefore H is positive definite if $2\omega^{-1}-1+\mu_i>0$, that is, if $0<\omega\leq\frac{2}{1-\mu_{min}}\leq\omega\leq\frac{2}{1-\mu_i}$. $\square$

Therefore $(D-\frac{\omega}{2}A)\geq 0$ for $\omega\in(0,2/(1-\mu_{min}))$ and hence

$$J(u^{(\nu)})-J(u^{(\nu+1)})\geq c\,\|\nabla J(u^{(\nu)})\|^2,$$

where $c=\lambda_{min}(R^\top(2\omega^{-1}D-A)R)/2$. It follows that the damped Jacobi iteration provides a minimizing sequence such that $\lim_{\nu\to+\infty}\nabla J(u^{(\nu)})=0$.

In a classical multigrid context, the criteria for choosing an iteration scheme is its ability to smooth errors. In an optimization context, it is required that the iterative scheme be a minimizer. Thus many other well-known iterative methods can be chosen like, for example, the steepest descent (gradient) method given by

$$R=R^{(\nu)}=\left(\frac{r^{(\nu)\top}r^{(\nu)}}{r^{(\nu)\top}A\,r^{(\nu)}}\right)I=\alpha_\nu\,I.$$

Also notice that $r^{(\nu)} = -J'(u^{(\nu)})$. Therefore we can write $u^{(\nu+1)} = u^{(\nu)} + \alpha_\nu r^{(\nu)}$. It follows that $J(u^{(\nu+1)}) = J(u^{(\nu)}) - \frac{\alpha_\nu}{2}\|\nabla J(u^{(\nu)})\|^2$.

The iterative schemes discussed above can be interpreted as the process of minimizing the functional J by optimizing successively with respect to each unknown variable (Gauss–Seidel scheme) or in parallel by updating all unknown variables at the same time (Jacobi scheme, steepest descent). In this sense these methods belong to the class of successive or parallel subspace correction (SSC or PSC) methods [334] and coordinate descent methods [341].

Convergence rates for SSC and PSC iterations applied to a convex functional $J(u)$ are proved in [334] assuming that $J : V \to \mathbb{R}$ is Gâteaux differentiable and that there exist constants K, $L > 0$, $p \geq q > 1$, such that

$$\langle J'(u) - J'(v), w - v\rangle \geq K\,\|u - v\|_V^p, \tag{5.14}$$

$$\|J'(u) - J'(v)\|_{V'} \leq L\,\|u - v\|_V^{q-1}, \tag{5.15}$$

for all $u, v \in V$, and $\langle\cdot,\cdot\rangle$ is the duality pairing between V and its dual space V'.

5.2.3 The Twogrid Scheme and the Approximation Property

After the application of ν_1 smoothing sweeps to the problem $A_h u_h = f_h$, we obtain an approximation $\tilde{u}_h$ whose error $\tilde{e}_h = u_h - \tilde{u}_h$ is smooth. Then $\tilde{e}_h$ can be approximated on a coarser space. We need to interpret this smooth error as the solution of a coarse problem whose matrix A_H and right-hand side have to be defined. For this purpose notice that in our model problem A_h is the second-order difference operator, approximating the one-dimensional minus Laplacian, and the *residual* $r_h = f_h - A_h\tilde{u}_h$ is a smooth function if $\tilde{e}_h$ is smooth. Obviously, because of linearity the original equation $A_h u_h = f_h$ and the residual equation $A_h\tilde{e}_h = r_h$ are equivalent. The difference is that $\tilde{e}_h$ and r_h are smooth; therefore we can think of representing them on a coarser grid with mesh size $H = 2h$. We define r_H as the restriction of the fine-grid residual to the coarse grid, that is, $r_H = I_h^H r_h$, where I_h^H is a suitable restriction operator (e.g., straight injection). This defines the right-hand side of the coarse problem. Since $\tilde{e}_h$ is the solution of a difference operator which can be represented analogously on the coarse discretization level, we define the following coarse problem

$$A_H\tilde{e}_H = r_H. \tag{5.16}$$

Here A_H represents the same discrete operator but relative to the grid with coarse mesh size H. Reasonably, one expects $\tilde{e}_H$ to be an approximation to $\tilde{e}_h$ on the coarse grid. Because of its smoothness, then we can apply a *prolongation* operator I_H^h to transfer $\tilde{e}_H$ to the fine grid. Therefore, since by definition $u_h = \tilde{u}_h + \tilde{e}_h$, we update the function $\tilde{u}_h$ applying the following coarse-grid correction step

$$\tilde{u}_h^{new} = \tilde{u}_h + I_H^h\tilde{e}_H.$$

Notice that $\tilde{e}_h$ was a smooth function and the last step has amended $\tilde{u}_h$ by its smooth error. In practice, also the interpolation procedure may introduce HF errors on the fine grid. Therefore it is convenient to complete the twogrid (TG) process by applying ν_2 postsmoothing sweeps after the coarse-grid correction.

We summarize the TG procedure with the following algorithm. To emphasize that the iteration $u_h^{(l)} = M u_h^{(l-1)} + N f_h$ is a smoothing procedure, we denote it by $u_h^{(l)} = S_h(u_h^{(l-1)}, f_h)$. When no confusion may arise, we also use S to denote the iteration matrix (in place of) M. Let u_h^0 be the starting approximation. A TG scheme is as follows.

ALGORITHM 5.1. TG scheme.

- Input: initial approx. $u_h^{(0)}$, V-cycle index $n = 0$, maximum n_{max}, tolerance tol.

1. While ($n < n_{max}$ && $\|r_h\| > tol$) do
2. Presmoothing steps: $u_h^{(l)} = S(u_h^{(l-1)}, f_h), l = 1, \dots, \nu_1$;
3. Computation of the residual: $r_h = f_h - A_h u_h^{(\nu_1)}$;
4. Restriction of the residual: $r_H = I_h^H r_h$;
5. Solution of the coarse-grid problem $e_H = (A_H)^{-1} r_H$;
6. Coarse-grid correction: $u_h^{(\nu_1+1)} = u^{(\nu_1)} + I_H^h e_H$;
7. Postsmoothing steps: $u_h^{(l)} = S(u_h^{(l-1)}, f_h), l = \nu_1 + 2, \dots, \nu_1 + \nu_2 + 1$;
8. Set $n = n + 1$;
9. End while

A TG scheme starts at the fine level with presmoothing, performs a coarse-grid correction solving a coarse-grid auxiliary problem, and ends with postsmoothing. This procedure defines one cycle. A pictorial representation of this process where "fine" is a high level and "coarse" is a low level looks like a "V" workflow. This is called the V-cycle. To solve the problem to a given tolerance, we have to apply the TG V-cycle iteratively. In fact, in Algorithm 5.1 a while loop is implemented that applies the TG cycle at most n_{max} times or until a given tolerance on the residual is achieved. Later, we shall define only the cycle and omit the repeated call of the cycle. The TG iterative scheme can be written in the form (5.1) as stated by the following lemma.

Lemma 5.4. *The TG iteration matrix is given by*

$$M_{TG} = S_h^{\nu_2}(I_h - I_H^h (A_H)^{-1} I_h^H A_h) S_h^{\nu_1}, \tag{5.17}$$

where I_h is the identity and S_h is the smoothing iteration matrix.

Proof. For the proof, notice that the coarse-grid correction gives $e_h^{(\nu_1+1)} = e^{(\nu_1)} - I_H^h e_H$. □

For the model problem considered here, it is possible to estimate the spectral radius of M_{TG}. Consider the damped Jacobi smoother with $\omega = 1/2$, assume that I_H^h is the piecewise linear interpolation given by

$$I_H^h = \frac{1}{2}\begin{pmatrix} 1 & 0 & 0 \\ 2 & 0 & 0 \\ 1 & 1 & 0 \\ 0 & 2 & 0 \\ 0 & 1 & 1 \\ 0 & 0 & 2 \\ 0 & 0 & 1 \end{pmatrix},$$

and let I_h^H be restriction by weighting such that $r_H(x_j) = (r_h(x_{j-1}) + 2r_h(x_j) + r_h(x_{j+1}))/4$, $j = 2,4,\ldots,n-1$. In stencil form we have

$$I_H^h = \frac{1}{4}\begin{pmatrix} 1 & 2 & 1 & 0 & 0 & 0 & 0 \\ 0 & 0 & 1 & 2 & 1 & 0 & 0 \\ 0 & 0 & 0 & 0 & 1 & 2 & 1 \end{pmatrix}.$$

With this setting the following theorem is proved [173] using discrete Fourier analysis.

Theorem 5.5. *Let the TG scheme* 5.1 *with* $\nu = \nu_1 + \nu_2 \geq 1$. *The spectral radius of the iteration matrix* M_{TG} *given by* (5.17) *is bounded by*

$$r(M_{TG}) \leq \max\{\chi(1-\chi)^\nu + (1-\chi)\chi^\nu \;\; : \;\; 0 \leq \chi \leq 1/2\} =: r_\nu < 1$$

uniformly with respect to the mesh size h. Hence (5.17) *is a convergent iteration.*

In the framework of local Fourier analysis, a simple and effective way to predict the convergence factor of the TG scheme, for usually moderate values of ν, is to assume that the coarse-grid correction step solves "exactly" the LF error components, and there is no interaction between HF and LF components. This can be considered an "ideal" situation. Then the reduction of a suitable norm of the error (e.g., discrete L^2-norm) by one V-cycle of the TG method is determined by the reduction of the HF components on the fine grid. For this reason the convergence factor, denoted by ρ, can be roughly estimated by

$$\rho_I = \mu^{\nu_1+\nu_2}. \tag{5.18}$$

A sharper bound can be computed by TG Fourier analysis [85]. For this purpose we need to construct the Fourier symbol of the TG coarse-grid correction operator

$$CG_h^H = [I_h - I_H^h (A_H)^{-1} I_h^H A_h].$$

We denote the corresponding symbol by

$$\widehat{CG}_h^H(\theta) = [\hat{I}_h - \hat{I}_H^h(\theta)(\widehat{A}_H(2\theta))^{-1}\, \hat{I}_h^H(\theta)\, \widehat{A}_h(\theta)].$$

(Recall that $e^{i\theta x/h} = e^{i(2\theta)x/H}$.) The symbol of the coarse-grid operator is

$$\widehat{A}_H(2\theta) = -\frac{2\cos(2\theta) - 2}{H^2},$$

and similarly one constructs $\widehat{A}_h(\theta)$ corresponding to the two harmonics, that is,

$$\widehat{A}_h(\theta) = \begin{bmatrix} -\frac{2\cos(\theta)-2}{h^2} & 0 \\ 0 & \frac{2\cos(\theta)+2}{h^2} \end{bmatrix}.$$

The symbol of the restriction operator is

$$\hat{I}_h^H(\theta) = \left[(1+\cos(\theta))/2 \qquad (1-\cos(\theta))/2 \right],$$

whereas for the injection operator we have $\hat{I}_k^{k-1}(\theta) = 1$. For the linear prolongation operator we have $\hat{I}_{k-1}^k(\theta) = \hat{I}_k^{k-1}(\theta)^\top$.

The symbol of the TG method is given by

$$\widehat{TG}_h^H(\theta) = \hat{S}_h(\theta)^{\nu_2}\, \widehat{CG}_h^H(\theta)\, \hat{S}_h(\theta)^{\nu_1}.$$

This is a 2×2 matrix corresponding to the two frequency components. In this framework the convergence factor is defined as follows

$$\rho(TG_h^H) = \sup \left\{ r(\widehat{TG}_h^H(\theta)) : \theta \in [-\pi/2, \pi/2) \right\}.$$

In Table 5.1, we report estimates of the convergence factor ρ_I as given by (5.18) and the estimates r_{TG} resulting from the TG convergence analysis. These are compared with the estimates of ρ_ν by Theorem 5.5. The estimated ρ_I approximates well the bound ρ_ν provided that $\nu_1 + \nu_2$ is small. For large ν, ρ_I has an exponential decay behavior, whereas ρ_ν and ρ_{TG} have a slower decay as observed in numerical experiments.

Table 5.1. *Comparison of error reduction factors.*

ν	ρ_I	ρ_{TG}	ρ_ν
1	0.5	0.4	0.5
2	0.25	0.19	0.25
3	0.12	0.12	0.12
4	0.06	0.08	0.08

Notice that measuring ρ requires the knowledge of the exact solution. Because this is usually not available, ρ is measured as the asymptotic value of the reduction of a suitable norm (usually the discrete L^2-norm) of the residuals after consecutive TG cycles.

Another theoretical approach to multigrid convergence analysis, related to the smoothing property (5.11), introduces the approximation property to measure how good the coarse-grid solution approximates the fine-grid solution. This is expressed by the following estimate:

$$\|A_h^{-1} - I_H^h A_H^{-1} I_h^H\| \le c_A h^\beta. \tag{5.19}$$

This estimate actually measures the accuracy between $u_h = A_h^{-1} f_h$ and $I_H^h u_H$, where $u_H = A_H^{-1} I_h^H f_h$. Standard accuracy estimates for our model problem result in $\beta = 2$. This is due to the second-order accuracy of the 3-point Laplacian in one dimension and the fact that the error in interpolation and restriction is of second order.

Using the estimates of the smoothing and approximation properties, one can prove convergence of the TG scheme as follows. Consider, for simplicity, $\nu_2 = 0$; i.e., only pre-smoothing is applied. Then for our model problem, we have

$$\begin{aligned}\|M_{TG}\| &= \|(I_h - I_H^h (A_H)^{-1} I_h^H A_h) S_h^\nu\| \\ &= \|(A_h^{-1} - I_H^h (A_H)^{-1} I_h^H) A_h S_h^\nu\| \\ &\leq \|A_h^{-1} - I_H^h (A_H)^{-1} I_h^H\| \|A_h S_h^\nu\| \\ &\leq c_A\, \eta(\nu), \end{aligned} \tag{5.20}$$

where $c_A\, \eta(\nu) < 1$ for sufficiently large ν. Notice that the coarse-grid correction without pre- and postsmoothing is not a convergent iteration, in general. In fact, I_h^H maps from a fine- to a coarse-dimensional space and $I_H^h (A_H)^{-1} I_h^H$ is not invertible. We may have $r(I_h - I_H^h (A_H)^{-1} I_h^H A_h) \geq 1$.

Next, we discuss the optimization properties of the coarse-grid correction step. In fact, we show that $u_h^{(\nu+1)} = u^{(\nu)} + I_H^h e_H$ with $e_H = (A_H)^{-1} r_H$ and $r_H = I_h^H r_h$ provides an update in the descent direction in the sense that

$$\nabla J_h(u_h^{(\nu)})^\top (I_H^h e_H) < 0,$$

unless $e_H = 0$, occurring at convergence. This means that the TG scheme results in an optimization iteration by choosing a smoothing scheme with minimization properties and by performing a globalization step along the coarse-grid correction.

We assume that

$$I_h^H = c_I\, (I_H^h)^\top \text{ for a constant } c_I > 0.$$

This assumption holds, for example, for I_h^H being full-weighting restriction and I_H^h bilinear interpolation. In this case we have $c_I = (h/H)^d$, with d the space dimension. It follows that

$$\begin{aligned}\nabla J_h(u_h^{(\nu)})^\top (I_H^h e_H) &= -{r_h}^\top (I_H^h e_H) = -\frac{1}{c_I} (I_h^H r_h)^\top e_H \\ &= -\frac{1}{c_I} r_H^\top e_H = -\frac{1}{c_I} (A e_H)^\top e_H < 0.\end{aligned}$$

In general, updating along a descent direction with a unitary steplength does not guarantee a reduction of the value of the objective. For this purpose a line search or an a priori choice of steplength α is required (globalization) such that

$$J(u^{(\nu)} + \alpha\, I_H^h e_H) < J(u^{(\nu)}).$$

This issue is discussed in detail in Section 5.4. In the case of our strictly convex quadratic minimization problem, we obtain

$$\alpha^* = \underset{\alpha}{\text{argmin}}\ J(u + \alpha\, I_H^h e_H) = 1.$$

That is, a steplength of one along the coarse-grid correction direction is optimal.

5.2.4 The Multigrid Scheme

In the TG scheme the size of the coarse problem may still be too large to be solved exactly. Therefore it is convenient to use recursively the TG method to solve the coarse problem (5.16) by introducing a further coarse-grid problem. This process can be repeated until a coarsest grid is reached where the corresponding residual equation is inexpensive to solve, e.g., by direct methods. This is a rough qualitative description of the multigrid method.

For a more detailed description, let us introduce a sequence of grids with mesh size $h_1 > h_2 > \cdots > h_L > 0$. Here $k = 1,2,\ldots,L$ is called the level number. With Ω_{h_k} we denote the set of grid points with grid spacing h_k. The number of interior grid points will be n_k. With V_k we denote the space of functions defined on Ω_{h_k}, which are vectors of $\mathbb{R}^{n_k}$. On each level k we define the problem $A_k u_k = f_k$. Here A_k is an $n_k \times n_k$ symmetric positive definite matrix, and u_k and f_k are vectors of size n_k. The transfer among levels is performed by two linear mappings: the restriction I_k^{k-1} and the prolongation I_{k-1}^k operators. With $u_k^{(l)} = S_k(u_k^{(l-1)}, f_k)$ we denote a smoothing iteration.

For variables defined on V_k we introduce the inner product $(\cdot,\cdot)_k$ with associated norm $\|u\|_k = (u,u)_k^{1/2}$. We denote with $|u|_k = (A_k u, u)^{1/2}$ the norm induced by A_k.

The following defines one cycle of the multigrid algorithm.

ALGORITHM 5.2. Multigrid scheme.

- Input: starting approx. $u_k^{(0)}$.

  1. If $k = 1$ solve $A_k u_k = f_k$ exactly.
  2. Presmoothing steps: $u_k^{(l)} = S(u_k^{(l-1)}, f_k)$, $l = 1,\ldots,\nu_1$;
  3. Computation of the residual: $r_k = f_k - A_k u_k^{(\nu_1)}$;
  4. Restriction of the residual: $r_{k-1} = I_k^{k-1} r_k$;
  5. Set $u_{k-1} = 0$;
  6. Call γ times the multigrid scheme to solve $A_{k-1} u_{k-1} = r_{k-1}$;
  7. Coarse-grid correction: $u_k^{(\nu_1+1)} = u_k^{(\nu_1)} + I_{k-1}^k u_{k-1}$;
  8. Postsmoothing steps: $u_k^{(l)} = S(u_k^{(l-1)}, f_k)$, $l = \nu_1+2,\ldots,\nu_1+\nu_2+1$;
  9. End.

The multigrid algorithm involves a new parameter, called the cycle index, γ, which is the number of times the multigrid procedure is applied to the coarse-level problem. Since this procedure converges fast, $\gamma = 1$ or $\gamma = 2$ is the typical value used. For $\gamma = 1$ we have a V-cycle, whereas $\gamma = 2$ is called the W-cycle. It turns out that with a reasonable γ, the coarse problem is solved almost exactly. Therefore in this case the convergence factor of a multigrid cycle equals that of the corresponding TG method, i.e., approximately $\rho = \mu^{\nu_1+\nu_2}$. Actually in many problems $\gamma = 2$ or even $\gamma = 1$ is sufficient to retain the TG convergence. A picture of the multigrid workflow is given in Figure 5.2.

Also the multigrid scheme can be expressed in the form (5.1) as stated by the following lemma.

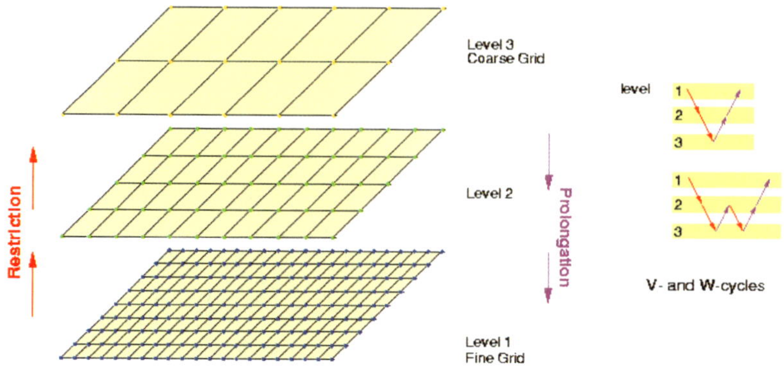

Figure 5.2. *Multigrid setting.*

Lemma 5.6. *The multigrid iteration matrix is given in recursive form by the following. For $k=1$, let $M_1=0$. For $k=2,\ldots,L$,*

$$M_k = S_k^{\nu_2}(I_k - I_{k-1}^k(I_{k-1} - M_{k-1}^{\gamma})A_{k-1}^{-1}I_k^{k-1}A_k)S_k^{\nu_1}, \tag{5.21}$$

where I_k is the identity, S_k is the smoothing iteration matrix, and M_k is the multigrid iteration matrix for the level k.

Proof. To derive (5.21) consider an initial error $e_k^{(0)}$. The action of ν_1 presmoothing steps gives $e_k = S_k^{\nu_1} e_k^{(0)}$ and the corresponding residual $r_k = A_k e_k$. On the coarse grid, this error is given by $e_{k-1} = A_{k-1}^{-1} I_k^{k-1} r_k$. However, in the multigrid algorithm we do not invert A_{k-1} (unless on the coarsest grid) and we apply γ multigrid cycles instead. That is, we denote with $v_{k-1}^{(\gamma)}$ the approximation to e_{k-1} obtained after γ application of M_{k-1}, and we have for the error (of the error) $\eta_{k-1}^{(\gamma)} = M_{k-1}^{\gamma}\eta_{k-1}^{(0)}$. That is,

$$e_{k-1} - v_{k-1}^{(\gamma)} = M_{k-1}^{\gamma}(e_{k-1} - v_{k-1}^{(0)}).$$

Following the multigrid Algorithm 5.2, we set $v_{k-1}^{(0)} = 0$ (step 5). Therefore we have $e_{k-1} - v_{k-1}^{(\gamma)} = M_{k-1}^{(\gamma)} e_{k-1}$, which can be rewritten as $v_{k-1}^{(\gamma)} = (I_{k-1} - M_{k-1}^{\gamma})e_{k-1}$. It follows that

$$\begin{aligned} v_{k-1}^{(\gamma)} &= (I_{k-1} - M_{k-1}^{\gamma})e_{k-1} = (I_{k-1} - M_{k-1}^{\gamma})\,A_{k-1}^{-1} I_k^{k-1}\, r_k \\ &= (I_{k-1} - M_{k-1}^{\gamma})\,A_{k-1}^{-1} I_k^{k-1}\, A_k e_k. \end{aligned}$$

Correspondingly, the coarse-grid correction is $u_k^{(\nu_1+1)} = u_k^{(\nu_1)} + I_{k-1}^k v_{k-1}^{(\gamma)}$. In terms of error functions this means that

$$e_k^{(\nu_1+1)} = e_k - I_{k-1}^k v_{k-1}^{(\gamma)}.$$

Substituting the expression for $v_{k-1}^{(\gamma)}$ given above, we obtain

$$e_k^{(\nu_1+1)} = [I_k - I_{k-1}^k(I_{k-1} - M_{k-1}^{\gamma})\,A_{k-1}^{-1} I_k^{k-1}\, A_k]e_k^{(\nu_1)}.$$

Finally, consideration of the pre- and postsmoothing sweeps proves the lemma. $\square$

We now illustrate a multigrid convergence theory due to Bramble, Pasciak, and Xu [82]. For this purpose, we consider a Poisson model problem and rewrite the multigrid iteration matrix above in the form of a classical iteration as

$$M_k = I_k - B_k A_k,$$

where I_k denotes the identity on V_k. Let $R_k : V_k \to V_k$ be an iteration operator such that $S_k = I_k - R_k A_k$ for $k > 1$. Consider

$$\begin{aligned} -\Delta u &= f \text{ in } \Omega, \\ u &= 0 \text{ on } \partial\Omega. \end{aligned} \tag{5.22}$$

The matrix form of this problem is

$$A_k u_k = f_k \qquad \text{in } V_k. \tag{5.23}$$

We introduce the following operators. We interpret $I_k^{k-1} : V_k \to V_{k-1}$ as the L_k^2 projection defined by

$$(I_k^{k-1} u, v)_{k-1} = (u, I_{k-1}^k v)_k$$

for all $u \in V_k$ and $v \in V_{k-1}$. Similarly, let $P_{k-1} : V_k \to V_{k-1}$ be the A_k projection defined by

$$(A_{k-1} P_{k-1} u, v)_{k-1} = (A_k u, I_{k-1}^k v)_k$$

for all $u \in V_k$ and $v \in V_{k-1}$.

The V-cycle multigrid algorithm to solve (5.23) in recursive form is given as follows.

ALGORITHM 5.3. Multigrid scheme: recursive form.

- Set $B_1 = A_1^{-1}$. For $k \geq 2$ define $B_k : V_k \to V_k$ in terms of B_{k-1} as follows. Let $f_k \in V_k$.

  1. Define $u^{(l)}$ for $l = 1, \ldots, \nu_1$ by
  $$u^{(l)} = u^{(l-1)} + R_k(f_k - A_k u^{(l-1)}).$$
  2. Set $u^{(\nu_1+1)} = u^{(\nu_1)} + I_{k-1}^k q$, where
  $$q = B_{k-1} I_k^{k-1}(f_k - A_k u^{(\nu_1)}).$$
  3. Set $B_k f_k = u^{(\nu_1+\nu_2+1)}$, where $u^{(\ell)}$ for $\ell = \nu_1 + 2, \ldots, \nu_1 + \nu_2 + 1$ is given by step 2 (with $R_k^\top$ instead of R_k for a symmetric multigrid scheme).

To simplify the analysis of this scheme, we choose $\nu_1 = 1$ and $\nu_2 = 0$, and take $u^{(0)} = 0$. From the definition of P_{k-1}, we see that

$$I_k^{k-1} A_k = A_{k-1} P_{k-1}. \tag{5.24}$$

Let $S_k u = S_k(u - u^{(0)}) = u - u^{(1)}$. Now for $u \in V_k$, $k = 2, \ldots, L$, we have

$$\begin{aligned}(I_k - B_k A_k)u &= u - u^{(1)} - I_{k-1}^k q \\ &= S_k u - I_{k-1}^k B_{k-1} A_{k-1} P_{k-1} S_k u \\ &= [I_k - I_{k-1}^k B_{k-1} A_{k-1} P_{k-1}] S_k u \\ &= [(I_k - I_{k-1}^k P_{k-1}) + I_{k-1}^k (I_{k-1} - B_{k-1} A_{k-1}) P_{k-1}] S_k u.\end{aligned} \tag{5.25}$$

It is immediate to see that this recurrence relation, including postsmoothing, can be written as

$$M_k = S_k [(I_k - I_{k-1}^k P_{k-1}) + I_{k-1}^k M_{k-1} P_{k-1}] S_k.$$

Clearly, it is equivalent to (5.21) with $\gamma = 1$. Starting from this recurrence relation, in [82] the following BPX multigrid convergence theorem is proved.

Theorem 5.7. *Let R_k satisfy* (5.26) *and* (5.27) *for $k > 1$. Then there exist positive constants $\delta_k < 1$ such that*

$$(A_k M_k u, u)_k \le \delta_k (A_k u, u)_k \quad \forall u \in V_k.$$

In the following, we sketch the proof of this theorem. The two conditions required in Theorem 5.7 are given below. The first condition concerns the smoothing operator R_k as follows. For simplicity, let R_k be symmetric and positive definite and S_k be nonnegative. We need $A_k S_k = S_k A_k$. There exist constants $C_R > 0$ and $c > 0$ independent of u and k such that

$$C_R \frac{\|u\|_k^2}{\lambda_k} \le (R u, u)_k \le c (A_k^{-1} u, u)_k \quad \forall u \in V_k, \tag{5.26}$$

where λ_k denotes the largest eigenvalue of A_k. In general, notice that if the spectrum $\sigma(S_k) = \sigma(I_k - R_k A_k) \in (-1, 1)$, then there exist positive constants a_0 and a_1 smaller than one such that

$$-a_0 (A_k u, u)_k \le (A_k (I_k - R_k A_k) u, u)_k \le a_1 (A_k u, u)_k.$$

This is the same as $(1 - a_1)(A_k^{-1} u, u)_k \le (R u, u)_k \le (1 + a_0)(A_k^{-1} u, u)_k$; see [75]. Notice that (5.26) corresponds to the smoothing property (5.11).

The second assumption is a regularity and approximation assumption. There exist $0 < \alpha \le 1$ and a constant $C_\alpha > 0$ independent of k such that

$$(A_k (I_k - I_{k-1}^k P_{k-1}) u, u)_k \le C_\alpha \left(\frac{\|A_k u\|_k^2}{\lambda_k} \right)^\alpha (A_k u, u)_k^{1-\alpha} \quad \forall u \in V_k. \tag{5.27}$$

The case $\alpha = 1$ corresponds to full elliptic regularity, i.e., $\|u\|_{H^2} \le c \|f\|_{L^2}$. Notice that (5.27) corresponds to the approximation property (5.19). In fact, we have

$$I_k - I_{k-1}^k P_{k-1} = \left(A_k^{-1} - I_{k-1}^k A_{k-1}^{-1} I_k^{k-1} \right) A_k,$$

and if $A_k u_k = f_k$, then $A_{k-1}(P_{k-1} u_k) = I_k^{k-1} f_k$.

The proof of Theorem 5.7 is now by induction. For $k = 1$ we have $M_1 = I_k - B_1 A_1 = I_k - A_1^{-1} A_1 = 0$ and the claim of the theorem is true. Now assume it is true for $k - 1$. We have

$$\begin{aligned}
(A_k M_k u,u)_k &= (A_k S_k (I_k - I_{k-1}^k P_{k-1}) S_k u,u)_k + (A_k S_k I_{k-1}^k M_{k-1} P_{k-1} S_k u,u)_k \\
&= (A_k (I_k - I_{k-1}^k P_{k-1}) z,z)_k + (A_k I_{k-1}^k M_{k-1} P_{k-1} z,z)_k \\
&= (A_k (I_k - I_{k-1}^k P_{k-1}) z,z)_k + (M_{k-1} P_{k-1} z, I_k^{k-1} A_k z)_{k-1} \\
&= (A_k (I_k - I_{k-1}^k P_{k-1}) z,z)_k + (M_{k-1} P_{k-1} z, A_{k-1} P_{k-1} z)_{k-1} \\
&= (A_k (I_k - I_{k-1}^k P_{k-1}) z,z)_k + (M_{k-1} v, A_{k-1} v)_{k-1} \\
&\le (A_k (I_k - I_{k-1}^k P_{k-1}) z,z)_k + \delta_{k-1} (A_{k-1} v, v)_{k-1} \\
&= (A_k (I_k - I_{k-1}^k P_{k-1}) z,z)_k + \delta_{k-1} (A_{k-1} P_{k-1} z, P_{k-1} z)_{k-1} \\
&= (A_k (I_k - I_{k-1}^k P_{k-1}) z,z)_k + \delta_{k-1} (A_k z, I_{k-1}^k P_{k-1} z)_k \\
&= (1-\delta_{k-1})(A_k (I_k - I_{k-1}^k P_{k-1}) z,z)_k + \delta_{k-1} (A_k z,z)_k,
\end{aligned}$$

where we let $z = S_k u$ (the case with ν pre- and postsmoothing sweeps requires $z = S_k^\nu u$) and $v = P_{k-1} z$. To complete the proof of the theorem, one considers the resulting inequality

$$(A_k M_k u,u)_k \le (1-\delta_{k-1})(A_k (I_k - I_{k-1}^k P_{k-1}) z,z)_k + \delta_{k-1} (A_k z,z)_k. \tag{5.28}$$

Using (5.27) with $\alpha = 1$ we have

$$(A_k (I_k - I_{k-1}^k P_{k-1}) z,z)_k \le C_1 \frac{\|A_k z\|_k^2}{\lambda_k}.$$

Next, we find a δ independent of k such that Theorem 5.7 holds for all k. We need the following lemma.

Lemma 5.8. *The following estimate holds*

$$(A_k (I_k - I_{k-1}^k P_{k-1}) S_k^\nu u, S_k^\nu u)_k \le \frac{C_1}{2\nu C_R} \left(|u|_k^2 - |S_k^\nu u|_k^2 \right).$$

Proof.

$$\begin{aligned}
(A_k (I_k - I_{k-1}^k P_{k-1}) S_k^\nu u, S_k^\nu u)_k &\le C_1 \lambda_k^{-1} \|A_k S_k^\nu u\|_k^2 \\
&= (C_1/C_R)(R_k A_k S_k^\nu u, A_k S_k^\nu u)_k \\
&= (C_1/C_R)((I - S_k) S_k^{2\nu} u, A_k u)_k \\
&\le \frac{C_1}{2\nu C_R} \left(|u|_k^2 - |S_k^\nu u|_k^2 \right).
\end{aligned}$$

The last inequality follows from

$$((I - S_k) S_k^{2\nu} u, A_k u)_k \le \frac{1}{2\nu} \sum_{j=0}^{2\nu-1} ((I - S_k) S_k^j u, A_k u)_k = \frac{1}{2\nu} \left(|u|_k^2 - |S_k^\nu u|_k^2 \right)$$

resulting from

$$(1-x)x^{2m} \leq \frac{1}{2m}(1-x)\sum_{j=0}^{2m-1} x^j = \frac{1}{2m}(1-x^{2m}) \text{ for } 0 \leq x \leq 1. \quad \square$$

With this lemma and (5.28) (with $\delta_{k-1} = \delta$) we obtain

$$(A_k M_k u, u)_k \leq (1-\delta)\frac{C_1}{2\nu C_R}\left(|u|_k^2 - |S_k^\nu u|_k^2\right) + \delta\,|S_k^\nu u|_k^2.$$

Now, choosing $\delta_k = \delta = C_1/(C_1 + 2\nu C_R)$ for all k, we have

$$(A_k M_k u, u)_k \leq \delta\,(A_k u, u)_k,$$

where $0 < \delta < 1$ for $\nu \geq 1$ and Theorem 5.7 is proved with δ independent of k.

Notice that since M_k is symmetric with respect to $(A_k \cdot, \cdot)_k$, it follows that $(A_k M_k^2 u, u)_k \leq \delta^2 (A_k u, u)_k$. This fact and the additional condition

$$(A_k I_{k-1}^k u, I_{k-1}^k u)_k \leq 2(A_{k-1} u, u)_{k-1} \quad \forall u \in V_k,$$

which characterizes the case of nested spaces, allow us to extend the theorem above to the case of W-cycles ($\gamma = 2$), and the same estimate of δ results [75].

5.2.5 The Algebraic Multigrid Method

Since the pioneering works in [86, 291, 324] there has been extensive research for the development of multigrid methods which resemble the geometric multigrid process and utilize only information contained in the algebraic equations resulting from the discretization of the differential problem at hand. The totality of these schemes defines the class of algebraic multigrid (AMG) methods; see [325] for an exhaustive review of AMG methods available.

A step towards the theoretical development of AMG involves the definition of smooth and rough components of the error in an algebraic rather than a geometrical sense. Given a classical iterative method, a smooth error is an error which is slowly reduced by such a method. If S is the smoothing scheme, then the smooth error e is characterized by $|Se|_k \approx |e|_k$. For the Gauss–Seidel method this condition roughly means that the error is such that the residual $|r_i| \ll |a_{ii}\, e_i|$, that is,

$$\left| a_{ii} e_i + \sum_{j \in \tilde{N}_i} a_{ij} e_j \right| \ll |a_{ii} e_i|, \tag{5.29}$$

where $A = (a_{ij})$ and $\tilde{N}_i = \{j \neq i \,:\, a_{ij} \neq 0\}$.

The next step in the development of AMG is the definition of a prolongation operator I_{k-1}^k, which has the smooth error vectors in its range. Notice that based on (5.29) it is possible to approximate the smooth error component e_i as a linear combination of its neighboring error components e_j, that is,

$$e_i \approx -\sum_{j \in \tilde{N}_i} \frac{a_{ij}}{a_{ii}} e_j. \tag{5.30}$$

Therefore, we can think of partitioning the whole set of unknown variables (also called points) in two subsets, C and F. The subset C has been selected as the subset of linearly independent components for the prolongation operator. It represents a coarse level of variables. The remaining variables belonging to the complement F of C are assumed to be expressed by means of an approximate version of (5.30). We have the following interpolation formula for the error function at the coarse level $k-1$ to the fine level k

$$(I_{k-1}^k e)_i = \begin{cases} \sum_{j \in P_i} w_{ij}\, e_j, & i \in F, \\ e_i, & i \in C, \end{cases} \tag{5.31}$$

where $P_i \subset \tilde{N}_i \cap C$ and $w_{ij} = -\alpha_i a_{ij}/a_{ii}$. The w_{ij}'s measure the coupling between variables. This formula is known as direct interpolation [325]. The leading coefficient α_i is introduced to take into account that, generally, $P_i \neq \tilde{N}_i$. It is selected by assuming that, for the limit case of zero row sum matrices, constants are interpolated exactly.

In order to have effective interpolation formulas, the set P_i should contain those indices j for which the absolute value of w_{ij} is larger, or at least some of them. In other words, the variables in P_i should have a certain amount of coupling for variable i. To this end the notion of strong coupling between variables is introduced. Assume, for simplicity, that $a_{ij} < 0$ whenever $i \neq j$ and that $a_{ii} > 0$. A point i is said to have strongly negatively coupling (or be n-coupled) to another point j if

$$-a_{ij} > \varepsilon_{str} \max_{a_{ik}<0} |a_{ik}| \quad \text{with } 0 < \varepsilon_{str} < 1. \tag{5.32}$$

Usually $\varepsilon_{str} = 0.2$, but this value is not critical. The set of points to which i is strongly n-coupled is

$$S_i = \left\{ j \in \tilde{N}_i \, : \, i \text{ is strongly n-coupled to } j \right\}. \tag{5.33}$$

Clearly the variables in S_i should be in P_i to have an effective interpolation formula. Consequently a point that has a large amount of points strongly n-coupled to it should be put in C. This amount is the number of elements in the set $S_i^T = \left\{ j \, : \, i \in S_j \right\}$. Note that $S_i^T \neq S_i$ even for symmetric matrices, since the relation of strong n-coupling is not symmetric. Once the point i has been put in C, the points in S_i^T that are not already in C are put in F. One could implement the following procedure to make the C/F splitting of points: Let U be the set of still undecided points. Then the next C point is the undecided variable for which the quantity $\lambda_i = \left| S_i^T \cap F \right|$ is higher (given a set X, $|X|$ denotes the number of elements it contains). Note that λ_i cannot be greater than the number of nonzero off-diagonal elements of the ith column of A.

This algorithm may have two drawbacks. First, if two or more unconnected points have a high value for $\left| S_i^T \right|$, then each of these points will be the root of unrelated C/F splittings whose merging can give rise to a nonuniform distribution of C and F variables, as shown in Figure 5.3. Second, it requires a computational burden which can be too high. In fact, a vector $\Lambda = (\lambda_1 \ldots \lambda_N)$ must be initialized and ordered. Then, each time a point becomes an F point, Λ needs to be updated and reordered. The ordering is needed in order to avoid a sequential search for the highest λ_i.

Therefore it is preferable to implement a different algorithm redefining the measure of importance as $\lambda_i = \left| S_i^T \cap F \right|$. With this modification the problem of unrelated splittings no longer arises, and the vector of nonzero λ_i is of a manageable size. The size $\left| S_i^T \cap U \right|$ is taken into account only when all λ_i are zero (e.g., at the start of the splitting). But instead

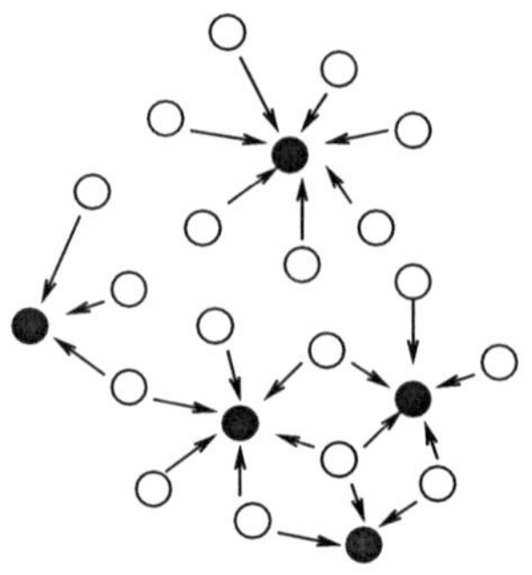

Figure 5.3. *Filled circles denote C points, empty circles denote F points, and the arrows denote strong n-coupling. Note that there is a double layer of F variables around the C point on the right. This figure first appeared in A. Borzì and G. Borzì, An algebraic multigrid method for a class of elliptic differential systems, SIAM J. Sci. Comput.,* 25(1) (2003), 302–323.

of searching for the maximal size $|S_i^T|$ among all variables, the search for the maximum is limited to a small set of randomly selected points. This does not cause much harm since $|S_i^T|$ is bounded by the number of nonzero elements in a column. These choices have been introduced in the algorithm to retain low complexity. It may happen that some points remain undecided, since they have strong n-connections only with F points. These undecided points are promoted to F points and interpolated from the neighboring F points. So, the interpolation process becomes a multipass interpolation (in this case a two-pass) [325].

Now, assume the sets C and F have been determined. The prolongation operator I_{k-1}^k is given by (5.31) with $P_i = S_i \cap C$. The restriction operator is given by the transpose of the prolongation operator, multiplied by the normalization coefficient as $I_k^{k-1} = \frac{N_{k-1}}{N_k}\left(I_{k-1}^k\right)^T$, where N_k is the number of variables at level k. The normalization coefficient is such that a restriction operator maps a constant to an almost constant vector with the same mean value.

The coarse matrix of coefficient A_{k-1} is defined by the Galerkin formula

$$A_{k-1} = I_k^{k-1} A_k I_{k-1}^k.$$

The coarsening strategy defined above is called *standard coarsening*. When standard coarsening is applied to matrices with small stencils, it may result in large memory complexity, since the reduction of the number of variables is not very high and the coarse matrix has bigger stencils. In order to maintain low memory complexity or to further reduce it, *aggressive coarsening* is introduced. This corresponds to a generalization of the notion of connectivity between points: the point i has a (p,l) long-range strong n-connection with the point j if there are at least p paths of length l, each given by a sequence of points $i_0, i_1, \ldots, i_l$ with $i_0 = i$ and $i_l = j$ such that $i_{q+1} \in S_{i_q}$ for $q = 0, 1, \ldots, l-1$ (see [325]). Aggressive coarsening is made by substituting the set $S_i^{p,l} = \{j \in \Omega : i \text{ strongly n-connected to } j \text{ w.r.t. } (p,l)\}$ to S_i in the algorithm previously described.

An alternative formulation of AMG schemes is based on smoothed aggregation [350]. In this approach, the prolongator is defined by a disaggregation followed by a smoothing.

Thus the coarsening process is determined by the selection of aggregates (disjoint sets), as opposed to the selection of C-points in classical AMG. The prolongator smoother usually has the form $\tilde{S}_k = I_k - \tilde{M}_k^{-1} A_k$.

5.3 Multigrid Methods for Nonlinear Problems

Two multigrid approaches to the solution of nonlinear problems are usually considered. The first is based on a generalization of the multigrid scheme described above, which is called the full approximation storage (FAS) scheme [83]. The second approach is based on the Newton method and uses the multigrid scheme as inner solver of the linearized equations defining the Jacobian in the Newton step.

An interesting comparison between the FAS scheme and the Newton-multigrid scheme is presented in [94]. First, it is shown that in terms of computing time, the exact Newton approach is not a viable method. Further, it is demonstrated that the inexact-Newton-multigrid scheme may provide efficiency similar to that of the FAS scheme. However, it remains an open issue how many inner multigrid cycles are possibly needed in the Newton-multigrid method to match the FAS efficiency. In this sense the FAS scheme is more robust, and we discuss this method in the following. We remark that the FAS scheme is a special instance of the class of nonlinear multigrid methods discussed in [173].

5.3.1 The FAS Multigrid Method

To illustrate the FAS method, consider the discrete nonlinear problem

$$A_k(u_k) = f_k, \tag{5.34}$$

where $A_k(\cdot)$ represents a nonlinear discrete operator on Ω_{h_k}.

The starting point for the FAS scheme is again to define a suitable smoothing process denoted by $u = S(u, f)$. This can be a Picard iteration or a local Newton–Gauss–Seidel iteration; see, e.g., [57]. Suppose now one applies this iterative method a few times to (5.34), obtaining some approximate solution $\tilde{u}_k$. The desired exact correction e_k is defined by $A_k(\tilde{u}_k + e_k) = f_k$. Here the coarse residual equation makes no sense because of the nonlinearity. Nevertheless the "correction" equation can instead be written in the form

$$A_k(\tilde{u}_k + e_k) - A_k(\tilde{u}_k) = r_k, \tag{5.35}$$

where $r_k = f_k - A_k(\tilde{u}_k)$. Now assume to represent $\tilde{u}_k + e_k$ on the coarse grid in terms of the coarse-grid variable

$$u_{k-1} := \hat{I}_k^{k-1} \tilde{u}_k + e_{k-1}. \tag{5.36}$$

Since $\hat{I}_k^{k-1} \tilde{u}_k$ and $\tilde{u}_k$ represent the same function but on different grids, the standard choice of the fine-to-coarse linear operator $\hat{I}_k^{k-1}$ is straight injection. The formulation of (5.35) on the coarse level is obtained by replacing $A_k(\cdot)$ by $A_{k-1}(\cdot)$, $\tilde{u}_k$ by $\hat{I}_k^{k-1}\tilde{u}_k$, and r_k by $I_k^{k-1} r_k = I_k^{k-1}(f_k - A_k(\tilde{u}_k))$; thus we get the FAS equation

$$A_{k-1}(u_{k-1}) = I_k^{k-1}(f_k - A_k(\tilde{u}_k)) + A_{k-1}(\hat{I}_k^{k-1} \tilde{u}_k). \tag{5.37}$$

This equation can also be written in the form

$$A_{k-1}(u_{k-1}) = I_k^{k-1} f_k + \tau_k^{k-1}, \tag{5.38}$$

where

$$\tau_k^{k-1} = A_{k-1}(\hat{I}_k^{k-1}\tilde{u}_k) - I_k^{k-1} A_k(\tilde{u}_k).$$

Observe that (5.38) without the τ_k^{k-1} term is the original equation represented on the coarse grid. At convergence we have $u_{k-1} = \hat{I}_k^{k-1} u_k$, because $f_k - A_k(u_k) = 0$ and $A_{k-1}(u_{k-1}) = A_{k-1}(\hat{I}_k^{k-1} u_k)$. The term τ_k^{k-1} is the fine-to-coarse *defect or residual correction*, that is, the correction to (5.38) such that its solution coincides with the fine-grid solution. This fact allows us to reverse the point of view of the multigrid approach [84]. Instead of regarding the coarse level as a device for accelerating convergence on the fine grid, one can view the fine grid as a device for calculating the correction τ_k^{k-1} to the coarse FAS equation which is computationally less expensive.

The direct use of u_{k-1} on fine grids, that is, the direct interpolation of this function by $I_{k-1}^k u_{k-1}$, cannot be used, since it introduces the interpolation errors of the full (possibly oscillatory) solution, instead of the interpolation errors of only the correction e_{k-1}, which is assumed smooth. For this reason the following coarse-grid correction is used

$$u_k = \tilde{u}_k + I_{k-1}^k (u_{k-1} - \hat{I}_k^{k-1} \tilde{u}_k). \tag{5.39}$$

A complete γ-cycle of the FAS scheme is summarized below.

ALGORITHM 5.4. FAS scheme.

- Input: starting approx. $u_k^{(0)}$.

  1. If $k = 1$ solve $A_k(u_k) = f_k$ exactly.
  2. Presmoothing steps: $u_k^{(l)} = S(u_k^{(l-1)}, f_k)$, $l = 1, \dots, \nu_1$;
  3. Computation of the residual: $r_k = f_k - A_k(u_k^{(\nu_1)})$;
  4. Restriction of the residual: $r_{k-1} = I_k^{k-1} r_k$;
  5. Set $u_{k-1} = \hat{I}_k^{k-1} u_k^{(\nu_1)}$;
  6. Set $f_{k-1} = r_{k-1} + A_{k-1}(u_{k-1})$
  7. Call γ times the FAS scheme to solve $A_{k-1}(u_{k-1}) = f_{k-1}$;
  8. Coarse-grid correction: $u_k^{(\nu_1+1)} = u_k^{(\nu_1)} + I_{k-1}^k (u_{k-1} - \hat{I}_k^{k-1} u_k^{(\nu_1)})$;
  9. Postsmoothing steps: $u_k^{(l)} = S(u_k^{(l-1)}, f_k)$, $l = \nu_1 + 2, \dots, \nu_1 + \nu_2 + 1$;
  10. End.

The action of one FAS scheme can be also expressed in terms of a (nonlinear) multigrid iteration operator B_k. Starting with an initial approximation $u_k^{(0)}$ the result of one FAS cycle is then denoted by $u_k = B_k(u_k^{(0)})\, f_k$.

ALGORITHM 5.5. FAS V-cycle scheme: recursive form.

- Set $B_1(u_1^{(0)}) \approx A_1^{-1}$ (e.g., iterating with S_1 starting with $u_1^{(0)}$). For $k = 2, \ldots, L$ define B_k in terms of B_{k-1} as follows:

  1. Set the starting approximation $u_k^{(0)}$.
  2. Presmoothing. Define $u_k^{(l)}$ for $l = 1, \ldots, \nu_1$ by
  $$u_k^{(l)} = S_k(u_k^{(l-1)}, f_k).$$
  3. Coarse-grid correction. Set
  $$u_k^{(\nu_1+1)} = u_k^{(\nu_1)} + I_{k-1}^k(u_{k-1} - \hat{I}_k^{k-1} u_k^{(\nu_1)}),$$
  where u_{k-1} is given by
  $$u_{k-1} = B_{k-1}(u_{k-1}^{(0)}) \left[I_k^{k-1}(f_k - A_k(u_k^{(\nu_1)})) + A_{k-1}(u_{k-1}^{(0)}) \right],$$
  and $u_{k-1}^{(0)} = \hat{I}_k^{k-1} u_k^{(\nu_1)}$.
  4. Postsmoothing. Define $u_k^{(l)}$ for $l = \nu_1 + 2, \ldots, \nu_1 + \nu_2 + 1$ by
  $$u_k^{(l)} = S_k(u_k^{(l-1)}, f_k).$$
  5. Set $B_k(u_k^{(0)}) f_k = u_k^{(\nu_1+\nu_2+1)}$.

Next, we discuss a multigrid approach to optimization which closely relates to the FAS scheme and provides a framework to investigate convergence of multigrid schemes for nonlinear problems.

5.3.2 The Full Multigrid Method

When dealing with nonlinear problems it may be essential to start the iterative procedure from a good initial approximation. The multigrid setting suggests a natural way to define this approximation. Suppose we start the solution process from a coarse working level $\ell = K < M$. At this level we solve the problem by multigrid iteration and then we interpolate the solution to the next working level to serve as initial approximation for the multigrid iterative process at this level; that is, we have

$$u_{\ell+1} = \tilde{I}_\ell^{\ell+1} u_\ell. \tag{5.40}$$

The approach of using a coarse-grid approximation as a first guess for the solution process on a finer grid is known as nested or cascadic iteration. The algorithm obtained by combining the multigrid scheme with nested iteration is called the full multigrid (FMG) method; see Figure 5.4. The interpolation operator (5.40) used in the FMG scheme is called the FMG interpolator. Because of the improvement on the initial solution at each starting level, the FMG scheme results in being more efficient than the iterative application of the multigrid cycle without FMG initialization. For Poisson model problems the com-

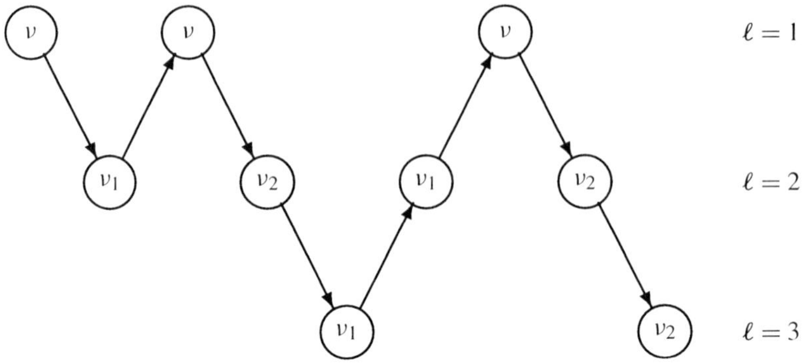

Figure 5.4. *The FMG scheme.*

putational complexity of the plain multigrid scheme is of order $\mathcal{O}(n \log n)$ while the FMG implementation provides $\mathcal{O}(n)$.

ALGORITHM 5.6. N-FMG scheme.

- Input: set $\ell = K$and the initial approximation $u_\ell^{(0)}$.

  1. Compute u_ℓ applying N-cycles of the multigrid scheme with $u_\ell^{(0)}$ as initial guess.
  2. If $k = L$ then stop.
  3. Else if $\ell < L$ then by interpolation set $u_{\ell+1}^{(0)} = \tilde{I}_\ell^{\ell+1} u_\ell$.
  4. Set $\ell = \ell + 1$, goto step 2.
  5. End

Within the N-FMG algorithm an estimate of the degree of accuracy of solutions can be obtained by comparing solutions at different levels. The norm of the solution error on level ℓ can be defined as

$$E_\ell = \|u_\ell - \hat{I}_{\ell+1}^\ell u_{\ell+1}\|, \tag{5.41}$$

where $\hat{I}_{\ell+1}^\ell$ is, e.g., the injection operator.

To discuss the efficiency of the FMG approach, consider the following three-dimensional Poisson problem with Dirichlet boundary conditions

$$\begin{cases} -(\frac{\partial^2 u}{\partial x^2} + \frac{\partial^2 u}{\partial y^2} + \frac{\partial^2 u}{\partial z^2}) = 3\sin(x+y+z) & \text{in} \quad \Omega = (0,2)^3, \\ u(x,y,z) = \sin(x+y+z) & \text{for} \quad (x,y,z) \in \partial\Omega. \end{cases} \tag{5.42}$$

On each level ℓ, we choose a uniform mesh with grid size $h_\ell = \frac{2}{n_\ell + 1}$, $n_\ell = 2^\ell - 1$, and finite difference discretization gives

$$-\left(\frac{u_{i+1jk}^\ell - 2u_{ijk}^\ell + u_{i-1jk}^\ell}{h_\ell^2} + \frac{u_{ij+1k}^\ell - 2u_{ijk}^\ell + u_{ij-1k}^\ell}{h_\ell^2} + \frac{u_{ijk+1}^\ell - 2u_{ijk}^\ell + u_{ijk-1}^\ell}{h_\ell^2} \right)$$

$$= 3\sin(ih_\ell, jh_\ell, kh_\ell), \quad i,j,k = 1,\ldots,n_\ell. \tag{5.43}$$

Thus we use a 7-point stencil approximation which is $O(h^2)$ accurate. To solve this problem we employ the FAS scheme with $\gamma = 1$ and $L = 7$ with Gauss–Seidel smoothing, $\nu_1 = 2$ and $\nu_2 = 1$. The smoothing factor estimated by local mode analysis is $\mu = 0.567$. Hence, by this analysis, the expected reduction factor is $\rho \approx \mu^{\nu_1+\nu_2} = 0.18$. The observed convergence factor is ≈ 0.20.

Here $\hat{I}_k^{k-1}$ is simple injection and I_k^{k-1} and I_{k-1}^k are the full weighting and (tri-)linear interpolation, respectively. Finally, the FMG operator $\tilde{I}_\ell^{\ell+1}$ is cubic interpolation [84, 173]. Now let us discuss the optimality of the FMG algorithm resulting with these components.

To show that the FMG scheme is able to solve the given discrete problem at a minimal cost, we compute E_ℓ (maximum norm) and show that it behaves like h_ℓ^2, demonstrating convergence. This means that the ratio $E_\ell/E_{\ell+1}$ at convergence should be a factor of $h_\ell^2/h_{\ell+1}^2 = 4$. Results are reported in Table 5.2. We find similar results with $N = 10$, $N = 3$, and $N = 1$ FMG schemes. Therefore the choice $N = 1$ in a FMG cycle is suitable to solve the problem to second-order accuracy.

Table 5.2. *The estimated solution error for various N-FMG cycles.*

Level	10-FMG	3-FMG	1-FMG
3	$6.75\,10^{-4}$	$6.79\ \ 10^{-4}$	9.44×10^{-4}
4	$1.73\,10^{-4}$	$1.75\ \ 10^{-4}$	2.34×10^{-4}
5	$4.36\,10^{-5}$	$4.40\ \ 10^{-5}$	5.92×10^{-5}
6	$1.09\,10^{-5}$	$1.10\ \ 10^{-5}$	1.48×10^{-5}

To estimate the amount of work invested in the FMG method, let us define the work unit (WU) [83], i.e., the computational work of one smoothing sweep at the finest level. On the level $\ell \le M$ the work involved is $(\frac{1}{2})^{3(M-\ell)}WU$, where the factor $\frac{1}{2}$ is given by the mesh size ratio $h_{\ell+1}/h_\ell$ and the exponent 3 is the number of spatial dimensions. Thus a multigrid cycle that uses $\nu = \nu_1 + \nu_2$ relaxation sweeps on each level requires

$$W_{cycle} = \nu \sum_{k=1}^{L} \left(\frac{1}{2}\right)^{3(L-k)} WU < \frac{8}{7}\nu WU,$$

ignoring transfer operations. Hence the computational work employed in an N-FMG method is roughly

$$W_{FMG} = N \sum_{\ell=K}^{L} \left(\frac{1}{2}\right)^{3(L-\ell)} W_{cycle},$$

ignoring the FMG interpolation and work on the coarsest grid. This means that, using the 1-FMG method, we solve the discrete three-dimensional Poisson problem to second-order accuracy with a number of computer operations which is proportional to the number of unknowns on the finest grid. In the present case we have an FMG work of approximately $4WU$.

5.4 The Multigrid Optimization Scheme

As optimization problems become more large-sized, the need of fast optimization schemes increases. In fact, most of these problems have complicated structure that make it difficult to implement sophisticated solution strategies, thus making the use of simple gradient-based optimization schemes the only viable choice. However, in the last few years, a multigrid optimization (MGOPT) strategy has been proposed [232, 260] that can be used to accelerate classical optimization schemes. The MGOPT multigrid approach to optimization problems resembles the FAS scheme [83] and the nonlinear multigrid methods discussed in [173]. The MGOPT scheme is designed for unconstrained optimization and constructs a hierarchy of objective functionals to be minimized. However, it is formally equivalent to the FAS scheme with damping of the coarse-grid correction step [178] applied to the first-order optimality condition equations. Further contributions to this topic can be found in, e.g., [45, 152, 267].

Some successful applications to bilinear optimal control problems [344, 345], to image reconstruction problems [267], and to hyperbolic problems [232] suggest that the MGOPT framework could provide the strategy of choice to construct fast optimization schemes. However, the successful application of MGOPT schemes requires the appropriate choice of the MGOPT components that are formally similar to the classical multigrid components. In particular, a hierarchy of optimization problems on a hierarchy of optimization spaces ought to be defined. Then interspace transfer operators between coarser and finer spaces are needed. A fast optimization scheme is obtained when the optimization step obtained with the coarsening-refining process is able to complement the classical optimization scheme.

In [45, 232, 260, 345], it is emphasized that under appropriate assumptions, the multigrid coarse-grid correction provides a descent direction and, therefore, combining this fact with a linesearch procedure and a minimizing classical iteration, a globally convergent algorithm is obtained. Numerical experiments, e.g., [260, 345], demonstrate that MGOPT greatly improves the efficiency of the underlying optimization scheme used as a "smoother," suggesting that the MGOPT scheme may be beneficial in combination with well-known optimization algorithms. This claim appears to be true as long as a line search along the coarse-level correction is performed and the problem has an underlying geometrical structure for which the construction of interspace transfer operators becomes evident.

Consider the following (locally) convex optimization problem

$$\min_{u_k} J_k(u_k), \tag{5.44}$$

where $k = 1, 2, \ldots, L$ is the resolution or discretization parameter, L denotes the finest resolution, and u_k is the (unconstrained) optimization variable in the space V_k. For variables defined on V_k, we introduce the inner product $(\cdot,\cdot)_k$ with associated norm $\|x\|_k = (x,x)_k^{1/2}$. Among spaces V_k, restriction operators $I_k^{k-1} : V_k \to V_{k-1}$ and prolongation operators $I_{k-1}^k : V_{k-1} \to V_k$ are defined. We require that $(I_k^{k-1}u, v)_{k-1} = (u, I_{k-1}^k v)_k$ for all $u \in V_k$ and $v \in V_{k-1}$, that is, $I_k^{k-1} = c_I\,(I_{k-1}^k)^\top$ for a constant $c_I > 0$. Notice that for optimization problems with PDEs, the definition of the hierarchy of spaces V_k and of the intergrid transfer operators follows the guidelines of geometrical/algebraical multigrid techniques. However, in principle the MGOPT framework is not restricted to PDE-based optimization problems, and therefore it could be applied to minimization problems without a geometric context. In this case, given J_L and V_L at the finest resolution it is an open issue of how to choose the hierarchy of J_k and V_k. We remark that the MGOPT scheme is an iterative gradient-based optimization method. Therefore it must be formulated in the same space where the gradient is defined. In the following, we assume an L^2 formulation of the gradient.

We denote with O_k an optimization algorithm (for example, the truncated Newton scheme used in [260]) and require that given an initial approximation $u_k^{(0)}$ to the solution of (5.44), the application of O_k results in sufficient reduction as follows

$$J_k(O_k(u_k^{(0)})) \leq J_k(u_k^{(0)}) - \eta \|\nabla J_k(u_k^{(0)})\|^2 \qquad \text{for some } \eta \in (0,1).$$

To define one cycle of the MGOPT method, it is convenient to consider the minimization problem $\min_{u_k}(J_k(u_k) - (f_k, u_k)_k)$, where $f_L = 0$. Let $u_k^{(0)}$ be the starting approximation at resolution k.

ALGORITHM 5.7. MGOPT scheme.

- Input: starting approx. $u_k^{(0)}$

  1. If $k = 1$ solve $\min_{u_k}(J_k(u_k) - (f_k, u_k)_k)$ exactly, i.e., solve $\nabla J_k(u_k) = f_k$.
  2. Preoptimization. Define $u_k^{(1)} = O_k(u_k^{(0)})$.
  3. Set up and solve a coarse-grid minimization problem. Define $u_{k-1}^{(1)} = I_k^{k-1} u_k^{(1)}$. Compute the fine-to-coarse gradient correction

     $$\tau_{k-1} = \nabla J_{k-1}(u_{k-1}^{(1)}) - I_k^{k-1} \nabla J_k(u_k^{(1)}), \qquad f_{k-1} = I_k^{k-1} f_k + \tau_{k-1}.$$

     The coarse-grid minimization problem is given by

     $$\min_{u_{k-1}} (J_{k-1}(u_{k-1}) - (f_{k-1}, u_{k-1})_{k-1}). \tag{5.45}$$

     Solve (5.45) with MGOPT to obtain u_{k-1}.
  4. Line search and coarse-grid correction. Perform a line search in the $I_{k-1}^k(u_{k-1} - I_k^{k-1} u_k^{(1)})$ direction to obtain α_k that minimizes J_k. The coarse-grid correction is given by

     $$u_k^{(2)} = u_k^{(1)} + \alpha_k I_{k-1}^k (u_{k-1} - I_k^{k-1} u_k^{(1)}).$$

  5. Postoptimization. Define $u_k^{(3)} = O_k(u_k^{(2)})$.
  6. End.

Roughly speaking, the essential guideline for constructing J_k on coarse levels is that it must sufficiently well approximate the convexity properties of the functional at finest resolution. This property and the following remark give an insight into the fact that the coarse-grid correction provides a descending direction; recall the discussion at the end of Section 5.2.3 and see also Lemma 5.13 below.

Remark 5.9. We have that

$$\nabla(J_{k-1}(u_{k-1})-(f_{k-1},u_{k-1})_{k-1})|_{u^{(1)}_{k-1}} = I_k^{k-1}\left(\nabla J_k(u_k^{(1)})-f_k\right).$$

That is, the gradient of the coarse-grid functional at the coarse approximation $u^{(1)}_{k-1} = I_k^{k-1}u_k^{(1)}$ equals the restriction of the gradient of the fine-grid functional at corresponding fine approximation $u_k^{(1)}$.

As remarked in [261], if the coarse optimization problem approximates the corresponding fine problem, the reduction obtained in the coarse problem should approximate the reduction in the fine problem obtained after the line search. In classical optimization terms, the coarse reduction represents the predicted reduction while the actual one is that obtained in the fine grid. We have the following estimates for the predicted and actual reductions of the value of the objective [261]. Denote $\hat{J}_k(u_k) = J_k(u_k)-(f_k,u_k)_k$ and $e_{k-1} = u_{k-1}-u^{(1)}_{k-1}$. The predicted reduction, up to $O(\|e_{k-1}\|^3)$ terms, is given as follows

$$\begin{aligned} R_p &= \hat{J}_{k-1}(u^{(1)}_{k-1})-\hat{J}_{k-1}(u_{k-1}) \\ &= -(e_{k-1},I_k^{k-1}\nabla J_k(u_k^{(1)}))_{k-1}-\frac{1}{2}(e_{k-1},\nabla^2\hat{J}_{k-1}(u^{(1)}_{k-1})e_{k-1})_{k-1}. \end{aligned}$$

The actual reduction, up to $O(\|e_{k-1}\|^3)$ terms, is given by the following

$$\begin{aligned} R_a &= \hat{J}_k(u_k^{(1)})-\hat{J}_k(u_k^{(2)}) \\ &= -(e_{k-1},I_k^{k-1}\nabla J_k(u_k^{(1)}))_{k-1}-\frac{1}{2}(e_{k-1},I_k^{k-1}\nabla^2\hat{J}_k(u_k^{(1)})I_{k-1}^k e_{k-1})_{k-1}. \end{aligned}$$

Therefore, we can state the following.

Remark 5.10. In the MGOPT process, the difference between the fine-grid (actual) reduction of the objective value and the corresponding coarse (predicted) reduction is estimated with

$$R_a-R_p = \frac{1}{2}\left(e_{k-1},\left[\nabla^2\hat{J}_{k-1}(u^{(1)}_{k-1})-I_k^{k-1}\nabla^2\hat{J}_k(u_k^{(1)})I_{k-1}^k\right]e_{k-1}\right)_{k-1}+O(\|e_{k-1}\|^3).$$

That is, this difference provides a measure of the discrepancy between the coarse Hessian and the projected fine Hessian.

5.4.1 Convergence of the MGOPT Method

Assume that for each k, J_k is twice Fréchet differentiable and $\nabla^2 J_k$ is strictly positive definite and satisfies the condition $(\nabla^2 J_k(u)v,v)_k \geq \beta\|v\|_k^2$ together with $\|\nabla^2 J_k(u)-\nabla^2 J_k(v)\|_k$

$\leq \lambda \|u-v\|_k$ uniformly for some positive constants β and λ. We use the expansion

$$J_k(u+z) = J_k(u) + (\nabla J_k(u), z)_k + \frac{1}{2}\int_0^1 (\nabla^2 J_k(u+tz)z, z)_k \, dt. \tag{5.46}$$

The main tool for our discussion is the following lemma [178].

Lemma 5.11. *For $u, v \in V_k$ assume $(\nabla J_k(u), v)_k \leq 0$ and let γ be such that*

$$0 < \gamma \leq -2\delta(\nabla J_k(u), v)_k \left[\int_0^1 (\nabla^2 J_k(u+t\gamma v)v, v)_k dt\right]^{-1}$$

for some $\delta \in (0,1]$. Then

$$-(1-\delta)\gamma(\nabla J_k(u), v)_k \leq J_k(u) - J_k(u+\gamma v) \leq -\gamma(\nabla J_k(u), v)_k. \tag{5.47}$$

Proof. Set $z = \gamma v$ in (5.46). The first inequality follows from the restriction to γ. The second inequality follows from the positivity of $\nabla^2 J_k$. □

The next lemma provides an explicit estimate for the steplength α_k for the coarse-grid gradient correction in step 4 of Algorithm 5.7.

Lemma 5.12. *For $u, v \in V_k$ assume $(\nabla J_k(u), v)_k \leq 0$ and let*

$$\alpha(u,v) = \min\left\{2, \frac{-(\nabla J_k(u), v)_k}{(\nabla^2 J_k(u)v, v)_k + \lambda \|v\|_k^3}\right\}. \tag{5.48}$$

Then

$$0 \leq -\frac{1}{2}\alpha(u,v)(\nabla J_k(u), v)_k \leq J_k(u) - J_k(u+\alpha(u,v)v). \tag{5.49}$$

Proof. For the proof it is enough to verify that Lemma 5.11 may be applied with $\gamma = \alpha(u,v)$ and $\delta = 1/2$. Notice that

$$\int_0^1 (\nabla^2 J_k(u+t\alpha\, v)v, v)_k \, dt \leq (\nabla^2 J_k(u)v, v)_k + \lambda \|v\|_k^3.$$

Therefore we have

$$\alpha(u,v) \leq \frac{-(\nabla J_k(u), v)_k}{(\nabla^2 J_k(u)v, v)_k + \lambda \|v\|_k^3} \leq \frac{-(\nabla J_k(u), v)_k}{\int_0^1 (\nabla^2 J_k(u+t\alpha\, v)v, v)_k dt}.$$

Hence α satisfies the condition of Lemma 5.11 with $\delta = 1/2$. □

The following lemma states that the MGOPT coarse-grid correction with steplength $0 < \alpha \leq 2$ given by Lemma 5.12 is a minimizing step. Notice that the above lemmas are formulated for a functional $J_k(u_k)$ and its gradient $\nabla J_k(u_k)$. They hold true considering $J_k(u_k) - (f_k, u_k)_k$ and $\nabla J_k(u_k) - f_k$.

Lemma 5.13. *Take $u_k^{(1)} \in V_k$ and define $u_{k-1}^{(1)} = I_k^{k-1} u_k^{(1)} \in V_{k-1}$. Denote $\hat{J}_{k-1}(u_{k-1}) = J_{k-1}(u_{k-1}) - (f_{k-1}, u_{k-1})_{k-1}$, where $f_{k-1} = I_k^{k-1} f_k + \tau_{k-1}$ and $\tau_{k-1} = \nabla J_{k-1}(u_{k-1}^{(1)}) -$*

$I_k^{k-1}\nabla J_k(u_k^{(1)})$. *Let* $u_{k-1} \in V_{k-1}$ *be such that* $\hat{J}_{k-1}(u_{k-1}) \leq \hat{J}_{k-1}(u_{k-1}^{(1)})$ *and define* $q = I_{k-1}^k(u_{k-1} - u_{k-1}^{(1)})$. *Then*

$$\hat{J}_k(u_k^{(1)} + \alpha(u_k^{(1)}, q)q) - \hat{J}_k(u_k^{(1)}) \leq \frac{1}{2}\alpha(u_k^{(1)}, q)(\nabla \hat{J}_k(u_k^{(1)}), q)_k, \tag{5.50}$$

where $\alpha(u_k^{(1)}, q)$ *is defined in Lemma* 5.12 *(strict inequality holds if* $\hat{J}_{k-1}(u_{k-1}) < \hat{J}_{k-1}(u_{k-1}^{(1)})$*).*

Proof. The proof follows from Lemma 5.12 after showing that $(\nabla \hat{J}_k(u_k^{(1)}), q)_k \leq 0$. From (5.46) we obtain

$$(\nabla \hat{J}_{k-1}(u_{k-1}^{(1)}), u_{k-1} - u_{k-1}^{(1)})_k \leq \hat{J}_{k-1}(u_{k-1}) - \hat{J}_{k-1}(u_{k-1}^{(1)}) \leq 0.$$

Now we have

$$\begin{aligned}(\nabla \hat{J}_k(u_k^{(1)}), q)_k &= (\nabla \hat{J}_k(u_k^{(1)}), I_{k-1}^k(u_{k-1} - u_{k-1}^{(1)}))_k \\ &= (I_k^{k-1}(\nabla \hat{J}_k(u_k^{(1)})), u_{k-1} - u_{k-1}^{(1)})_{k-1} \\ &= (\nabla \hat{J}_{k-1}(u_{k-1}^{(1)}), u_{k-1} - u_{k-1}^{(1)})_{k-1} \leq 0.\end{aligned} \tag{5.51}$$

For the last equality recall Remark 5.9 and the discussion at the end of Section 5.2.3. ☐

Notice that in Lemma 5.13 it is not required to solve exactly the coarse minimization problem: find $u \in V_{k-1}$ such that $\hat{J}_{k-1}(u) = \min_{u_{k-1}} \hat{J}_{k-1}(u_{k-1})$. This is (formally) required only on the coarsest grid. The following theorem states convergence of the MGOPT method.

Theorem 5.14. *The MGOPT method described above provides a minimizing iteration, and if* J *is strictly convex, then (the index* L *of the finest level is omitted)*

$$\lim_{i\to\infty} \|u^{(i)} - u\| = 0,$$

where $J(u) = \min_v J(v)$ *and* (i) *is the MGOPT cycle index.*

Proof. The proof of the first part is by induction. For $k = 2$ we have u, where $\hat{J}_1(u) = \min_{u_1} \hat{J}_1(u_1)$, and from Lemma 5.13 it follows that

$$\begin{aligned}\hat{J}_2(u_2^3) = \hat{J}_2(O_2(u_2^2)) \leq \hat{J}_2(u_2^2) &= \hat{J}_2(u_2^1 + \alpha\, I_{k-1}^k(u - I_k^{k-1}u_2^1)) \\ &\leq \hat{J}_2(u_2^1) = \hat{J}_2(O_2(u_2^0)) \leq \hat{J}_2(u_2^0).\end{aligned} \tag{5.52}$$

If $\hat{J}_2(u_2^0) > \min_{u_2} \hat{J}_2(u_2)$, then (5.52) holds with strict inequality.

For $k > 2$, due to the induction hypothesis and because of Lemma 5.13 the theorem holds.

The sequence $\{u_L^{(i)}\}_{i\geq 1}$ is in the compact set $A = \{v \in V_L : J_L(v) \leq J_L(u_L^0)\}$ and $\{J_L(u_L^{(i)})\}_{i\geq 1}$ is a nonincreasing sequence in the compact set $V = \{J_L(v) : v \in A\}$, so this sequence converges. We can write $\lim_{i\to\infty}(J_L(u_L^{(i)}) - J_L(u_L)) = 0$. Strict convexity and (5.46) give that $\lim_{i\to\infty} \|u_L^{(i)} - u_L\|_L = 0$. ☐

Linear multigrid schemes including AMG methods [51, 291, 325] can be interpreted as MGOPT schemes for the quadratic functional

$$J_k(u) = \frac{1}{2}(u, A_k u)_k - (u, b_k)_k, \quad u \in V_k,$$

where $V_k = \mathbb{R}^{n_k}$ and A_k is an $n_k \times n_k$ symmetric positive definite matrix. Consider $n_{k-1} < n_k$ and take I_{k-1}^k to be full rank. Then, with the Galerkin formula $A_{k-1} = I_k^{k-1} A_k I_{k-1}^k$ and $b_{k-1} = I_k^{k-1} b_k$, one obtains a suitable coarse functional $J_{k-1}(u) = \frac{1}{2}(u, A_{k-1}u)_{k-1} - (u, b_{k-1})_{k-1}$. A TG analysis of the MGOPT scheme applied to this problem reveals that we have convergence for $\alpha = 1$. In fact, considering (5.48) with $v = I_H^h e_H$, we have

$$\alpha = -\frac{(\nabla J_h, I_H^h e_H)_h}{(A_h I_H^h e_H, I_H^h e_H)_h} = \frac{(I_h^H r_h, e_H)_H}{(A_h I_H^h e_H, I_H^h e_H)_h} = \frac{(A_H e_H, e_H)_H}{(I_h^H A_h I_H^h e_H, e_H)_H} = 1.$$

(Notice that in the linear case $\lambda = 0$.)

The MGOPT convergence theory given above applies also to analysis of the FAS scheme. For this purpose we assume the existence of a functional J_k such that $\nabla J_k(u_k) = A_k(u_k) - f_k$. Then, subject to the conditions given above, mesh-independent convergence of the FAS scheme is proved if one proves that α given by (5.48) is always greater than or equal to one.

5.4.2 The Construction of the MGOPT Components

Experience suggests that for optimization problems with an underlying geometrical and/or differential structure, strategies similar to that of geometrical multigrid methods applied to PDE problems can be followed for the construction of a hierarchy of objectives. Thus the objective J_k at level k is the approximation of the functional J at the k level. Also AMG methods provide guidelines for the implementation of a hierarchy of functionals

$$J_k(x) = \frac{1}{2}(x, A_k x)_k - (x, b_k)_k, \quad x \in V_k,$$

where $V_k = \mathbb{R}^{n_k}$ with Euclidean inner product $(\cdot,\cdot)_k$ normalized by n_k, and A_k is an $n_k \times n_k$ symmetric positive definite matrix. Consider $n_{k-1} < n_k$ and take $I_{k-1}^k : V_{k-1} \to V_k$ to be full rank. Let $I_k^{k-1} = \left(\frac{n_{k-1}}{n_k}\right)\left(I_{k-1}^k\right)^T$. Then, with the choice $A_{k-1} = I_k^{k-1} A_k I_{k-1}^k$ and $b_{k-1} = I_k^{k-1} b_k$, one obtains a suitable coarse functional

$$J_{k-1}(y) = \frac{1}{2}(y, A_{k-1} y)_{k-1} - (y, b_{k-1})_{k-1}, \quad y \in V_{k-1}.$$

Now, we discuss the construction of the MGOPT components in the general case where only the function $J : \mathbb{R}^n \to R$ is available. We focus on the case where J is twice Fréchet differentiable and $\nabla^2 J$ is strictly positive definite. We denote $n_k = n$ and $V_k = \mathbb{R}^{n_k}$.

Consider the following quadratic model of the objective. We have

$$J_k(x) = J_k(x^0) + (\nabla J_k(x^0), x - x^0) + \frac{1}{2}(x - x^0, \nabla^2 J_k(x^0)(x - x^0)), \tag{5.53}$$

where $x^0 \in V_k$ is sufficiently close to x^*, the J-minimizer.

Associated to the Hessian, we have the following eigenvalue problem

$$\nabla^2 J_k(x^0)\, v_j = \mu_j v_j, \qquad j = 1,\dots,n,$$

where the eigenvalues μ_j result in being real and positive and the eigenvectors can be taken orthonormal. We assume that the eigenvalues have increasing value such that $\mu_i < \mu_j$ if $i < j$. Therefore, the set of vectors $(v_j)_{j=1,n_k}$ provides a basis for V_k, and we have the following

$$x^* - x^0 = c_1 v_1 + \cdots + c_n v_n.$$

Next, we assume that the iterative optimization scheme denoted by O is more effective in minimizing along some of the eigenvectors and less effective with respect to the remaining ones. Specifically, we assume that the optimization scheme is fast in solving the components in the subspace span$\{v_{m+1},\dots,v_n\}$, $1 < m < n_k$. Hence, after the optimization step, $x^1 = O(x^0)$, we have

$$x^* - x^1 \approx c_1 v_1 + \cdots + c_m v_m. \tag{5.54}$$

Notice that (5.54) can also be written as follows

$$x^* - x^1 \approx [v_1, \dots, v_m] \begin{pmatrix} c_1 \\ \vdots \\ c_m \end{pmatrix}.$$

This result suggests that the vectors v_j, $j = 1,\dots,m$, provide the columns of a matrix which can play the role of the prolongation operator. Therefore, we write

$$I_{k-1}^k = [v_1, \dots, v_m]. \tag{5.55}$$

By doing this, we identify the coarse space as the space of the coefficients $c = (c_1,\dots,c_m)^T$. (This corresponds to the space of the error in the coarse grids of a linear multigrid method.) We set $n_{k-1} = m$ and $V_{k-1} = \mathbb{R}^{n_{k-1}}$.

In this space, an appropriate coarse function is given by the following quadratic model

$$J_{k-1}(c) = (I_k^{k-1} \nabla J_k(x^1), c) + \frac{1}{2}(c, I_k^{k-1} \nabla^2 J_k(x^1)\, I_{k-1}^k c). \tag{5.56}$$

This model has a minimum at $c^* \in V_{k-1}$ given by the solution of the following

$$\left(I_k^{k-1} \nabla^2 J_k(x^1)\, I_{k-1}^k \right) c = -I_k^{k-1} \nabla J_k(x^1),$$

which corresponds to a Newton step defined on the coarse space with a Hessian obtained by a Galerkin projection. Therefore, the solution to the coarse minimization problem defined by (5.56) provides a coarse-grid correction $I_{k-1}^k c^*$ that updates x^1 towards the minimizer x^*, that is,

$$x^* \approx x^2 = x^1 + I_{k-1}^k c^*.$$

Notice that no gradient correction has been introduced. In fact, in the coarse model (5.56) the variable c represents corrections for the minimizer, that is, differences between a given approximation and the minimizer sought. This corresponds to the classical linear multigrid approach to PDE problems. In a nonlinear approach, the coarse model is a function of a coarse representation of x, which we denote by x^c, in the sense that

$$x^* - x^1 \approx I_{k-1}^k (x^c - I_k^{k-1} x^1).$$

This results in the coarse quadratic model

$$F(x^c) = J_{k-1}(x^c) - (I_k^{k-1} \nabla^2 J(x^1) I_{k-1}^k I_k^{k-1} x^1, x^c), \tag{5.57}$$

obtained neglecting terms depending only on x^1. Further, from (5.56) we have

$$\left(I_k^{k-1} \nabla^2 J_k(x^1) I_{k-1}^k \right) x^c = \nabla J_{k-1}(x^c) - I_k^{k-1} \nabla J_k(x^1),$$

which we use in (5.57) to obtain the following

$$F(x^c) = J_{k-1}(x^c) - (\nabla J_{k-1}(I_k^{k-1} x^1) - I_k^{k-1} \nabla J_k(x^1), x^c). \tag{5.58}$$

Thus we obtain the classical MGOPT coarse-space minimization problem; compare with (5.45). We remark that the coarse quadratic model is defined based on x^1 while the classical optimization step was analyzed based on the eigenvectors of the Hessian defined on x^0. This inconsistency becomes relevant for highly nonlinear optimization problems for which the quadratic model starts losing its validity. This occurrence could be controlled in part with the help of robust globalization procedures.

However, the open challenge is how to construct an approximation to the set of eigenvectors defining the prolongation operator without having to solve the eigenvalue problem for the Hessian. Indeed, we can easily construct a set of orthonormal vectors spanning the space where the optimization procedure effectively acts, e.g., applying the Gram–Schmidt procedure to the set of approximation increments. But what we need is the subspace orthogonal to this set.

Another, less explored possibility to construct appropriate transfer operators is to pursue the approach of AMG smoothed aggregation starting from simple disaggregation operators and using the classical optimization scheme for smoothing.

5.5 Multigrid and Reduced SQP for Parameter Optimization

Optimization problems of the form

$$\min J(y,u), \tag{5.59}$$

$$c(y,u) = 0 \tag{5.60}$$

are called parameter optimization problems if the space U for the influence variables is finite-dimensional, i.e., $u \in U = \mathbb{R}^{n_u}$, and there is no further multigrid structure conceived within U. However, the constraint (5.60) is thought of as a PDE. This situation happens frequently in simulation-based optimization problems such as, e.g., parameter identification problems or shape optimization with an a priori defined shape parametrization. Typically, the number of parameters is low—often below 100. In this framework, the method of choice is a reduced SQP approach as demonstrated in [310, 309, 118]. Some details have to be discussed if the PDE in (5.60) is solved by a multigrid method. A straightforward implementation of a *reduced SQP method* within the separability framework applied to problem (5.59)–(5.60) iterates over the following steps.

ALGORITHM 5.8. Reduced SQP method.

- Initialize $\ell = 0, y_0, u_0$.
  1. Solve with multigrid the adjoint problem $c_y^*(y_\ell, u_\ell)p_\ell = -J_y(y_\ell, u_\ell)$ and build the reduced gradient $\nabla_u J_\ell = J_u^* + c_u^*(y_\ell, u_\ell)p_\ell$
  2. Build some approximation $B_\ell \approx H(y_\ell, u_\ell, p_\ell)$ of the reduced Hessian, e.g., by quasi-Newton update formula
  3. Solve
$$\Delta u = \arg \min_{u \in \mathcal{L}U(u_\ell)} \frac{1}{2}(u, B_\ell u) + (\nabla_u J_\ell, u),$$
     where $\mathcal{L}U(u_\ell)$ denotes the linearization of U in u_ℓ
  4. Solve with multigrid the linear problem
$$c_y(y_\ell, u_\ell)\Delta y = -(c_u(y_\ell, u_\ell)\Delta u + c(y_\ell, u_\ell))$$
  5. Update $y_{\ell+1} = y_\ell + \tau\,\Delta y$ and $u_{\ell+1} = u_\ell + \tau\,\Delta u$, where τ is some linesearch updating factor in the early iterations.

The linesearch factor τ in step 5 can be determined by the use of classical merit functions as discussed in [265]. Alternatively, efficient trust region approaches are discussed in [187]. In steps 1 and 4, linear systems have to be solved in our context by application of multigrid methods. Since these systems are adjoint to each other, the natural question arises of whether this fact should be reflected in the respective multigrid solvers, as well. This fact is particularly important if the linear systems are solved up to some tolerance. Classical reduced SQP convergence theory can guarantee convergence only if we know that the reduced gradient γ can be interpreted as a derivative, i.e., we need the consistency condition
$$\nabla_u J_\ell^* = \frac{\partial}{\partial u} J(y_\ell - Ac_u(y_\ell, u_\ell)u, u_\ell + u), \tag{5.61}$$
where A is an approximation to $c_y(y_\ell, u_\ell)^{-1}$ defined by the multigrid algorithm for the linearized forward problem. If we do not satisfy this consistency condition, the necessary condition $\gamma_\ell \approx 0$ for optimality cannot be used as an indication for the optimal solution and therefore as a stopping rule for the iterations if we want to use only a comparatively coarse accuracy in the forward solver. Then, this inconsistent reduced gradient might even give a direction which is not a descent direction.

In [309] a proof is given for the fact that the condition (5.61) leads to the following requirements for the construction of grid transfer operators and the smoothing operators
$${}_{\mathcal{A}}\mathcal{I}_k^{k-1} = \left({}_{\mathcal{F}}\mathcal{I}_{k-1}^k\right)^*, \quad {}_{\mathcal{A}}\mathcal{S} = ({}_{\mathcal{F}}\mathcal{S})^*, \quad {}_{\mathcal{A}}\mathcal{I}_{k-1}^k = \left({}_{\mathcal{F}}\mathcal{I}_k^{k-1}\right)^*,$$
where ${}_{\mathcal{A}}\mathcal{I}$ and ${}_{\mathcal{F}}\mathcal{I}$ with indices mean the transfer operators for the adjoint and forward problems, respectively. Here ${}_{\mathcal{A}}\mathcal{S}$ and ${}_{\mathcal{F}}\mathcal{S}$ represent the respective smoothing operators. For efficiency reasons, the accuracy in the respective linear systems can be adapted in the style of inexact reduced SQP methods as in [187], where the accuracy of the linear subproblems is continuously increased, when zooming in to the solution. This is not required when using approximate reduced SQP methods, as in [310], where the optimization problem is reformulated so that a stagnation point of the resulting approximate algorithm is always an

optimal solution, regardless of whether the accuracy of the linear subproblems is increased during the nonlinear iterations or not.

Often, additional constraints are to be satisfied, either formulated in the description of the set U or as a finite number of state constraints. These additional constraints can be efficiently taken care of within the setup above in so-called partially reduced SQP methods, as introduced in [310].

5.6 Schur-Complement-Based Multigrid Smoothers

The earliest multigrid optimization approaches and many later ones have been based on a smoothing concept which can be interpreted as a Schur-complement splitting of the KKT matrix. Considering a Newton iteration for the necessary conditions to problem (5.59)–(5.60), we obtain an incremental iteration where the increments in all variables

$$\mathbf{w} = \begin{pmatrix} \Delta y \\ \Delta u \\ \Delta p \end{pmatrix}$$

are the solution to the equation

$$\mathcal{A}\mathbf{w} = \begin{pmatrix} -\nabla_y L(y,u,p) \\ -\nabla_u L(y,u,p) \\ -c(y,u) \end{pmatrix} =: \mathbf{f}. \tag{5.62}$$

The function $L(y,u,p)$ is the Lagrangian of the optimization problem, and the operator matrix

$$\mathcal{A} = \begin{bmatrix} L_{yy} & L_{yu} & c_y^* \\ L_{uy} & L_{uu} & c_u^* \\ c_y & c_u & 0 \end{bmatrix} \tag{5.63}$$

is the KKT matrix, i.e., the matrix of second-order derivatives of the Lagrangian of the optimization problem. All variants of SQP methods for nonlinear problems play with variable approximations of the matrix $\mathcal{A}$ above, since the system (5.62) can be viewed as a linear-quadratic optimization problem.

For the linear-quadratic problem of Example 2.16(c), equation (2.8), the operator matrix is constant,

$$\mathcal{A} = \begin{bmatrix} I & 0 & -\Delta \\ 0 & \nu I & -I \\ -\Delta & -I & 0 \end{bmatrix},$$

where I is the identity operator in the interior of the domain Ω, and Δ represents the Laplacian with homogeneous Dirichlet boundary conditions.

Schur-complement smoothing approaches that can still be written in the form

$$\mathbf{w}^{(l)} = \mathbf{w}^{(l-1)} + \mathcal{R}(\mathbf{f} - \mathcal{A}\mathbf{w}^{(l-1)})$$

aim at maintaining a high degree of modularity in the implementation of a multigrid optimization method. If, for example, one has a fast Poisson solver for inverting $-\Delta$, one aims at iterative methods which use this fast solver and thus at a modular method. This is the starting point of the early multigrid optimization methods in [168]. Before going into more detail, we briefly refer to the basic Schur-complement approach.

A Schur decomposition of a general 2×2-block matrix

$$K = \begin{bmatrix} A & B^\top \\ B & D \end{bmatrix},$$

with symmetric blocks A and D, and A invertible, is an explicit reformulation of a block Gauss decomposition, i.e.,

$$K \begin{bmatrix} I & -A^{-1}B^\top \\ 0 & I \end{bmatrix} = \begin{bmatrix} A & 0 \\ B & S \end{bmatrix},$$

where $S = D - BA^{-1}B^\top$ is the so-called Schur complement. Obviously, in Schur-complement approaches, one needs the inverses of the blocks A and S or at least approximations of them, thus defining iterative methods rather than factorization methods. Iterative Schur-complement solvers are based on the scheme

$$\mathbf{w}^{(l)} = \mathbf{w}^{(l-1)} + \begin{bmatrix} I & -\tilde{A}^{-1}B^\top \\ 0 & I \end{bmatrix} \begin{bmatrix} \tilde{A} & 0 \\ B & \tilde{S} \end{bmatrix}^{-1} (\mathbf{f} - K\,\mathbf{w}^{(l-1)}), \qquad (5.64)$$

where $\tilde{A}$ and $\tilde{S}$ are approximations to A and S.

If we want to employ this technique, we first have to match the blocks in (5.63) with the blocks in the matrix K. A possible approach is the identification

$$A = \begin{bmatrix} L_{yy} & L_{yu} \\ L_{uy} & L_{uu} \end{bmatrix},$$

and B and D are chosen accordingly. The factorization is a so-called range space factorization. In many cases, the A-block thus defined may not be invertible, which is a limiting factor for the method. Therefore, this arrangement is not well suited for PDE-constrained optimization problems, unlike for variational problems like Stokes or Navier–Stokes problems [74, 362].

Interchanging the second and third rows and columns in the matrix $\mathcal{A}$ and identifying

$$A = \begin{bmatrix} L_{yy} & c_y^* \\ c_y & 0 \end{bmatrix}, \qquad B = \begin{bmatrix} L_{uy} & c_u^* \end{bmatrix}, \qquad D = L_{uu}$$

leads to a so-called nullspace decomposition. In the iterative version of this approach, the A and S blocks are inverted only approximatively. With this decomposition the Schur complement reads as

$$S = L_{uu} - L_{uy}c_y^{-1}c_u - c_u^* c_y^{*-1} L_{yu} + c_u^* c_y^{*-1} L_{yy} c_y^{-1} c_u,$$

which is also the otherwise-called reduced Hessian that characterizes the optimization problem. Recall that coercivity of the reduced Hessian guarantees well-posedness of the overall optimization problem. For the purpose of illustration, we elaborate on the above expression in the case of Example 2.16(c). The reduced Hessian is

$$S = \nu I - 0\Delta^{-1} I - I\Delta^{-1} 0 + I\Delta^{-1} I \Delta^{-1} I = \nu I + (\Delta^{-1})^2,$$

which is the compact operator $(\Delta^{-1})^2$ perturbed by $\nu \cdot I$. In [168], Hackbusch uses this insight in proposing a multigrid smoother for integral Fredholm operators of the second kind, operating on the controls u only:

$$u^{(l)} = \frac{1}{\nu}(\gamma - (\Delta^{-1})^2 u^{(l-1)}) = u^{(l-1)} + \frac{1}{\nu}\left(\gamma - S u^{(l-1)}\right),$$

where $\gamma = \Delta^{-1} z$ is the reduced gradient at zero. In the nullspace Schur-complement setting, this corresponds to choosing

$$\tilde{A} = \begin{bmatrix} 0 & c_y^* \\ c_y & 0 \end{bmatrix}, \qquad \tilde{S} = \nu I.$$

Notice that because in Example 2.16(c) we have $L_{uy} = 0$ and $L_{yu} = 0$, certain terms in the iteration (5.64) vanish, so that only one exact solution with c_y and one exact solution with c_y^* are to be performed. For this purpose a fast Poisson solver is used. But still, the forward system and the adjoint system are solved exactly. Similar ideas lead to the generalization to parabolic optimal control problems in [171, 2].

If one wants to save effort and so performs a full solution of the forward and adjoint systems not in each smoothing step but rather in successive smoothing steps for the forward and adjoint systems, the resulting iteration (5.64) is no longer a smoothing step of a multigrid method of the second kind. In [311, 240, 314, 118] iteration (5.64) is interpreted as a transforming smoothing iteration and successfully applied in various practical problems. The choices for the algorithmic blocks are

$$\tilde{A} = \begin{bmatrix} L_{yy} & \tilde{c}_y^* \\ \tilde{c}_y & 0 \end{bmatrix} \qquad \text{and} \qquad \tilde{S} = \nu I,$$

where $\tilde{c}_y$ is some approximation to c_y useful for smoothing, e.g., just the diagonal of c_y in Jacobi smoothing. The block $\tilde{S}$ mostly consists of the regularizing part, but in numerical experiments it has been shown that a deterioration of the algorithmic performance for $\nu \to 0$ can be avoided by a small number of conjugate gradient iterations for the Schur-complement system. The same iteration is used as an iterative solver in [200] and in [163] as a preconditioner for Krylov methods for the optimality conditions.

Each smoothing step of the approximate nullspace iterations for the solution of system (5.62) runs through the following steps:

(1) compute defects $\begin{pmatrix} d_a \\ d_d \\ d_c \end{pmatrix} := \mathcal{A}\mathbf{w}^{(l)} - \mathbf{f}$

(2) $\tilde{d}_c := \tilde{c}_y^{-1} d_c$

(3) $\tilde{d}_a := \tilde{c}_y^{-*}(d_a - L_{yy}\tilde{d}_c)$

(4) $\Delta u := -\tilde{S}^{-1}(d_d + L_{uy}^* \tilde{d}_c + c_u^* \tilde{d}_a)$

(5) $\Delta y := \tilde{d}_c + \tilde{c}_y^{-1} c_u \Delta u$

(6) $\Delta\lambda := \tilde{d}_a + \tilde{c}_y^{-*}(L_{yu}\Delta u - L_{yy}\Delta y)$

(7) $\mathbf{w}^{(l+1)} = \mathbf{w}^{(l)} + \begin{pmatrix} \Delta y \\ \Delta u \\ \Delta\lambda \end{pmatrix}$

This iteration shows good smoothing properties in practical applications as demonstrated in [311, 240, 314, 118]. The convergence theory is based not on Fourier analysis as later discussed for other smoothing concepts but rather on transforming smoothers and is discussed in [315]. The Schur-complement-based smoothing approaches discussed above decouple the smoothing of the forward and adjoint equations from the smoothing of the design equation or Schur-complement equation. In this way, the smoothing algorithm for the overall optimization system is still a highly modular algorithm. The price for this high degree of modularity is the necessity to deal with the nullspace Schur complement. Typically, only the easily accessible part stemming from regularization is used as an approximation, i.e., $\tilde{S} = L_{uu}$. If the regularization parameter ν tends towards zero, this strategy runs into trouble. In [2], this problem is resolved by a more refined analysis of the reduced Hessian for inverse problems [117]. In Example 2.16(c), additional accuracy with respect to the Schur-complement system

$$\bar{S}u^l = -\gamma, \qquad \text{where} \qquad \bar{S} = \nu I + c_u^*(\tilde{c}_y^*)^{-1}\tilde{c}_y^{-1}c_u,$$

can be achieved by a small number of conjugate gradient steps as demonstrated in [315]. It should be noted that this approximate Schur complement is formed with the approximations $\tilde{c}_y \approx c_y$ which are cheaply inverted because they are used for smoothing the forward and the adjoint systems.

A variation of the nullspace Schur-complement iteration is presented in [218, 184], where the $D = L_{uu}$-block is used as a pivot instead of the A-block. The resulting Schur complement is then

$$S = \begin{bmatrix} L_{yy} & c_y^* \\ c_y & 0 \end{bmatrix} - \begin{bmatrix} L_{yu}L_{uu}^{-1}L_{uy} & L_{yu}L_{uu}^{-1}c_u^* \\ c_uL_{uu}^{-1}L_{uy} & c_uL_{uu}^{-1}c_u^* \end{bmatrix}.$$

In particular in cases similar to Example 2.16(c), where $L_{yu} = 0$ and $L_{uu} = \nu \cdot I$, we see that

$$S = \begin{bmatrix} L_{yy} & c_y^* \\ c_y & 0 \end{bmatrix} - \begin{bmatrix} 0 & 0 \\ 0 & \frac{1}{\nu}c_uc_u^* \end{bmatrix}.$$

This means that

$$\begin{bmatrix} L_{yy} & c_y^* \\ c_y & 0 \end{bmatrix}^{-1} S = I - \text{compact operator},$$

which makes the above-mentioned multigrid methods of the second kind applicable. So far, this has been used only in the form of two-level cascadic methods.

Schur-complement multigrid approaches for parabolic problems have not been widely used. They can be found, e.g., in [172], where the forward and the adjoint systems are solved exactly. Schur-complement approaches are advantageous in hyperbolic optimization problems because of the nondiffusive nature of the forward problem, where an exact solver can be provided. This is shown in more detail in the next section in the form of an application to optical flow problems.

A Multigrid Solver and a Hyperbolic Optimal Control Problem

Multigrid methods for optimal control problems governed by hyperbolic systems are a much less investigated field of research, partly because multigrid methods are not considered as a natural choice for the solution of hyperbolic equations, even though there are many examples of successful application of these methods to hyperbolic problems. Therefore, it is more natural to use multigrid schemes in a decoupled form as discussed in the previous section on Schur-complement approaches. In fact, in many cases an elliptic nullspace Schur complement results, making the use of multigrid methods obvious.

In this context, an interesting application problem results from the optimal control formulation of the optical flow problem as proposed in [55, 56]. An optical flow is the field of apparent velocities in a sequence of images; see [24, 201, 359]. From the knowledge of the optical flow, information about the spatial arrangement of objects and the rate of change of this arrangement ought to be obtained.

The forward model is based on the assumptions that the image brightness of an object point remains constant in the images when the object moves. That is, the total time derivative of the brightness at each point (x_1, x_2) at time t is zero. This approach leads to the following *optical flow constraint* (OFC) equation

$$\frac{\partial y}{\partial t} + u\frac{\partial y}{\partial x_1} + v\frac{\partial y}{\partial x_2} = 0, \tag{5.65}$$

where $y = y(x_1, x_2, t)$ denotes the image brightness at (x_1, x_2) and t, and $\vec{w} = (u, v)$ represents the optical flow vector.

Now consider a sequence of image frames $\{Y_k\}_{k=0,N}$ on Ω sampled at increasing time steps, $t_k \in [0, T]$, $k = 0, 2, \ldots, N$, where $t_0 = 0$ and $t_N = T$. In the optimal control formulation we require estimation of $\vec{w}$ such that the resulting $y(\cdot, t_k, \vec{w})$ approximates Y_k at the sampling times. This means solving

$$\begin{cases} y_t + \vec{w}\cdot\nabla y = 0 & \text{in } Q = \Omega \times (0, T], \\ y(\cdot, 0) = Y_0 \end{cases} \tag{5.66}$$

and minimizing the cost functional

$$\begin{aligned} J(y, \vec{w}) = {} & \frac{1}{2}\int_\Omega \sum_{k=1}^{N} |y(x_1, x_2, t_k) - Y_k|^2 d\Omega \\ & + \frac{\alpha}{2}\int_Q \Phi\left(\left|\frac{\partial \vec{w}}{\partial t}\right|^2\right) dq + \frac{\beta}{2}\int_Q \Psi(|\nabla u|^2 + |\nabla v|^2) dq + \frac{\gamma}{2}\int_Q |\nabla\cdot\vec{w}|^2 dq. \end{aligned} \tag{5.67}$$

Here, α, β, and γ are predefined nonnegative weights. The term with Φ provides bounded variation type regularization across edges and corners of images, where $\nabla\vec{w}$ is large; see, e.g., [209]. The last term in (5.67) improves the filling-in properties of the optimal control solution; see [55, 56] for details.

The optimal solution is characterized by the following optimality system

$$\begin{aligned} y_t + \vec{w}\cdot\nabla y = 0, \quad & \text{with } y(\cdot, 0) = Y_0, \\ p_t + \nabla\cdot(\vec{w}p) = \sum_{k=1}^{N-1} \left[\delta(t - t_k)(y(\cdot, t_k) - Y_k)\right], \quad & \text{with } p(\cdot, T) = -(y(\cdot, T) - Y_N), \\ \alpha\frac{\partial^2 u}{\partial t^2} + \beta\nabla\cdot[\Psi'(|\nabla u|^2 + |\nabla v|^2)\nabla u] + \gamma\frac{\partial}{\partial x_1}(\nabla\cdot\vec{w}) = p\frac{\partial y}{\partial x_1}, & \\ \alpha\frac{\partial^2 v}{\partial t^2} + \beta\nabla\cdot[\Psi'(|\nabla u|^2 + |\nabla v|^2)\nabla v] + \gamma\frac{\partial}{\partial x_2}(\nabla\cdot\vec{w}) = p\frac{\partial y}{\partial x_2}, & \end{aligned} \tag{5.68}$$

where δ denotes the Dirac δ-function. The interpretation of the second equation in (5.68) is

$$p_t + \nabla \cdot (\vec{w} p) = 0 \text{ on } t \in (t_{k-1}, t_k) \quad \text{for } k = 1, \ldots, N, \tag{5.69}$$

$$p(\cdot, t_k^+) - p(\cdot, t_k^-) = y(\cdot, t_k) - Y_k \quad \text{for } k = 1, \ldots, N-1. \tag{5.70}$$

The last two equations are nonlinear elliptic equations representing the optimality condition. As boundary conditions for $\vec{w}$ one can choose homogeneous Dirichlet boundary conditions on the spatial boundary and natural boundary conditions at the temporal boundaries of Q.

To solve (5.68), an explicit time-marching second-order TVD scheme for the forward-backward hyperbolic subsystem and a FAS multigrid method for the elliptic control equations is proposed in [55, 56]. The proposed method is summarized as follows.

ALGORITHM 5.9. Loop for solving the optical flow problem.

1. Apply the Horn and Schunck method [201] for a starting approximation to the optical flow.
2. Solve the OFC equation to obtain y.
3. Solve (backward) the adjoint OFC equation to obtain p.
4. Update the right-hand sides of the elliptic system.
5. Apply a few cycles of multigrid to solve the control equations.
6. Go to 2 and repeat I_{loop} times.

Results presented in [55, 56] show that the optimal control approach allows accurate and robust determination of optical flow also in the limit case where only two image frames are given.

A known benchmark for verification of optical flow solvers is the "Hamburg taxi sequence"; see [24]. It consists of a sequence of frames of a taxi coming from the right in the main road and turning right into a side street in Hamburg (Germany). One photo of the sequence and the corresponding brightness pattern are depicted in Figure 5.5.

We consider a sequence of five photos of the moving taxi taken at regular intervals ($T = 4$). The space-time computational domain is a $128 \times 96 \times 128$ grid, where 128 time subdivisions are taken in the time direction. This grid can be obtained from a coarse $4 \times 3 \times 4$ mesh by halving the mesh size six times. Our algorithm is applied with $\alpha = 5.0$, $\beta = 0.25$, $\gamma = 0.25$, and $I_{loop} = 10$. In Figure 5.5 the optical flow computed with the optimal control approach at $t = 2$ is presented. Comparing with the solution obtained with the Horn and Schunck method [24], the optimal control approach provides a better reconstruction optical flow for the taxi sequence.

5.7 The Collective Smoothing Multigrid Approach

A collective smoothing multigrid (CSMG) approach means solving the optimality system for the state, the adjoint, and the control variables simultaneously in the multigrid process by using collective smoothers for the optimizations variables. The CSMG approach is in contrast to the sequential solving of the state, adjoint, and control equations. The last approach requires that the uncontrolled state equations be solvable, and thus it cannot be applied to singular optimal control problems [236], where the uncontrolled system may not

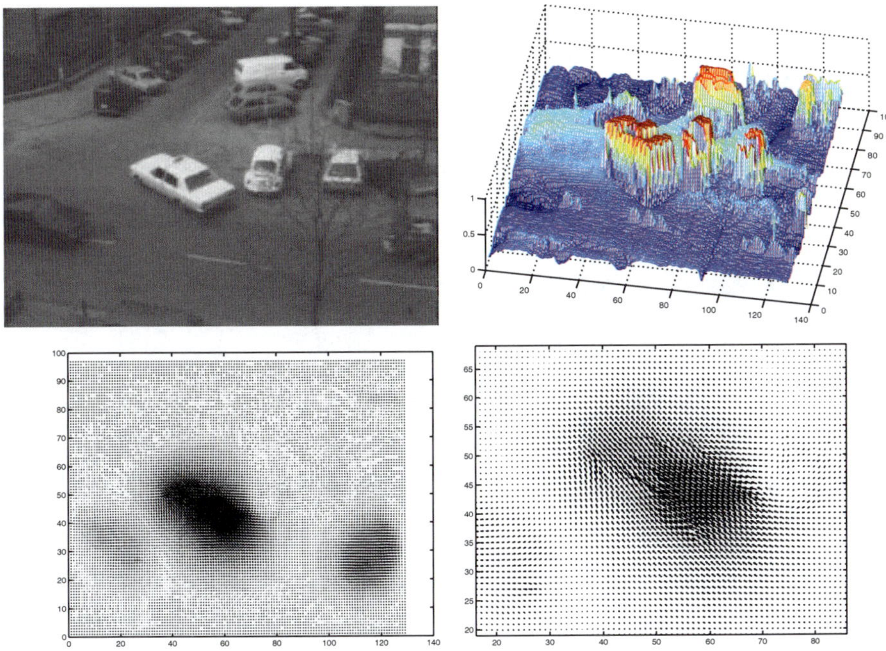

Figure 5.5. *First frame of the taxi sequence (top left); the corresponding brightness distribution (top right). Optical flow for the taxi sequence (bottom left). Close-ups of the solution containing the region of the taxi (bottom right). This figure first appeared in A. Borzì, K. Ito, and K. Kunisch, Optimal control formulation for determining optical flow, SIAM J. Sci. Comput.*, 24(3) (2002), 818–847.

have a solution or blowup in finite time. A CSMG-based scheme aims at realizing the tight coupling in /the optimality system along the hierarchy of grids. By employing collective smoothing, that is, by realizing the coupling in the optimality system at the smoothing step level, robustness and typical multigrid efficiency is achieved; see, e.g., [64].

Strategies of development of collective smoothers for optimality systems appear well established in an AMG context [51, 52] that also provide an example of application of the CSMG approach to convection-diffusion problems. For these problems, another recent contribution in a geometric multigrid context can be found in [203]. Further extensions of the CSMG strategy to problems with control or state constraints are given in [47, 48, 59]. The CSMG approach has also been successfully applied to parameter identification problems [12, 345]. In the control-constrained case, the CSMG approach allows one to construct robust multigrid schemes that apply also in the case $\nu = 0$, thus allowing the investigation of bang-bang control problems. In particular, using the multigrid scheme in [48, 59] it is possible to show the phenomenon of "chattering control" [35] for elliptic systems, which appears to be a less investigated problem. Further extension of the CSMG method can be found in [44], where this method is combined with sparse-grid techniques for solving elliptic optimal control problems with random coefficients.

The results above concern multigrid methods for linear and nonlinear elliptic optimality systems with linear and bilinear control mechanisms. Early works concerning multigrid solution of parabolic optimal control problems are [171, 172]. Within the CSMG framework, recent contributions are given in [46, 49, 50, 53, 58, 67, 150]. The starting point for these recent developments is represented by space-time parabolic multigrid methods [170, 347] and also the approach presented in [169]. Based on the CSMG strategy it is also possible to solve bang-bang parabolic control problems [150]. In [6] the CSMG approach is extended to the solution of integral Fredholm optimal control problems.

In the following sections we discuss the development and convergence properties of CSMG schemes for control problems governed by PDEs.

5.7.1 CSMG Schemes for Elliptic Control Problems

We discuss in detail the design of a collective smoothing iteration for an elliptic optimal control problem with control constraints [48, 59]. This procedure appears to be robust with respect to changes of the value of the weight and, in particular, it allows the choice $\nu = 0$. This fact makes the CSMG scheme a useful tool to investigate bang-bang control phenomena [147].

Consider the following basic elliptic distributed optimal control problem

$$\left\{ \begin{array}{rcll} \min J(y,u) & := & \frac{1}{2}\|y-z\|^2_{L^2(\Omega)} + \frac{\nu}{2}\|u\|^2_{L^2(\Omega)}, & \\ -\Delta y & = & u+g & \text{in } \Omega, \\ y & = & 0 & \text{on } \partial\Omega, \end{array} \right. \tag{5.71}$$

where we require that $u \in U_{ad}$ and the set of admissible controls is the closed convex subset of $L^2(\Omega)$ given by

$$U_{ad} = \{u \in L^2(\Omega) \,|\, \underline{u}(\mathbf{x}) \leq u(\mathbf{x}) \leq \overline{u}(\mathbf{x}) \text{ a.e. in } \Omega\}, \tag{5.72}$$

where $\underline{u}$ and $\overline{u}$ are elements of $L^\infty(\Omega)$.

As illustrated in Chapter 2, the solution to (5.71) is characterized by the following optimality system

$$\begin{array}{rll} -\Delta y = & u+g & \text{in } \Omega, \\ y = & 0 & \text{on } \partial\Omega, \\ -\Delta p = & -(y-z) & \text{in } \Omega, \\ p = & 0 & \text{on } \partial\Omega, \\ (\nu u - p, v - u) \geq & 0 & \forall v \in U_{ad}. \end{array} \tag{5.73}$$

Notice that the last equation in (5.73) giving the optimality condition is equivalent to (see [235, 245])

$$u = \max\left\{\underline{u}, \min\left\{\overline{u}, \frac{1}{\nu}p(u)\right\}\right\} \text{ in } \Omega \quad \text{if } \nu > 0. \tag{5.74}$$

The unique solution u to (5.71)–(5.72) with $\nu = 0$ corresponds to the optimality condition given by [59]

$$p = \min\{0, p+u-\underline{u}\} + \max\{0, p+u-\overline{u}\} \quad \text{in } \Omega. \tag{5.75}$$

We illustrate the construction of a smoothing iteration for the constrained-control problem with $\nu > 0$ in the framework of finite differences. Recall the discretization setting introduced in Chapter 3, and consider the discrete optimality system at $\mathbf{x} \in \Omega_h$, where $\mathbf{x} = (ih, jh)$ and i, j index the grid points, e.g., lexicographically. We have

$$-(y_{i-1\,j} + y_{i+1\,j} + y_{i\,j-1} + y_{i\,j+1}) + 4y_{ij} - h^2 u_{ij} = h^2 g_{ij}, \tag{5.76}$$

$$-(p_{i-1\,j} + p_{i+1\,j} + p_{i\,j-1} + p_{i\,j+1}) + 4p_{ij} + h^2 y_{ij} = h^2 z_h, \tag{5.77}$$

$$(\nu u_{ij} - p_{ij})(v_{ij} - u_{ij}) \geq 0 \quad \forall v_h \in U_{adh}. \tag{5.78}$$

A collective smoothing step at $\mathbf{x}$ consists in updating the values y_{ij} and p_{ij} such that the resulting residuals of the two equations at that point are zero. The neighboring variables are considered constant during this process. Therefore, replacing these two constants in (5.76) and (5.77), we obtain y_{ij} and p_{ij} as functions of u_{ij} as follows

$$y_{ij} = (A_{ij} + h^2 u_{ij})/4 \tag{5.79}$$

and

$$p_{ij} = (4\,B_{ij} - h^2 A_{ij} - h^4 u_{ij})/16. \tag{5.80}$$

Now to obtain the u_{ij} update, replace the expression for p_{ij}, as a function of u_{ij}, in the inequality constraint and define the auxiliary variable

$$\tilde{u}_{ij} = \frac{1}{16\nu + h^4}(4\,B_{ij} - h^2 A_{ij}). \tag{5.81}$$

Here $\tilde{u}_{ij}$ is defined as the solution to the optimality condition equation without constraints, i.e., $\nabla \hat{J}(u) = \nu u - p(u) = 0$, and therefore (5.81) defines the u_{ij} update in the case of no constraints. In the presence of constraints, the new value for u_{ij} resulting from the smoothing step is given by projection of $\tilde{u}_{ij}$ as follows

$$u_{ij} = \begin{cases} \overline{u}_{ij} & \text{if } \tilde{u}_{ij} > \overline{u}_{ij}, \\ \tilde{u}_{ij} & \text{if } \underline{u}_{ij} \leq \tilde{u}_{ij} \leq \overline{u}_{ij}, \\ \underline{u}_{ij} & \text{if } \tilde{u}_{ij} < \underline{u}_{ij} \end{cases} \tag{5.82}$$

for all $\mathbf{x} = (ih, jh) \in \Omega_h$. With the new value of u_{ij} given, new values for y_{ij} and p_{ij} are obtained. This completes the description of the collective smoothing step. It satisfies the inequality constraint; see [59]. Further, in case $\nu = 0$ the smoothing iteration defined above satisfies (5.75). Because of (5.82) we can consider that the present iteration belongs to the class of projected iterative schemes [88].

We report results of experiments with the following objective function

$$z(x_1, x_2) = \sin(2\pi x_1)\sin(\pi x_2).$$

We choose the following constraints: $\underline{u} = -30$ and $\overline{u} = 30$.

We obtain that the constraints are active in large portions of the domain for the three choices of $\nu = \{10^{-4}, 10^{-6}, 10^{-8}\}$ considered here. For $\nu = 10^{-6}$ this can be seen in Figure 5.6. From the results of numerical experiments reported in Table 5.3 we observe that for $\nu = 10^{-4}$ the multigrid convergence behavior is similar to that observed in the unconstrained case. Reducing the value of ν results in steeper gradients of the adjoint and control variables, particularly close to the boundary where p and u are required to be zero. Furthermore, decreasing ν results in an increasingly more complex switching structure of

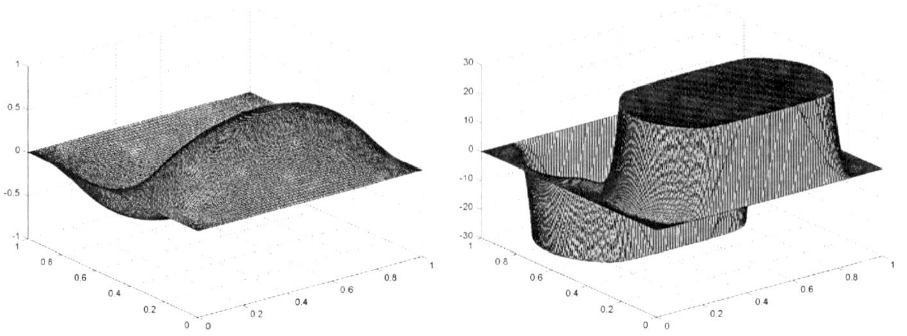

Figure 5.6. *Numerical solutions for the state (left) and the control (right);* $\nu = 10^{-6}$ *and* 513×513 *mesh. Reprinted with permission from A. Borzì and K. Kunisch, A multigrid scheme for elliptic constrained optimal control problems, Comput. Optim. Appl.,* 31(3) (2005), 309–333.

Table 5.3. *Results of experiments.*

$\nu = 10^{-4}$			
Mesh	$\rho(y), \rho(p)$	$\lvert y-z\rvert_0$	$\lvert r(y)\rvert_0, \lvert r(p)\rvert_0$
129×129	0.04, 0.04	$1.11\,10^{-1}$	$3.1\,10^{-10}, 1.2\,10^{-13}$
257×257	0.03, 0.04	$1.11\,10^{-1}$	$6.8\,10^{-10}, 7.1\,10^{-14}$
513×513	0.03, 0.04	$1.11\,10^{-1}$	$4.9\,10^{-10}, 1.5\,10^{-13}$
1025×1025	0.03, 0.03	$1.11\,10^{-1}$	$3.2\,10^{-10}, 7.2\,10^{-13}$
$\nu = 10^{-6}$			
Mesh	$\rho(y), \rho(p)$	$\lvert y-z\rvert_0$	$\lvert r(y)\rvert_0, \lvert r(p)\rvert_0$
129×129	0.56, 0.56	$5.30\,10^{-2}$	$1.3\,10^{-6}, 2.2\,10^{-10}$
257×257	0.52, 0.51	$5.30\,10^{-2}$	$1.5\,10^{-7}, 1.3\,10^{-11}$
513×513	0.03, 0.03	$5.30\,10^{-2}$	$3.5\,10^{-10}, 5.3\,10^{-14}$
1025×1025	0.03, 0.03	$5.30\,10^{-2}$	$2.2\,10^{-10}, 2.2\,10^{-13}$
$\nu = 10^{-8}$			
Mesh	$\rho(y), \rho(p)$	$\lvert y-z\rvert_0$	$\lvert r(y)\rvert_0, \lvert r(p)\rvert_0$
129×129	0.63, 0.63	$5.28\,10^{-2}$	$1.6\,10^{-3}, 8.3\,10^{-8}$
257×257	0.54, 0.54	$5.28\,10^{-2}$	$2.4\,10^{-6}, 7.4\,10^{-11}$
513×513	0.64, 0.60	$5.28\,10^{-2}$	$2.5\,10^{-7}, 3.7\,10^{-12}$
1025×1025	0.68, 0.66	$5.28\,10^{-2}$	$2.7\,10^{-7}, 2.1\,10^{-12}$
2049×2049	0.74, 0.71	$5.28\,10^{-2}$	$7.8\,10^{-7}, 3.5\,10^{-12}$
4097×4097	0.76, 0.70	$5.28\,10^{-2}$	$7.4\,10^{-8}, 2.9\,10^{-12}$

the control between upper and lower bounds; see Figure 5.7. The results for $\nu = 10^{-6}$ in Table 5.3 suggest that once the mesh size is sufficiently fine to resolve completely the switching structure the typical multigrid convergence rate is obtained. They further indicate that the multigrid convergence factor depends only weakly on the mesh size provided the problem is sufficiently well resolved on the mesh.

The ability of the multigrid scheme in solving constrained control problems with very small value of ν allows us to investigate the occurrence of bang-bang control for the

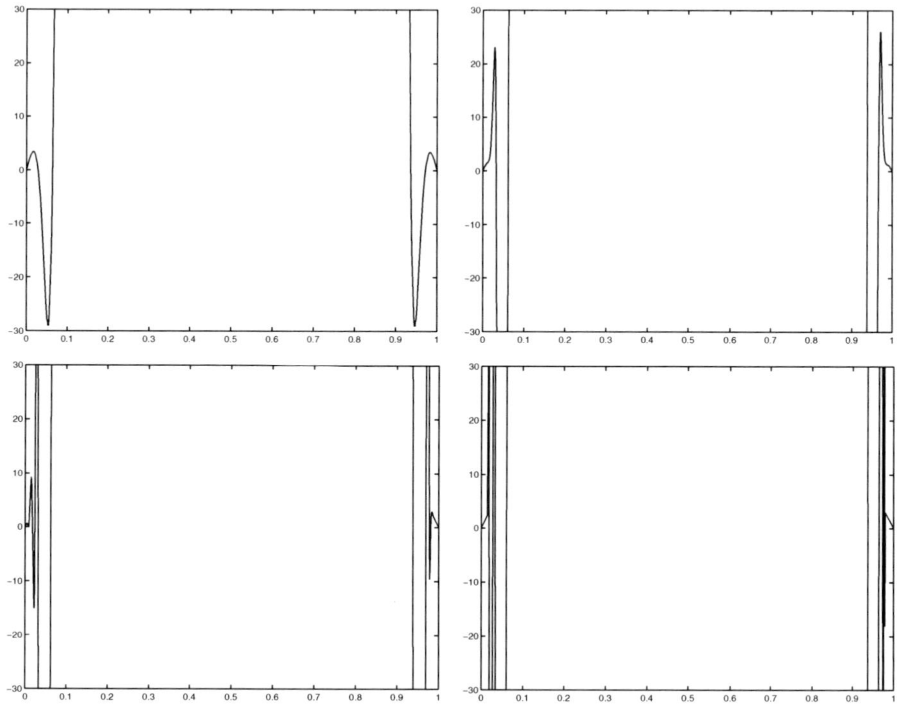

Figure 5.7. *The control function for* $x_1 = 3/4$ *and* $x_2 \in [0,1]$ *obtained with* $\nu = 10^{-8}$ *(top left),* $\nu = 10^{-10}$ *(top right),* $\nu = 10^{-12}$ *(bottom left), and* $\nu = 0$ *(bottom right);* 2049×2049 *mesh. Reprinted with permission from A. Borzì and K. Kunisch, A multigrid scheme for elliptic constrained optimal control problems, Comput. Optim. Appl.,* 31(3) (2005), 309–333.

present class of problems. In particular, with the choice of z given above we can observe fast switching of the control function in the x_2 direction as depicted in Figure 5.7. In this figure we give plots of the control function for $x_1 = 3/4$ and $x_2 \in [0,1]$ for the following choices $\nu \in \{10^{-8}, 10^{-10}, 10^{-12}, 0\}$. We can see that as the value of ν is reduced the number of switching points increases.

The solution obtained for $\nu = 0$ is interesting. In this case, by further refining the mesh size, additional switching points can be seen closer to the boundary while the existing switching points obtained at the previous coarser grids are retained.

We complete this discussion considering another desired state given by

$$z_1(x_1,x_2) = \sin(4\pi x_1)\sin(2\pi x_2).$$

The difference between this objective function and the previous one is that the gradient of z_1 is larger close to the boundary. For the choice $\nu = 0$ the constraints are everywhere active; i.e., the control is bang-bang. Moreover, no fast switching of the control occurs. In Figure 5.8 the optimal control and the corresponding state for $\nu = 0$ are depicted. The numerical results in Table 5.4 document the convergence factors.

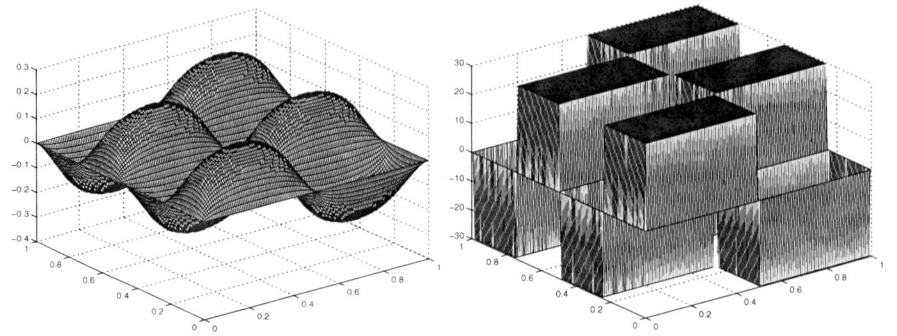

Figure 5.8. *Numerical solutions with z_1 and $\nu = 0$. The state (left) and the control (right); 257×257 mesh. Reprinted with permission from A. Borzì and K. Kunisch, A multigrid scheme for elliptic constrained optimal control problems, Comput. Optim. Appl.,* 31(3) (2005), 309–333.

Table 5.4. *Results of experiments with z_1 and $\nu = 0$.*

Mesh	$\rho(y), \rho(p)$	$\lvert y-z\rvert_0$	$\lvert r(y)\rvert_0, \lvert r(p)\rvert_0$
513×513	0.12, 0.13	$3.70\,10^{-1}$	$2.9\,10^{-8}, 1.3\,10^{-13}$
1025×1025	0.12, 0.13	$3.70\,10^{-1}$	$2.5\,10^{-8}, 4.2\,10^{-13}$
2049×2049	0.12, 0.16	$3.70\,10^{-1}$	$1.9\,10^{-8}, 1.6\,10^{-12}$

The approach described above can be extended to the case of boundary optimal control problems with constraints. Consider

$$\left\{\begin{array}{rcll} \min J(y,u) & := & \frac{1}{2}\|y-z\|^2_{L^2(\Omega)} + \frac{\nu}{2}\|u\|^2_{L^2(\partial\Omega)}, & \\ -\Delta y + y & = & g & \text{in } \Omega, \\ \frac{\partial y}{\partial n} & = & u & \text{on } \partial\Omega, \end{array}\right. \tag{5.83}$$

where Ω is an open bounded set of $\mathbb{R}^2$, $g \in L^2(\Omega)$, and $z \in L^2(\Omega)$ is the objective function. The set of admissible controls is given in this case by

$$U_{ad} = \{u \in L^2(\partial\Omega) \,|\, \underline{u}(\mathbf{x}) \le u(\mathbf{x}) \le \overline{u}(\mathbf{x}) \text{ a.e. in } \partial\Omega\}, \tag{5.84}$$

where $\underline{u}$ and $\overline{u}$ are functions of $L^\infty(\partial\Omega)$.

For the existence of a unique solution to (5.83)–(5.84) we refer the reader to [235, 339]. The solution is characterized by the following optimality system

$$\begin{array}{rcll} -\Delta y + y & = & g & \text{in } \Omega, \\ \frac{\partial y}{\partial n} & = & u & \text{on } \partial\Omega, \\ -\Delta p + p & = & -(y-z) & \text{in } \Omega, \\ \frac{\partial p}{\partial n} & = & 0 & \text{on } \partial\Omega, \\ (\nu u - p, v - u) & \ge & 0 & \forall v \in U_{ad}. \end{array} \tag{5.85}$$

To implement the control on the boundary, one can approximate $\frac{\partial}{\partial n}$ by centered difference quotient and combine the Neumann boundary conditions with the stencil of the

Laplace operator considered at the boundary. We discuss this approach explicitly for one lateral boundary of $\Omega = (0,1) \times (0,1)$.

Let $\mathbf{x} = (ih, jh)$ be a boundary grid point on the side $x = 0$. Based on the finite differences framework of Chapter 3, we obtain the following discretized state and adjoint equations

$$\begin{aligned} -(y_{i-1\,j} + y_{i+1\,j} + y_{i\,j-1} + y_{i\,j+1}) + (4+h^2)y_{ij} &= h^2 g_{ij}, \\ y_{i-1\,j} - y_{i+1\,j} &= 2hu_{ij}, \\ -(p_{i-1\,j} + p_{i+1\,j} + p_{i\,j-1} + p_{i\,j+1}) + (4+h^2)p_{ij} + h^2\, y_{ij} &= h^2\, z_{ij}, \\ p_{i-1\,j} - p_{i+1\,j} &= 0. \end{aligned}$$

Summing up the minus Laplacian stencil with the normal derivative, the ghost variables outside of Ω are eliminated. On the boundary, we obtain

$$\begin{aligned} -(2y_{i+1\,j} + y_{i\,j-1} + y_{i\,j+1}) + (4+h^2)y_{ij} - 2hu_{ij} &= h^2 g_{ij}, \\ -(2p_{i+1\,j} + p_{i\,j-1} + p_{i\,j+1}) + (4+h^2)p_{ij} + h^2\, y_{ij} &= h^2\, z_{ij}, \\ (\nu u_{ij} - p_{ij})(v_{ij} - u_{ij}) \geq 0 \quad & \forall v_h \in U_{adh}. \end{aligned}$$

On the corners we consider the minus Laplacian stencil with the normal derivatives in both directions. The equations obtained in this way have the same structure as (5.76) and (5.77), and therefore the development of the collective smoothing iteration proceeds along the same lines as described above; see [59].

An example of application of the CSMG for solving boundary optimal control problems is the following. Consider the desired target given by

$$z(x_1, x_2) = (x_1^2 - x_2^2)\sin(\pi x_1)\sin(\pi x_2),$$

and let $g = 0$. We choose constraints given by $\underline{u} = -1$ and $\overline{u} = 1$ which are active in part of the boundary for $\nu \leq 10^{-6}$. The multigrid setting is the same as in previous sections. Results for this case are reported in Table 5.5.

Table 5.5. *Results of experiments with a boundary control problem;* 1025×1025 *mesh.*

ν	$\rho(y), \rho(p)$	$\lvert y - z \rvert_0$	$\lvert r(y) \rvert_0, \lvert r(p) \rvert_0$
10^{-6}	0.05, 0.05	$8.09\,10^{-2}$	$1.7\,10^{-10}, 2.9\,10^{-13}$
10^{-8}	0.14, 0.12	$8.09\,10^{-2}$	$3.7\,10^{-8}, 2.9\,10^{-13}$
10^{-10}	0.28, 0.28	$8.09\,10^{-2}$	$4.7\,10^{-5}, 9.9\,10^{-11}$
0	0.25, 0.26	$8.09\,10^{-2}$	$3.5\,10^{-5}, 4.8\,10^{-11}$

In the presence of nonlinearities in the state equation it is not possible to solve explicitly the optimality system at the grid point level. Instead, a collective local Newton–Gauss–Seidel iteration can be applied, resulting in a CSMG scheme that also in this case provides convergence factors that are almost independent of ν and of the mesh size [48, 47, 57].

Consider the following nonlinear optimal control problem

$$\left\{ \begin{array}{rcll} \min J(y,u) & := & \frac{1}{2}\|y - z\|^2_{L^2(\Omega)} + \frac{\nu}{2}\|u\|^2_{L^2(\Omega)}, & \\ -\Delta y + G(y) & = & u + f & \text{in } \Omega, \\ y & = & 0 & \text{on } \partial\Omega, \end{array} \right. \tag{5.86}$$

where U_{ad} is as in (5.72). Existence of solutions to (5.86) can be established under suitable conditions for various forms of the nonlinearity; see, e.g., [57, 148, 236]. We take $G \in C^\infty$ monotonically increasing function with $G(0) = 0$, but the following procedure remains effective for other nonlinearities as well.

Optimal solutions are characterized by the following optimality system

$$\begin{aligned} -\Delta y + G(y) = & \quad u + g && \text{in } \Omega, \\ y = & \quad 0 && \text{on } \partial\Omega, \\ -\Delta p + G'(y)\, p = & \quad -(y - z) && \text{in } \Omega, \\ p = & \quad 0 && \text{on } \partial\Omega, \\ (\nu u - p, v - u) \geq & \quad 0 && \forall v \in U_{ad}. \end{aligned} \tag{5.87}$$

Similar to the linear case, we consider the discrete version of this optimality system at the grid point i, j and thus obtain the following system for the three scalar variables y_{ij}, p_{ij}, and u_{ij}. We have

$$-A_{ij} + 4\, y_{ij} + h^2\, G(y_{ij}) - h^2\, u_{ij} = 0, \tag{5.88}$$

$$-B_{ij} + 4\, p_{ij} + h^2\, G'(y_{ij})\, p_{ij} + h^2\, y_{ij} = 0, \tag{5.89}$$

$$(\nu u_{ij} - p_{ij})(v_{ij} - u_{ij}) \geq 0 \quad \forall v_h \in U_{adh}. \tag{5.90}$$

The inverse of the Jacobian for (5.88)–(5.89) is given by

$$J_{ij}^{-1} = \frac{1}{\det J_{ij}} \begin{pmatrix} 4 + h^2 G'(y_{ij}) & 0 \\ -h^2(1 + G''(y_{ij})\, p_{ij}) & 4 + h^2 G'(y_{ij}) \end{pmatrix}, \tag{5.91}$$

where $\det J_{ij} = (4 + h^2 G'(y_{ij}))(4 + h^2 G'(y_{ij}))$. Notice that also in the case of nonmonotone nonlinearities such that $G' < 0$, we can choose h sufficiently small to guarantee that $\det J_{ij} \neq 0$. Also notice that second-order necessary conditions for a minimum require that $(1 + G''(y_{ij})\, p_{ij}) \geq 0$. Hence, given u_{ij}, a local Newton update for the state and the adjoint variables $\hat{y}_{ij}$ and $\hat{p}_{ij}$ at i, j can be performed, and it is given by

$$\begin{pmatrix} \hat{y}_{ij} \\ \hat{p}_{ij} \end{pmatrix} = \begin{pmatrix} y_{ij} \\ p_{ij} \end{pmatrix} + J_{ij}^{-1} \begin{pmatrix} r_{ij}^y \\ r_{ij}^p \end{pmatrix}, \tag{5.92}$$

where $r_{ij}^y = A_{ij} - 4\, y_{ij} - h^2\, G(y_{ij}) + h^2\, u_{ij}$ and $r_{ij}^p = B_{ij} - 4\, p_{ij} - h^2\, G'(y_{ij})\, p_{ij} - h^2\, y_{ij}$ denote the residual of (5.88) and (5.89), respectively. Notice that r_{ij}^y depends explicitly on u_{ij}. This fact allows us to write $\hat{p}_{ij}$ as a function of u_{ij} as follows

$$\begin{aligned} \hat{p}_{ij}(u_{ij}) = p_{ij} &+ \frac{1}{\det J_{ij}} \left(-h^2(1 + G''(y_{ij})\, p_{ij})(A_{ij} - 4\, y_{ij} - h^2\, G(y_{ij}))\right) \\ &+ \frac{1}{\det J_{ij}} \left((4 + h^2 G'(y_{ij}))(B_{ij} - 4\, p_{ij} - h^2\, G'(y_{ij})\, p_{ij} - h^2\, y_{ij})\right) \\ &- \frac{1}{\det J_{ij}} (1 + G''(y_{ij})\, p_{ij}) h^4\, u_{ij}. \end{aligned} \tag{5.93}$$

Now to obtain first the update for u_{ij}, replace $\hat{p}_{ij}$ in the inequality constraint. From $\nu\tilde{u} - \hat{p}(\tilde{u}) = 0$ at i, j, we obtain the auxiliary variable

$$\tilde{u}_{ij} = \left(\nu + \frac{(1+G''(y_{ij})\,p_{ij})h^4}{\det J_{ij}}\right)^{-1}$$
$$\times\left[p_{ij} + \frac{1}{\det J_{ij}}\left(-h^2(1+G''(y_{ij})\,p_{ij})(A_{ij} - 4\,y_{ij} + h^2\,G(y_{ij}))\right)\right.$$
$$\left. + \frac{1}{\det J_{ij}}\left((4+h^2G'(y_{ij}))(B_{ij} - 4\,p_{ij} - h^2\,G'(y_{ij})\,p_{ij} - h^2\,y_{ij})\right)\right].$$

This $\tilde{u}$ provides the update for the control in the case of no constraints. In the presence of constraints we apply the projection given by (5.82).

It should be clear that the CSMG multigrid procedure just described can be easily extended to solve the optimality system (5.87) with the fourth-order discretization discussed in Section 3.2. See [48].

A CSMG Scheme for State-Constrained Elliptic Control Problems

In state-constrained optimal control problems, bounds are given to the admissible set of values of the state variable. Consider the following

$$\left\{\begin{array}{rclll} \min J(y,u) & := & \frac{1}{2}\|y-z\|^2_{L^2(\Omega)} + \frac{\nu}{2}\|u\|^2_{L^2(\Omega)}, & & \\ \Delta y + G(y) & = & u+g & \text{in } \Omega, & \\ y & = & 0 & \text{on } \partial\Omega, & \\ y_L \;\le\; y & \le & y_H & \text{a.e. in } \Omega, & \end{array}\right. \tag{5.94}$$

where y_L and y_H are continuous functions. Existence and uniqueness of solutions to state-constrained semilinear elliptic optimal control problems depend on the given constraints and on the nonlinearity. For a nonempty solution set, uniqueness can be proved for sufficiently regular G such that the state operator is monotone; see, e.g., [207, 255, 339].

The solution approach to state-constrained optimal control problems through Lagrange multipliers associated with the state constraints leads to difficulties [254]. In particular, the fact that the Lagrange multipliers associated with the state constraints are only regular Borel measures prevents us from using classical approximation techniques. The remedy is to introduce appropriate regularization; see [207, 254] and the references given therein.

In the following, we consider the Lavrentiev regularization approach because it elegantly accommodates our framework. The Lavrentiev-type regularization consists in approximating the pointwise state constraints $y_L(x) \le y(x) \le y_H(x)$ with the following

$$y_L(x) \le y(x) - \lambda u(x) \le y_H(x) \quad \text{a.e. in } \Omega,$$

where $\lambda > 0$ is a small parameter. As a result, the associated Lagrange multipliers can be assumed to be functions in $L^2(\Omega)$; see, e.g., [254]. The following regularized state-constrained optimal control problem results

$$\left\{\begin{array}{rcl} \min J(y,u) & := & \frac{1}{2}\|y-z\|^2_{L^2(\Omega)} + \frac{\nu}{2}\|u\|^2_{L^2(\Omega)}, \\ -\Delta y + G(y) & = & u+g, \\ y & = & 0, \\ y_L \;\le\; y-\lambda u & \le & y_H. \end{array}\right. \tag{5.95}$$

Now, introduce the auxiliary variable $v = y - \lambda u$ (use $v = y + \lambda u$ if you have the minus Laplacian) and express the control function u in terms of $v \in L^2(\Omega)$. The regularized state-constrained optimal control problem becomes

$$\left\{ \begin{array}{rcl} \min J(y,v) & := & \frac{1}{2}\|y-z\|^2_{L^2(\Omega)} + \frac{\nu}{2\lambda^2}\|y-v\|^2_{L^2(\Omega)}, \\ -\Delta y + G(y) - y/\lambda + v/\lambda & = & g, \\ y & = & 0, \\ y_L \quad \le v & \le & y_H. \end{array} \right. \tag{5.96}$$

Notice that after the transformation, an optimal control problem is obtained having a "control-constrained" structure. The solution to (5.96) is characterized by the following optimality system

$$\begin{array}{rcll} -\Delta y + G(y) - y/\lambda + v/\lambda & = & g & \text{in } \Omega, \\ y & = & 0 & \text{on } \partial\Omega, \\ -\Delta p + G'(y)p - p/\lambda + (y-z) + \gamma(y-v) & = & 0 & \text{in } \Omega, \\ p & = & 0 & \text{on } \partial\Omega, \\ (p/\lambda - \gamma(y-v), t-v) & \ge & 0, & \end{array} \tag{5.97}$$

where $\gamma = \nu/\lambda^2$ and the inequality must hold for all $t \in V_{ad}$, and V_{ad} is defined by

$$V_{ad} = \{v \in L^2(\Omega) \,|\, y_L(x) \le v(x) \le y_H(x) \text{ a.e. in } \Omega\}.$$

Next, the construction of the CSMG smoother for the regularized state-constrained optimality system given by (5.97) is discussed. We use the finite difference framework of Chapter 3 and define the two constants

$$A_{ij} = -(y_{i-1\,j} + y_{i+1\,j} + y_{i\,j-1} + y_{i\,j+1}) - h^2 g_{ij}$$

and

$$B_{ij} = -(p_{i-1\,j} + p_{i+1\,j} + p_{i\,j-1} + p_{i\,j+1}) - h^2 z_{ij}.$$

Consider the following finite difference discretized state and adjoint equations:

$$A_{ij} - \alpha\, y_{ij} + h^2 G(y_{ij}) + (h^2/\lambda)\, v_{ij} = 0, \tag{5.98}$$

$$B_{ij} - \alpha\, p_{ij} + h^2 G'(y_{ij})\, p_{ij} + (1+\gamma)h^2 y_{ij} - \gamma\, h^2 v_{ij} = 0, \tag{5.99}$$

where $\alpha = (4 + h^2/\lambda)$. In addition, we have the inequality

$$(p_{ij}/\lambda - \gamma\, y_{ij} + \gamma\, v_{ij}) \cdot (t_{ij} - v_{ij}) \ge 0$$

for all $t \in V_{adh} = \{v \in L^2_h(\Omega_h) \,|\, y_L \le v \le y_H \text{ a.e. in } \Omega_h\}$.

Now consider the Jacobian of the system (5.98)–(5.99) with respect to y_{ij}, p_{ij}, that is,

$$J_{ij} = \begin{pmatrix} -\alpha + h^2 G'(y_{ij}) & 0 \\ (1+\gamma)h^2 + h^2 G''(y_{ij})\, p_{ij} & -\alpha + h^2 G'(y_{ij}) \end{pmatrix}.$$

Hence, the following local Newton update for y_{ij} and p_{ij} at (i,j) results

$$\begin{pmatrix} \hat{y}_{ij} \\ \hat{p}_{ij} \end{pmatrix} = \begin{pmatrix} y_{ij} \\ p_{ij} \end{pmatrix} + J_{ij}^{-1} \begin{pmatrix} r_{ij}(v_{ij}) \\ s_{ij}(v_{ij}) \end{pmatrix}, \tag{5.100}$$

where $r_{ij}(v_{ij})$ and $s_{ij}(v_{ij})$ denote the residuals of (5.98) and (5.99), respectively. Both residuals depend explicitly on the control variable v_{ij}. Therefore the update above defines the values $\hat{y}_{ij}(v_{ij})$ and $\hat{p}_{ij}(v_{ij})$ as functions of v_{ij}.

Similarly to the control-constrained case, we now denote with $\tilde{v}_{ij}$ the solution to the unconstrained optimality condition equation

$$p_{ij}(v_{ij})/\lambda - \gamma\, y_{ij}(v_{ij}) + \gamma\, v_{ij} = 0.$$

It is given by $\tilde{v}_{ij} = N_{ij}/D_{ij}$, where

$$\begin{aligned} N_{ij} = &-(\lambda(\alpha B_{ij} + (1+\gamma)A_{ij}h^2 - A_{ij}\alpha\gamma\lambda - h^2(B_{ij} - A_{ij}\gamma\lambda + ((1+\gamma)h^2 - \alpha\gamma\lambda)y_{ij})G'(y_{ij}) \\ &- h^4\gamma\lambda y_{ij}G'(y_{ij})^2 + h^2 A_{ij}\, p_{ij}G''(y_{ij}) - \alpha h^2 p_{ij}\, y_{ij}G''(y_{ij}) + h^2 G(y_{ij})(h^2(1+\gamma) \\ &- \alpha\gamma\lambda + h^2\gamma\lambda G'(y_{ij}) + h^2 p_{ij}G''(y_{ij})))) \end{aligned}$$

and

$$\begin{aligned} D_{ij} = &\, h^4(1+\gamma) - 2\alpha h^2\gamma\lambda + \alpha^2\gamma\lambda^2 + 2h^2\gamma\lambda(h^2 - \alpha\lambda)G'(y_{ij}) \\ &+ h^4\gamma\lambda^2 G'(y_{ij})^2 + h^4\, p_{ij}\, G''(y_{ij}). \end{aligned}$$

Now recall that the update to v_{ij} must satisfy the constraints $y_L(x_{ij}) \le v_{ij} \le y_H(x_{ij})$. Therefore a feasible update is given by

$$v_{ij} = \begin{cases} y_{Hij} & \text{if} \quad \tilde{v}_{ij} \ge y_{Hij}, \\ \tilde{v}_{ij} & \text{if} \quad y_{Lij} < \tilde{v}_{ij} < y_{Hij}, \\ y_{Lij} & \text{if} \quad \tilde{v}_{ij} \le y_{Lij}. \end{cases} \tag{5.101}$$

Updates for the adjoint and state variables are obtained using (5.100), which defines the mappings $p_{ij} = \hat{p}_{ij}(v_{ij})$ and $y_{ij} = \hat{y}_{ij}(v_{ij})$, respectively.

The effectiveness of the resulting CSMG scheme can be seen in the following computation of a state-constrained optimal control problem. Take $z(x_1, x_2) = \sin(2\pi x_1)\sin(\pi x_2)$ and $y_L(x) = -1/2$ and $y_H(x) = 1/2$. Results for the choice $\nu = 10^{-7}$ and $\lambda = 10^{-3}$ are reported in Figures 5.9 and 5.10. In Figure 5.9, the constrained state solution is depicted. Convergence history is reported in Figure 5.10. We notice an increase of the value of the sum of the L^2-norm of the residuals during the first few iterations of the smoothing scheme. This behavior results from an increase of the residual of the state equation, while the residuals of the adjoint and control equations decrease monotonically. On the other hand, we observe typical convergence behavior of the multigrid scheme based on the proposed smoother.

The results reported in Table 5.6 show typical multigrid convergence factors that are mesh independent. These values are obtained choosing $\nu \approx \lambda^2$. With ν held fixed and decreasing λ the resulting convergence factors worsen.

Local Fourier Analysis: Linear Elliptic Case

In this section, we discuss the extension of the TG local Fourier analysis [85, 361, 340] in the case of the optimality system (5.73) assuming no constraints on the control; see [48, 61].

Consider a sequence of (infinite) grids, $G_k = \{(j_1 h_k, j_2 h_k),\ \boldsymbol{j} = (j_1, j_2) \in \mathbb{Z}^2\}$, and on these grids define the Fourier components

$$\phi_k(\boldsymbol{\theta}, \boldsymbol{j}) = e^{i j_1 \theta_1}\, e^{i j_2 \theta_2}.$$

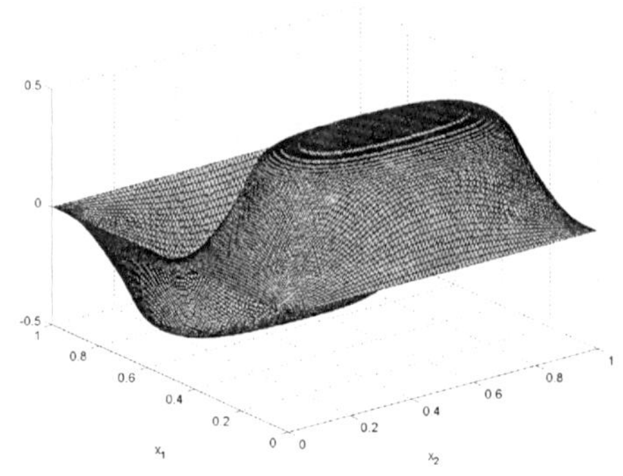

Figure 5.9. *State-constrained case. The optimal state for $\nu = 10^{-7}$ and $\lambda = 10^{-3}$. Reprinted with permission from A. Borzì, Smoothers for control- and state-constrained optimal control problems, Comput. Vis. Sci., 11(1) (2008), 59–66.*

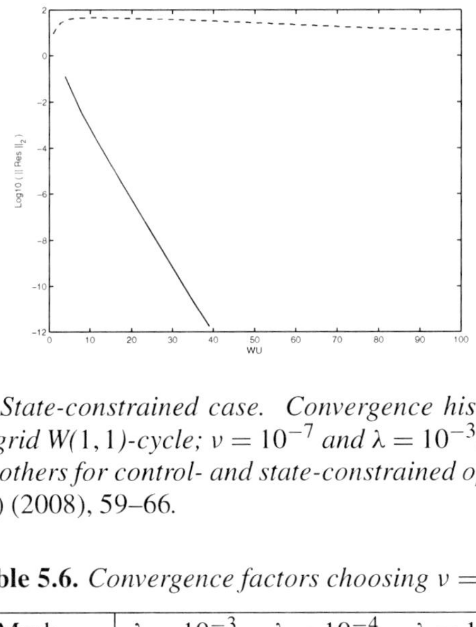

Figure 5.10. *State-constrained case. Convergence history for smoothing only (dashed line) and multigrid W(1,1)-cycle; $\nu = 10^{-7}$ and $\lambda = 10^{-3}$. Reprinted with permission from A. Borzì, Smoothers for control- and state-constrained optimal control problems, Comput. Vis. Sci., 11(1) (2008), 59–66.*

Table 5.6. *Convergence factors choosing $\nu = \lambda^2$.*

Mesh	$\lambda = 10^{-3}$	$\lambda = 10^{-4}$	$\lambda = 10^{-5}$
257×257	0.06	0.06	0.07
513×513	0.07	0.07	0.08
1025×1025	0.07	0.07	0.07

For any LF $\boldsymbol{\theta} = (\theta_1, \theta_2) \in [-\pi/2, \pi/2)^2$, consider

$$\boldsymbol{\theta}^{(0,0)} := (\theta_1, \theta_2), \qquad \boldsymbol{\theta}^{(1,1)} := (\overline{\theta_1}, \overline{\theta_2}),$$
$$\boldsymbol{\theta}^{(1,0)} := (\overline{\theta_1}, \theta_2), \qquad \boldsymbol{\theta}^{(0,1)} := (\theta_1, \overline{\theta_2}),$$

where

$$\overline{\theta_j} = \begin{cases} \theta_j + \pi \text{ if } \theta_j < 0, \\ \theta_j - \pi \text{ if } \theta_j \geq 0. \end{cases}$$

We have $\phi(\boldsymbol{\theta}^{(0,0)},\cdot) = \phi(\boldsymbol{\theta}^{(1,1)},\cdot) = \phi(\boldsymbol{\theta}^{(1,0)},\cdot) = \phi(\boldsymbol{\theta}^{(0,1)},\cdot)$ for $\boldsymbol{\theta}^{(0,0)} \in [-\pi/2,\pi/2)^2$ and $(x_1,x_2) \in G_{k-1}$. That is, we have a quadruple of distinct Fourier components that coincide (aliases) on G_{k-1} with the LF component $\phi(\boldsymbol{\theta}^{(0,0)},\cdot)$.

Denote with $\boldsymbol{\alpha} = (\alpha_1,\alpha_2)$ and consider $\boldsymbol{\alpha} \in \{(0,0),(1,1),(1,0),(0,1)\}$; then on G_{k-1} we have $\phi_k(\boldsymbol{\theta}^{\boldsymbol{\alpha}},\mathbf{x}) = \phi_{k-1}(2\boldsymbol{\theta}^{(0,0)},\mathbf{x})$. The four components $\phi_k(\boldsymbol{\theta}^{\boldsymbol{\alpha}},\cdot)$ are called harmonics. Their span is denoted with

$$E_k^{\theta} = \text{span}[\phi_k(\boldsymbol{\theta}^{\boldsymbol{\alpha}},\cdot) : \boldsymbol{\alpha} \in \{(0,0),(1,1),(1,0),(0,1)\}].$$

The purpose of this analysis is to investigate the action of the smoothing and coarse-grid correction operators on couples (ϵ_y,ϵ_p) defined by

$$\epsilon_y(\boldsymbol{j}) = \sum_{\boldsymbol{\alpha},\boldsymbol{\theta}} Y_{\boldsymbol{\alpha},\boldsymbol{\theta}}\, \phi_k(\boldsymbol{\theta}^{\boldsymbol{\alpha}},\boldsymbol{j}) \quad \text{and} \quad \epsilon_p(\boldsymbol{j}) = \sum_{\boldsymbol{\alpha},\boldsymbol{\theta}} P_{\boldsymbol{\alpha},\boldsymbol{\theta}}\, \phi_k(\boldsymbol{\theta}^{\boldsymbol{\alpha}},\boldsymbol{j}).$$

Here (ϵ_y,ϵ_p) represent the error functions for y_h and p_h and $W_{\boldsymbol{\alpha},\boldsymbol{\theta}} = (Y_{\boldsymbol{\alpha},\boldsymbol{\theta}}, P_{\boldsymbol{\alpha},\boldsymbol{\theta}})$ denote the corresponding Fourier coefficients. With this decomposition of the error, the action of one smoothing step can be expressed as $W^{(1)}_{\boldsymbol{\alpha},\boldsymbol{\theta}} = \hat{S}(\boldsymbol{\theta})\, W^{(0)}_{\boldsymbol{\alpha},\boldsymbol{\theta}}$, where $\hat{S}(\boldsymbol{\alpha},\boldsymbol{\theta})$ is the Fourier symbol [340] of the smoothing operator. To determine $\hat{S}(\boldsymbol{\alpha},\boldsymbol{\theta})$, recall that the functions $\phi_k(\boldsymbol{\theta}^{\boldsymbol{\alpha}},\mathbf{x})$ are eigenfunctions of any discrete operator described by a difference stencil on the G_k grid. Therefore we have $S_k\phi_k(\boldsymbol{\theta}^{\boldsymbol{\alpha}},\boldsymbol{j}) = \hat{S}_k(\boldsymbol{\theta})\phi_k(\boldsymbol{\theta},\boldsymbol{j})$; that is, the symbol of S_k is its (formal) eigenvalue.

Now, consider the collective Gauss–Seidel (CGS) step applied to the following optimality system obtained with the replacement $u_h = p_h/\nu$, that is,

$$-\Delta_h y_h - p_h/\nu = g_h, \tag{5.102}$$

$$-\Delta_h p_h + y_h = z_h. \tag{5.103}$$

In this case, one smoothing step at $\mathbf{x}$ corresponds to an update which sets the residuals at $\mathbf{x}$ equal to zero. In terms of Fourier modes $\boldsymbol{\theta}$, $\hat{S}_k(\boldsymbol{\theta})$ is given by

$$\hat{S}_k(\boldsymbol{\theta}) = \begin{bmatrix} -(e^{-i\theta_1}+e^{-i\theta_2}-4) & -h_k^2/\nu \\ h_k^2 & -(e^{-i\theta_1}+e^{-i\theta_2}-4) \end{bmatrix}^{-1} \times \begin{bmatrix} (e^{i\theta_1}+e^{i\theta_2}) & 0 \\ 0 & (e^{i\theta_1}+e^{i\theta_2}) \end{bmatrix}, \tag{5.104}$$

which represents the Fourier symbol of the smoothing operator.

The smoothing factor of S_k measures the action of this iteration on the HF error components and can be defined as follows

$$\mu(S_k) = \sup\left\{ |r(\hat{S}_k(\boldsymbol{\theta}))| \,:\, \boldsymbol{\theta} \in \{\boldsymbol{\theta}^{(1,1)},\boldsymbol{\theta}^{(1,0)},\boldsymbol{\theta}^{(0,1)}\} \right\}, \tag{5.105}$$

where r denotes the spectral radius. Notice that by local Fourier analysis the problem of computing the smoothing factor is reduced to that of determining the spectral radius of $\hat{S}(\boldsymbol{\theta})$, a 2×2 matrix. This task may be performed using any symbolic package. Notice that later, alternative equivalent formulation of the smoothing factor will be given.

The next step is to construct the Fourier symbol of the TG coarse-grid correction operator given by

$$\widehat{CG}_k^{k-1}(\boldsymbol{\theta}) = [\hat{I}_k - \hat{I}_{k-1}^k(\boldsymbol{\theta})(\widehat{A}_{k-1}(2\boldsymbol{\theta}))^{-1}\,\hat{I}_k^{k-1}(\boldsymbol{\theta})\,\widehat{A}_k(\boldsymbol{\theta})].$$

The Fourier symbol of the coarse-grid operator $\widehat{A}_{k-1}(\boldsymbol{\theta})$ is

$$\begin{bmatrix} -\frac{2(\cos(2\theta_1)+\cos(2\theta_2))-4}{h_{k-1}^2} & -1/\nu \\ 1 & -\frac{2(\cos(2\theta_1)+\cos(2\theta_2))-4}{h_{k-1}^2} \end{bmatrix},$$

and similarly one constructs $\widehat{A}_k(\boldsymbol{\theta})$ corresponding to the four harmonics, that is,

$$\begin{bmatrix}
l(\boldsymbol{\theta}^{(0,0)}) & 0 & 0 & 0 & -1/\nu & 0 & 0 & 0 \\
0 & l(\boldsymbol{\theta}^{(1,1)}) & 0 & 0 & 0 & -1/\nu & 0 & 0 \\
0 & 0 & l(\boldsymbol{\theta}^{(1,0)}) & 0 & 0 & 0 & -1/\nu & 0 \\
0 & 0 & 0 & l(\boldsymbol{\theta}^{(0,1)}) & 0 & 0 & 0 & -1/\nu \\
1 & 0 & 0 & 0 & l(\boldsymbol{\theta}^{(0,0)}) & 0 & 0 & 0 \\
0 & 1 & 0 & 0 & 0 & l(\boldsymbol{\theta}^{(1,1)}) & 0 & 0 \\
0 & 0 & 1 & 0 & 0 & 0 & l(\boldsymbol{\theta}^{(1,0)}) & 0 \\
0 & 0 & 0 & 1 & 0 & 0 & 0 & l(\boldsymbol{\theta}^{(0,1)})
\end{bmatrix},$$

where

$$l(\boldsymbol{\theta}^{\alpha}) = -\frac{2(\cos(\theta_1^{\alpha_1}) + \cos(\theta_2^{\alpha_2})) - 4}{h_k^2}.$$

The Fourier symbol of restriction operator is

$$\hat{I}_k^{k-1}(\boldsymbol{\theta}) = \begin{bmatrix} I(\boldsymbol{\theta}^{(0,0)}) & I(\boldsymbol{\theta}^{(1,1)}) & I(\boldsymbol{\theta}^{(1,0)}) & I(\boldsymbol{\theta}^{(0,1)}) & 0 & 0 & 0 & 0 \\ 0 & 0 & 0 & 0 & I(\boldsymbol{\theta}^{(0,0)}) & I(\boldsymbol{\theta}^{(1,1)}) & I(\boldsymbol{\theta}^{(1,0)}) & I(\boldsymbol{\theta}^{(0,1)}) \end{bmatrix},$$

where

$$I(\boldsymbol{\theta}^{\alpha}) = \frac{1}{4}(1 + \cos(\theta_1^{\alpha_1}))(1 + \cos(\theta_2^{\alpha_2})).$$

For the prolongation operator we have $\hat{I}_{k-1}^k(\boldsymbol{\theta}) = \hat{I}_k^{k-1}(\boldsymbol{\theta})^\top$. Finally, the symbol of the TG method with ν_1 presmoothing steps and ν_2 postsmoothing steps is given by

$$\widehat{TG}_k^{k-1}(\boldsymbol{\theta}) = \hat{S}_k(\boldsymbol{\theta})^{\nu_2}\,\widehat{CG}_k^{k-1}(\boldsymbol{\theta})\,\hat{S}_k(\boldsymbol{\theta})^{\nu_1}.$$

This 8×8 matrix corresponds to the pairs $(Y_{\boldsymbol{\alpha},\boldsymbol{\theta}}, P_{\boldsymbol{\alpha},\boldsymbol{\theta}})$, $\boldsymbol{\alpha} \in \{(0,0),(1,1),(1,0),(0,1)\}$. In this framework the TG convergence factor is defined as follows

$$\rho(TG_k^{k-1}) = \sup\left\{r\left(\widehat{TG}_k^{k-1}(\boldsymbol{\theta})\right) : \boldsymbol{\theta} \in [-\pi/2, \pi/2)^2\right\}. \tag{5.106}$$

In Table 5.7, values of $\mu(S_k)$ and of $\rho(TG_k^{k-1})$ corresponding to the setting $h_k = 1/64$ and $\nu \in \{10^{-4}, 10^{-8}\}$ are reported. These results show robustness of the multigrid solver with respect to values of ν and suggest mesh independence and typical multigrid efficiency.

We report in Table 5.8 the values of $\rho(TG_k^{k-1})$ and that of $\mu(S_k)^{m_1+m_2}$ obtained with the TG analysis with the forward Gauss–Seidel smoother. For comparison, the observed

Table 5.7. *Convergence factors and smoothing factors obtained with local Fourier analysis; $h = 1/64$ and different ν.*

(m_1,m_2)	ν	μ	ρ
(1,1)	10^{-4}	0.50	0.20
(2,2)	10^{-4}	0.50	0.08
(1,1)	10^{-8}	0.55	0.27
(2,2)	10^{-8}	0.55	0.12

Table 5.8. *Estimated and observed convergence factors (averages).*

(m_1,m_2)	$\mu^{m_1+m_2}$	ρ	ρ_{exp}
(1,1)	0.25	0.25	0.30
(2,1)	0.125	0.12	0.12
(2,2)	0.06	0.08	0.08
(3,2)	0.03	0.06	0.06
(3,3)	0.01	0.05	0.05

values ρ_{exp} of convergence factor defined as the "asymptotic" value of the ratio between the discrete L^2-norms of residuals resulting from two successive multigrid cycles on the finest mesh are reported. Notice that the values reported in Table 5.8 are typical of the standard Poisson model problem. These values have been obtained considering the mesh size value h ranging in the interval $[0.01, 0.25]$ corresponding to the interval of mesh sizes used in the multigrid code. The value of the weight ν has been taken in the interval $[10^{-6}, 1]$.

CSMG Convergence Theory

TG local Fourier analysis of multigrid schemes applied to optimality systems provides sharp convergence estimates at the cost of simplifying assumptions. On the other hand, the multigrid theory provided in this section does not require special assumptions on the boundary; it applies to polygonal domains and guarantees convergence of the multigrid method to weak solutions of the optimality system.

We discuss convergence of the CSMG scheme in the framework of [61, 78, 82]. For a related theoretical framework see [308].

Consider the following optimality system

$$-\Delta_h y_h - p_h/\nu = g_h, \tag{5.107}$$

$$-\Delta_h p_h + y_h = z_h. \tag{5.108}$$

The starting point for this analysis is given in Section 5.2.4, where we have presented the theoretical framework of [82] for the scalar Poisson problem

$$\begin{aligned} -\Delta y &= f \text{ in } \Omega, \\ y &= 0 \text{ on } \partial\Omega. \end{aligned} \tag{5.109}$$

After discretization, this problem becomes

$$\hat{A}_k y_k = f_k. \tag{5.110}$$

Let $\hat{P}_{k-1} : V_k \to V_{k-1}$ (resp., $I_k^{k-1} : V_k \to V_{k-1}$) be the $\hat{A}_k$ (resp., L_k^2) projections defined by $(\hat{A}_{k-1}\hat{P}_{k-1}u, v)_{k-1} = (\hat{A}_k u, I_{k-1}^k v)_k$ (resp., $(I_k^{k-1}u, v)_{k-1} = (u, I_{k-1}^k v)_k$) for all $u \in V_k$ and $v \in V_{k-1}$. Let $\hat{R}_k : V_k \to V_k$ be an iteration operator. Then the V-cycle multigrid algorithm to solve (5.110) in recursive form is given by Algorithm 5.3. For this algorithm we proved the following convergence result (Theorem 5.7). (As in the scalar case, for the purpose of a simplified analysis, we take $\nu_1 = 1$ and $\nu_2 = 0$.)

Theorem 5.15. *Let $\hat{R}_k$ satisfy* (5.26) *and* (5.27) *for $k > 1$. Then there exists a positive constant $\delta < 1$ such that*

$$(\hat{A}_k \hat{M}_k u, u)_k \le \hat{\delta}(\hat{A}_k u, u)_k \quad \forall u \in V_k,$$

where $\hat{M}_k = I_k - \hat{B}_k\hat{A}_k$.

To investigate convergence of the multigrid scheme applied to the optimality system, we first consider the decoupled symmetric system as follows

$$\begin{aligned} -\nu\Delta y &= \nu g \text{ in } \Omega, \\ y &= 0 \text{ on } \partial\Omega, \\ -\Delta p &= z \text{ in } \Omega, \\ p &= 0 \text{ on } \partial\Omega. \end{aligned} \tag{5.111}$$

This system is exactly two copies of the Poisson problem; hence the multigrid convergence theory for this system inherits the properties of the scalar case. In fact, if we define

$$\hat{\mathcal{A}}_k = \begin{pmatrix} \nu\hat{A}_k & 0 \\ 0 & \hat{A}_k \end{pmatrix}, \tag{5.112}$$

and analogously $\hat{\mathcal{B}}_k$, $\hat{\mathcal{A}}_k$, etc., as the system counterparts of $\hat{B}_k$, $\hat{A}_k$, etc., then the multigrid algorithm has exactly the same form as Algorithm 5.3 with $\hat{\mathcal{B}}_k$, $\hat{\mathcal{A}}_k$, etc., replacing $\hat{B}_k$, $\hat{A}_k$, etc. As a consequence we have the following theorem. Let $\mathbf{w}_k = (y_k, p_k) \in V_k \times V_k =: \mathcal{V}_k$.

Theorem 5.16. *Under the assumption of Theorem* 5.7, *there exists a positive constant $\delta < 1$ such that*

$$(\hat{\mathcal{A}}_k\hat{\mathcal{M}}_k\mathbf{w}, \mathbf{w})_k \le \delta(\hat{\mathcal{A}}_k\mathbf{w}, \mathbf{w})_k, \tag{5.113}$$

where δ has the same form as in Theorem 5.7.

To analyze the optimality system, we define

$$\mathcal{A}_k = \hat{\mathcal{A}}_k + d_k,$$

where

$$d_k = \begin{pmatrix} 0 & -I_k \\ I_k & 0 \end{pmatrix}.$$

We note that

$$|(d_k\mathbf{w}, \mathbf{w}')| \le C\,|\mathbf{w}|_0\,|\mathbf{w}'|_0 \tag{5.114}$$

for some constant C. Now, the multigrid algorithm corresponding to this nonsymmetric problem has exactly the same recursive form as (5.25) with $\mathcal{B}_k$, $\mathcal{A}_k$, etc., replacing B_k, A_k, etc., and thus

$$\mathcal{M}_k = \mathcal{I}_k - \mathcal{B}_k\mathcal{A}_k = [\mathcal{I}_k - \mathcal{I}_{k-1}^k\mathcal{P}_{k-1} + \mathcal{I}_{k-1}^k(\mathcal{I}_{k-1} - \mathcal{B}_{k-1}\mathcal{A}_{k-1})\mathcal{P}_{k-1}]\mathcal{S}_k, \tag{5.115}$$

where $\mathcal{I}_k$ is the identity operator on $\mathcal{V}_k$.

Next, let $\mathcal{S}_k$ and $\hat{\mathcal{S}}_k$ represent the CGS smoothing and two copies of the scalar Gauss–Seidel smoothing, respectively. Based on a subspace decomposition of $\mathcal{V}_k = \sum_{i=1}^{\ell} \mathcal{V}_k^i$ one can prove the following two lemmas [61].

Lemma 5.17. *There exists some constant $C_{\mathcal{S}}$ independent of k such that*

$$|(\hat{\mathcal{A}}_k(\mathcal{S}_k - \hat{\mathcal{S}}_k)\mathbf{w}, \mathbf{v})_k| \le C_{\mathcal{S}}\, h_k\, |\mathbf{w}|_1\, |\mathbf{v}|_1 \tag{5.116}$$

for all $\mathbf{w}, \mathbf{v} \in \mathcal{V}_k$.

Lemma 5.18. *The following inequalities hold:*

$$|(\hat{\mathcal{A}}_{k-1}(\hat{\mathcal{P}}_{k-1} - \mathcal{P}_{k-1})\mathbf{w}, \mathbf{v})_{k-1}| \le C_{\mathcal{P}}\, h_{k-1}\, |\mathbf{w}|_1 |\mathbf{v}|_1 \; for\; \mathbf{w} \in \mathcal{V}_k,\, \mathbf{v} \in \mathcal{V}_{k-1} \tag{5.117}$$

and

$$|(\hat{\mathcal{A}}_k(\mathcal{I}_k - \mathcal{I}_{k-1}^k\mathcal{P}_{k-1})\mathbf{w}, \mathbf{v})_k| \le C_I\, h_k\, |\mathbf{w}|_1\, |\mathbf{v}|_1 \; for\; \mathbf{w} \in \mathcal{V}_k,\, \mathbf{v} \in \mathcal{V}_k, \tag{5.118}$$

where $C_{\mathcal{P}}$ and C_I are some constants independent of k.

We assume that I_{k-1}^k and I_k^{k-1} represent the bilinear interpolation and full-weighting restriction operators. The prolongation operator satisfies the following conditions [82]:

$$(\hat{A}_k I_{k-1}^k u_{k-1}, I_{k-1}^k u_{k-1})_k \le (\hat{A}_{k-1} u_{k-1}, u_{k-1})_{k-1} \qquad \forall u_{k-1} \in V_{k-1}, \tag{5.119}$$

$$(I_{k-1}^k u_{k-1}, I_{k-1}^k u_{k-1})_k \le (u_{k-1}, u_{k-1})_{k-1} \qquad \forall u_{k-1} \in V_{k-1}. \tag{5.120}$$

As a consequence of this lemma we have the following

$$(\mathcal{A}_k \mathcal{I}_{k-1}^k \mathbf{w}_{k-1}, \mathcal{I}_{k-1}^k \mathbf{w}_{k-1})_k \le (\mathcal{A}_{k-1}\mathbf{w}_{k-1}, \mathbf{w}_{k-1})_{k-1} \tag{5.121}$$

for all $\mathbf{w}_{k-1} = (u_{k-1}, v_{k-1}) \in \mathcal{V}_{k-1}$. Now we prove the following theorem.

Theorem 5.19. *There exist positive constants h_0 and $\tilde{\delta} < 1$ such that for all $h_1 < h_0$ we have*

$$(\hat{\mathcal{A}}_k\mathcal{M}_k\mathbf{w}, \mathbf{w})_k \le \tilde{\delta}\,(\hat{\mathcal{A}}_k\mathbf{w}, \mathbf{w})_k \quad \forall \mathbf{w} \in \mathcal{V}_k,$$

where $\tilde{\delta} = \delta + Ch_1$ and δ is as in Theorem 5.16.

Proof. Denoting the operator norm $\|\cdot\|_{\hat{\mathcal{A}}_k}$ by $\|\cdot\|$, we show that $\|\mathcal{M}_k - \hat{\mathcal{M}}_k\| \le c_k h_1$, where c_k is uniformly bounded. The error operator $\mathcal{M}_k$ can be written as

$$\mathcal{M}_k = (\mathcal{I}_k - \mathcal{I}_{k-1}^k\mathcal{B}_{k-1}\mathcal{A}_{k-1}\mathcal{P}_{k-1})\mathcal{S}_k,$$

and $\hat{\mathcal{M}}_k$ has similar representation. We compare the error operators and write their difference as

$$\begin{aligned}\mathcal{M}_k - \hat{\mathcal{M}}_k &= (\mathcal{I}_k - \mathcal{I}_{k-1}^k \mathcal{B}_{k-1}\mathcal{A}_{k-1}\mathcal{P}_{k-1})(\mathcal{S}_k - \hat{\mathcal{S}}_k) \\ &\quad - \mathcal{I}_{k-1}^k \mathcal{B}_{k-1}\mathcal{A}_{k-1}(\mathcal{P}_{k-1} - \hat{\mathcal{P}}_{k-1})\hat{\mathcal{S}}_k + \mathcal{I}_{k-1}^k(\mathcal{M}_{k-1} - \hat{\mathcal{M}}_{k-1})\hat{\mathcal{P}}_{k-1}\hat{\mathcal{S}}_k.\end{aligned}$$

Thus in terms of the operator norm, we have

$$\begin{aligned}\|\mathcal{M}_k - \hat{\mathcal{M}}_k\| &\le \|\mathcal{I}_k - \mathcal{I}_{k-1}^k \mathcal{B}_{k-1}\mathcal{A}_{k-1}\mathcal{P}_{k-1}\|\,\|\mathcal{S}_k - \hat{\mathcal{S}}_k\| \\ &\quad + \|\mathcal{B}_{k-1}\mathcal{A}_{k-1}\|\,\|\mathcal{P}_{k-1} - \hat{\mathcal{P}}_{k-1}\|\,\|\hat{\mathcal{S}}_k\| \\ &\quad + \|\mathcal{M}_{k-1} - \hat{\mathcal{M}}_{k-1}\|\,\|\hat{\mathcal{P}}_{k-1}\hat{\mathcal{S}}_k\|.\end{aligned} \tag{5.122}$$

Let us make the induction hypothesis: $\|\mathcal{M}_{k-1} - \hat{\mathcal{M}}_{k-1}\| \le c_{k-1}h_1$, where c_{k-1} is a constant to be defined below. By the triangle inequality and Theorem 5.16,

$$\|\mathcal{M}_{k-1}\| \le \delta + c_{k-1}h_1 \tag{5.123}$$

and

$$\|\mathcal{B}_{k-1}\mathcal{A}_{k-1}\| \le 1 + \delta + c_{k-1}h_1. \tag{5.124}$$

Using the induction hypothesis, (5.119), Lemma 5.17, and Lemma 5.18, we have

$$\begin{aligned}&\|\mathcal{I}_k - \mathcal{I}_{k-1}^k \mathcal{B}_{k-1}\mathcal{A}_{k-1}\mathcal{P}_{k-1}\| \\ &\quad \le \|\mathcal{I}_k - \mathcal{I}_{k-1}^k\mathcal{P}_{k-1}\| + \|\mathcal{I}_{k-1} - \mathcal{B}_{k-1}\mathcal{A}_{k-1}\|\,\|\mathcal{P}_{k-1}\| &(5.125)\\ &\quad \le C_I h_{k-1} + \|\mathcal{M}_{k-1}\|(1 + C_\mathcal{P} h_{k-1}) &(5.126)\\ &\quad \le C_I(h_{k-1} + \delta + c_{k-1}h_1), &(5.127)\end{aligned}$$

where we assumed C_I sufficiently large so that $1 + C_\mathcal{P}h_{k-1} \le C_I$. To prove the second inequality (5.126) we used the fact that $\|\hat{\mathcal{P}}_{k-1}\| \le 1$ and the chain of inequalities $\|\mathcal{P}_{k-1}\| \le \|\hat{\mathcal{P}}_{k-1}\| + \|\mathcal{P}_{k-1} - \hat{\mathcal{P}}_{k-1}\| \le 1 + C_\mathcal{P}h_{k-1}$.

Collecting (5.122)–(5.124), and using (5.119), Lemma 5.17, Lemma 5.18, and (5.125)–(5.127), we see that

$$\begin{aligned}\|\mathcal{M}_k - \hat{\mathcal{M}}_k\| &\le C_I C_\mathcal{S}(h_{k-1} + \delta + c_{k-1}h_1)h_k \\ &\quad + C_\mathcal{P}(1 + \delta + c_{k-1}h_1)h_{k-1} + c_{k-1}h_1 \\ &\le \left(\frac{C_I C_\mathcal{S}}{2} + C_\mathcal{P}\right) h_{k-1}(1 + \delta + c_{k-1}h_1) + c_{k-1}h_1\end{aligned}$$

for all k.

Now let $\hat{C} := \frac{C_I C_\mathcal{S}}{2} + C_\mathcal{P}$ and define

$$c_k := c_{k-1} + \hat{C}h_1^{-1}h_{k-1}(1 + \delta + c_{k-1}h_1). \tag{5.128}$$

To see that the sequence c_k is uniformly bounded in k, one notes that $c_j \le c_k$ for $j \le k$ and hence

$$\begin{aligned} c_k &= c_{k-1} + \hat{C} h_1^{-1}(1+\delta+c_{k-1}h_1)h_{k-1} \\ &= c_1 + \hat{C} h_1^{-1} \sum_{j=2}^{k} (1+\delta+c_{j-1}h_1)h_{j-1} \\ &\leq c_1 + \hat{C} h_1^{-1} \sum_{j=2}^{k} (1+\delta+c_k h_1)h_{j-1} \\ &\leq c_1 + 2\hat{C}(1+\delta) + 2\hat{C} h_1 c_k. \end{aligned}$$

Now move the c_k term to the left to get

$$c_k \leq (c_1 + 2\hat{C}(1+\delta))/(1-2\hat{C}h_1),$$

provided that h_1 is small enough. Therefore, if the coarsest grid is sufficiently fine, we have $\tilde{\delta} = \delta + C h_1 < 1$. □

The constants in (5.116), (5.117), and (5.118) depend on the features of the optimality system such as, for example, nonsymmetry. They account for the induction hypothesis where the coarsest mesh size, h_1, enters the analysis and results in the estimate $\tilde{\delta} = \delta + C h_1$. The requirement for a sufficiently small h_1 has no correspondence to our numerical experience (using CGS). However, the estimate of Theorem 5.19 states that, for sufficiently small h_1, we have $\tilde{\delta} \approx \delta$; that is, the convergence factor of the multigrid method applied to the optimality system is close to the convergence factor of the multigrid scheme applied to the scalar Poisson problem. This fact agrees with our numerical experience.

5.7.2 Algebraic Multigrid Methods for Optimality Systems

In this section, we describe the extension of the AMG scheme discussed in Section 5.2.5 in order to solve optimality systems. In particular, we discuss the AMG solution of the following system of n_e linear elliptic equations

$$-\sum_{k=1}^{d} \frac{\partial}{\partial x_k}\left(d_k \frac{\partial u^{(l)}}{\partial x_k}\right) + \sum_{k=1}^{d} c_k \frac{\partial u^{(l)}}{\partial x_k} + \sum_{p=1}^{n_e} b^{(l,p)} u^{(p)} = f^{(l)}, \qquad l = 1, \ldots, n_e, \tag{5.129}$$

where the functions $d_k = d_k(\mathbf{x})$, $c_k = c_k(\mathbf{x})$, and $b^{(l,p)} = b^{(l,p)}(\mathbf{x})$ are in $L^\infty(\Omega)$ and represent the diffusion coefficients, the convection coefficients, and the reaction coefficients, respectively. The right-hand side is given by $f^{(l)} \in L_2(\Omega)$. We assume that on the boundary $\partial\Omega$, the solution is subject to general Robin boundary conditions.

Let us denote with $\hat{A}$ the differential part of (5.129) and denote with B the coupling operator. After discretization, the following system of algebraic equations is obtained

$$\hat{A}_k u_k^{(l)} + \sum_{p=1}^{n_e} B_k^{(l,p)} u_k^{(p)} = F_k^{(l)}, \qquad l = 1, \ldots, n_e, \tag{5.130}$$

where the boundary conditions enter the definition of $\hat{A}_k$ and of $F_k^{(l)}$. For $k = 1$, system (5.130) denotes the problem to be solved, which is also the "finest" algebraic problem in the AMG solution process. In order to solve (5.130), the algebraic multigrid algorithm constructs a hierarchy of coarser problems denoted by (5.130) with $k = 2, \dots, L$, where L is the index of the coarsest level (opposite to the geometrical multigrid case).

It is convenient to represent (5.130) in block form as follows

$$\hat{\mathcal{A}}_k \mathbf{u}_k + \mathcal{B}_k \mathbf{u}_k = \mathcal{F}_k, \tag{5.131}$$

where $\hat{\mathcal{A}}_k$ and $\mathcal{B}_k$ have $\hat{A}_k$ and B_k as constitutive blocks. Here, we denote with $\mathbf{u}_k = (u_k^{(l)})_{l=1,n_e}$, with $\mathcal{F}_k = (\mathcal{F}_k^{(l)})_{l=1,n_e}$, and with N_k the total number of (variables) points at level k.

As a smoother we use the CGS method based on a blockwise ordering of the unknowns and block splitting of $\mathcal{A}_k = \hat{\mathcal{A}}_k + \mathcal{B}_k$, that is, $\mathcal{A}_k = \mathcal{D}_k - \mathcal{L}_k - \mathcal{U}_k$, where $\mathcal{D}_k$ is the block-diagonal matrix, $\mathcal{L}_k$ is the lower block-triangular matrix, and $\mathcal{U}_k$ is the upper block-triangular matrix. Then, the CGS scheme can be written as

$$\mathbf{u}_{k,i}^{new} = \mathbf{u}_{k,i}^{old} + \mathcal{D}_{k,i}^{-1} \mathcal{R}_{k,i}, \quad i = 1, 2, \dots, N_k, \tag{5.132}$$

where $\mathcal{R}_{k,i} = (\mathcal{R}_{k,i}^{(l)})_{l=1,n_e}$ denotes the (dynamic) residuals of the n_e equations at the point i immediately before the relaxation step and the block $\mathcal{D}_{k,i}$ in the case of $n_e = 2$ is given by

$$\mathcal{D}_i = \begin{pmatrix} a_{ii} + b_{ii}^{1,1} & b_{ii}^{1,2} \\ b_{ii}^{2,1} & a_{ii} + b_{ii}^{2,2} \end{pmatrix}. \tag{5.133}$$

For the coarse-grid correction a system of coarse algebraic problems is constructed at level $k+1$,

$$\hat{\mathcal{A}}_{k+1} \mathbf{e}_{k+1} + \mathcal{B}_{k+1} \mathbf{e}_{k+1} = \mathcal{I}_k^{k+1} \mathcal{R}_k, \tag{5.134}$$

where $\mathbf{e}_{k+1}$ aims to represent, on the coarse level, the error $\mathbf{e}_k$ on the next finer level. The operator $\mathcal{I}_k^{k+1}$ restricts the residual computed at level k to the level $k+1$. It represents the action of the AMG restriction operator I_k^{k+1} applied n_e times. The coarse matrix of coefficient $\hat{\mathcal{A}}_{k+1}$ and the mass matrix associated with any of the $\mathcal{B}_{k+1}^{(l,p)}$ terms are defined by the Galerkin formula

$$\hat{\mathcal{A}}_{k+1} = \mathcal{I}_k^{k+1} \hat{\mathcal{A}}_k \mathcal{I}_{k+1}^k \quad \text{and} \quad \mathcal{B}_{k+1} = \mathcal{I}_k^{k+1} \mathcal{B}_k \mathcal{I}_{k+1}^k.$$

Once the coarse grid problem is solved, a new approximation to the error at level k is obtained, $\mathcal{I}_{k+1}^k \mathbf{e}_{k+1}$, and the coarse-grid correction follows:

$$\mathbf{u}_k^{new} = \mathbf{u}_k + \mathcal{I}_{k+1}^k \mathbf{e}_{k+1}, \tag{5.135}$$

where $\mathcal{I}_{k+1}^k$ is an interpolation operator. It represents the action of I_{k+1}^k applied n_e times. Here $\mathbf{u}_k$ represents the current approximation at level k as it was obtained by the smoothing process and before coarsening.

Now, notice that the AMG scheme for differential systems illustrated above can be efficiently implemented in the case where the blocks $\hat{A}_k$ are all equal to a "master" block $\tilde{A}_k$ or some of them are the transpose of this block, $\hat{A}_{k'} = \tilde{A}_k^\top$. In particular, this configuration occurs in the case of optimality systems where the discretization of the state and adjoint equations may result in blocks that are the transpose of each other. It is clear that in these cases an AMG scheme can be developed that requires one to store only the master block and the restriction operator I_k^{k+1}, constructed based only on $\hat{A}_k$. This is done in [51] for the case of elliptic control problems with jumping coefficients and in [52], where the following convection-diffusion optimality system is considered. We have

$$-\sum_{k=1}^{d} \frac{\partial}{\partial x_k}\left(d_k \frac{\partial y}{\partial x_k}\right) + \sum_{k=1}^{d} c_k \frac{\partial y}{\partial x_k} - \frac{1}{\nu} p = g \quad \text{in } \Omega, \tag{5.136}$$

$$\alpha \frac{\partial y}{\partial \nu} + \beta y = \gamma \quad \text{on } \partial\Omega, \tag{5.137}$$

$$-\sum_{k=1}^{d} \frac{\partial}{\partial x_k}\left(d_k \frac{\partial p}{\partial x_k}\right) - \sum_{k=1}^{d} c_k \frac{\partial p}{\partial x_k} + y = z \quad \text{in } \Omega, \tag{5.138}$$

$$\alpha \frac{\partial p}{\partial \nu} + \beta p = 0 \quad \text{on } \partial\Omega, \tag{5.139}$$

where the control equation $\nu u - p = 0$ has been used to eliminate the control function. It is assumed that $\sum_{k=1}^{d} \partial c_k / \partial x_k = 0$. The differential operator in (5.138) is the adjoint of the differential operator in (5.136): notice the change of sign of the convection term.

We discuss results of experiments where the desired target is given by

$$z(x, y, z) = \sin(3\pi x)\cos(3\pi y)\sin(\pi z).$$

The boundary conditions result from the following choice of the boundary parameters

$$\begin{array}{llll} \text{planes } x = 0, x = 1: & \alpha = 1, & \beta = 0, & \gamma = 0; \\ \text{planes } y = 0, y = 1: & \alpha = 1, & \beta = 0, & \gamma = 0; \\ \text{planes } z = 0, z = 1: & \alpha = 0, & \beta = 1, & \gamma = 0. \end{array} \tag{5.140}$$

We solve the optimal control problem (5.136)–(5.140), with $d_k(\mathbf{x}) = 1$, and the convecting recirculating flow given by

$$\begin{array}{rcl} c_1 & = & -\sin \pi x \cos \pi y, \\ c_2 & = & \sin \pi y \cos \pi x, \\ c_3 & = & 0. \end{array} \tag{5.141}$$

Two techniques are considered: standard coarsening and aggressive coarsening. The number of coarse points obtained for the present test case is reported in Table 5.9. Observe that the reduction factor of the number of points from one level to the next coarser one is almost independent of the number of initial points. Aggressive coarsening is used only to pass from the finest to the next coarser level. In the first coarsening step, standard coarsening is only approximately halving the number of variables while aggressive coarsening reduces this number by a factor of approximately eight.

In Table 5.10 we report the convergence behavior of AMG solving the optimal control problem. We notice a weak dependency of the convergence factor ρ on the size of the problem. This behavior is observed when solving the optimal control problem with recirculating convection which appears to be the "worst case" for the AMG solver.

Table 5.9. *Number of variables at various levels of the coarsening process.*

	Standard coarsening		Aggressive coarsening	
k	N_i	N_i	N_i	N_i
1	120000	499200	120000	499200
2	60000	249600	14198	58067
3	10000	41608	6259	25454
4	1255	5225	991	3985
5	272	1210	161	639
6	41	193	31	98
7	-	34	-	18

Table 5.10. *Convergence properties depending on* N_i $(\nu = 10^{-6})$.

	Standard coarsening				Aggressive coarsening			
N_i	ρ	No. iter.	c_i/c_r	$\|\|u-z\|\|$	ρ	No. iter.	c_i/c_r	$\|\|u-z\|\|$
120000	0.14	15	2.65/2.68	$2.79\,10^{-2}$	0.40	33	1.70/1.69	$2.79\,10^{-2}$
499200	0.19	21	2.69/2.73	$2.61\,10^{-2}$	0.51	42	1.72/1.72	$2.60\,10^{-2}$
712800	0.23	25	2.70/2.74	$2.59\,10^{-2}$	0.60	47	1.72/1.72	$2.60\,10^{-2}$

Table 5.11. *Tracking properties depending on* ν; $N_i = 120000$, *standard coarsening.*

ν	$\|\|u-z\|\|$	ρ	No. iter.
10^{-4}	$2.61\,10^{-1}$	0.08	12
10^{-6}	$2.79\,10^{-2}$	0.14	15
10^{-8}	$3.89\,10^{-3}$	0.03	8

Also in Table 5.10 we compare the performance of AMG when using standard coarsening and aggressive coarsening. The clear advantage of aggressive coarsening is smaller values of the complexity factors compared with those obtained by standard coarsening. However, as may be expected [325], the convergence factor obtained using aggressive coarsening is larger than that obtained by standard coarsening.

In Table 5.11 tracking errors and convergence factors depending on ν are reported. In all experiments we observed AMG convergence that does not deteriorate as ν tends to be small. Notice that the computational performance of gradient methods, applied to optimal control problems belonging to the class considered here, worsen as ν tends to zero; see the discussion in [36].

A CSMG Scheme for a Shape Optimization Problem

We now describe a CSMG multigrid scheme for the shape optimization problem (2.9)–(2.10), that is, a simplified shape optimization problem formulated as an elliptic boundary control problem. Here the shape should be optimized so that a flow along the boundary approaches a certain pressure distribution.

Let us assume that $\Omega = (0,1)^2$ and Γ_1 consists of the boundary points with $x_2 = 0$, the bottom boundary. The optimality system is given by

$$\begin{array}{rcll} -\Delta y &=& 0 & \text{in } \Omega, \\ \frac{\partial y}{\partial \vec{n}} &=& \frac{\partial u}{\partial x} & \text{on } \Gamma_1, \\ y &=& 0 & \text{on } \partial\Omega\backslash\Gamma_1, \\ -\Delta p &=& 0 & \text{in } \Omega, \\ \frac{\partial p}{\partial \vec{n}} &=& -\frac{\partial}{\partial x}(\frac{\partial y}{\partial x} - P(x)) & \text{on } \Gamma_1, \\ p &=& 0 & \text{on } \partial\Omega\backslash\Gamma_1. \end{array} \tag{5.142}$$

As shown in [331], an appropriate choice for determining the control is to set $u = \frac{\partial p}{\partial x}$.

Denote first-order backward and forward partial derivatives of v_h in the x_i direction by ∂_i^- and ∂_i^+, respectively. They are given by

$$\partial_i^- v_h(\mathbf{x}) = \frac{v_h(\mathbf{x}) - v_h(\mathbf{x} - \hat{i}\,h)}{h} \qquad \text{and} \qquad \partial_i^+ v_h(\mathbf{x}) = \frac{v_h(\mathbf{x} + \hat{i}\,h) - v_h(\mathbf{x})}{h},$$

where $\hat{i}$ denotes the i coordinate direction vector and v_h is extended by 0 on grid points outside of Ω; see [174]. We have the five-point Laplacian $\Delta_h = \partial_1^+\partial_1^- + \partial_2^+\partial_2^-$.

In this framework, the discretization of (5.142) gives

$$\begin{array}{rcll} -\Delta_h y &=& 0 & \text{in } \Omega, \\ \partial_h^n y &=& \partial_1^+\partial_1^- p_h & \text{on } \Gamma_1, \\ y &=& 0 & \text{on } \partial\Omega\backslash\Gamma_1, \\ -\Delta_h p &=& 0 & \text{in } \Omega, \\ \partial_h^n p &=& -\partial_1^+\partial_1^- y_h + P_x & \text{on } \Gamma_1, \\ p &=& 0 & \text{on } \partial\Omega\backslash\Gamma_1, \end{array} \tag{5.143}$$

where $P_x = \frac{\partial P}{\partial x}$, and $\partial_h^n v_h = -(v_{i\,j+1} - v_{i\,j-1})/2h$ at Γ_1.

Notice that to solve (5.143) we need to realize the coupling on Γ_1 while in the interior of the domain we have two Laplace equations. For this purpose we combine the Neumann boundary conditions with the stencil of the discrete elliptic operators considered at the boundary. For $\mathbf{x} = (ih, jh)$ being a boundary grid point on the side $x_2 = 0$ we have

$$-(2y_{i+1\,j} + y_{i\,j-1} + y_{i\,j+1} - 4y_{i\,j}) - \frac{2}{h}(p_{i+1\,j} - 2p_{i\,j} + p_{i-1\,j}) = 0,$$

$$-(2p_{i+1\,j} + p_{i\,j-1} + p_{i\,j+1} - 4p_{i\,j}) + \frac{2}{h}(y_{i+1\,j} - 2y_{i\,j} + y_{i-1\,j}) = 2hP_x.$$

The equations obtained in this way have the same structure as (5.76) and (5.77) (without constraints) and the application of the CGS iteration follows along the same lines as above. Hence we can apply the CSMG multigrid method previously described. Notice that bilinear prolongation and full-weighting restriction have to be used to guarantee the right scaling for the coarse-grid problem formulation at the boundary; see the discussion in [340]. Clearly, on the boundary these transfer operators are mirrored.

To numerically validate the FAS multigrid algorithm for solving (5.143), consider the desired profile given by $P_x = 1$. Results for this case are reported in Table 5.12 for a V-cycle with two pre- and postsmoothing steps. These results show typical multigrid efficiency.

Table 5.12. *Results of experiments for a shape optimization problem.*

Mesh	129×129	257×257	517×517
ρ	0.08	0.08	0.08

CSMG Schemes for Bilinear Optimization Problems

An important class of optimization problems is represented by bilinear optimal control problems and inverse problems. Bilinear optimal control problems with PDEs is a less investigated subject representing a nonlinear control strategy with the aim of obtaining better system response than possible with linear control. Bilinear structures typically arise in quantum control problems and in parameter identification problems as we discuss in the applications chapter. A multigrid scheme for solving parameter identification problems of the form $\nabla \cdot (e^u \nabla y) = q$ is presented in [12]. Here we illustrate a CSMG scheme and a MGOPT scheme discussed in [345] to solve the following bilinear optimal control problem

$$\left\{ \begin{array}{rll} \min & J(y,u) := \frac{1}{2}\|y-z\|^2_{L^2(\Omega)} + \frac{\nu}{2}\|u\|^2_{L^2(\Omega)}, & \\ -\Delta y - u\, y & = g \quad \text{in} \quad \Omega, & \\ y & = 0 \quad \text{on} \quad \partial\Omega, & \end{array} \right. \tag{5.144}$$

where $z \in L^2(\Omega)$ is the target function. Similar to the linear case, we choose $f \in L^2(\Omega)$ and $U = L^2(\Omega)$. The solution to problem (5.144) is characterized by the following optimality system

$$\begin{array}{rcll} -\Delta y - uy & = & g & \text{in } \Omega, \\ y & = & 0 & \text{on } \partial\Omega, \\ -\Delta p + y - up & = & z & \text{in } \Omega, \\ p & = & 0 & \text{on } \partial\Omega, \\ \nu u - yp & = & 0 & \text{in } \Omega. \end{array} \tag{5.145}$$

For a given u, $y(u)$ and $p(u)$ are the solutions of the state and adjoint equations with homogeneous Dirichlet boundary conditions. Their existence requires that the operator $(\Delta + u)$ be invertible. Now notice that u cannot be constant on Ω since it satisfies $u = yp/\nu$ and it inherits the homogeneous boundary conditions as the state and the adjoint variables. Then we can use Lemma 3.2 from [158] based on results given in [139].

We can now derive the reduced gradient and the reduced Hessian as follows

$$\nabla \hat{J}(u) = \nu u - y\, p \tag{5.146}$$

and

$$\nabla^2 \hat{J}(u) = \nu \mathrm{I} + y(\Delta + u)^{-2} y + p(\Delta + u)^{-1} y + y(\Delta + u)^{-1} p. \tag{5.147}$$

In this case, we cannot state positivity of the reduced Hessian, and it is difficult to find the ellipticity and Lipschitz constants, unlike in the linear case. However, since $p = (\Delta + u)^{-1}(y - z)$ we can expect that for sufficiently accurate tracking, i.e., small $\|y - z\|$, and moderate values of ν the reduced Hessian is a positive definite operator. This situation may take place whenever z is (almost) attainable. That is, there exists a u such that $y(u) \approx z$.

Next, we define the collective smoothing iteration for the finite difference discretization of the bilinear elliptic optimal control problem given above. We have

$$
\begin{aligned}
-\Delta_k y_k - u_k y_k &= g_k, \\
-\Delta_k p_k + y_k - u_k p_k &= z_k, \\
\nu u_k - y_k p_k &= 0.
\end{aligned}
\tag{5.148}
$$

In this case, we have

$$
\begin{aligned}
-(y_{i-1,j} + y_{i+1,j} + y_{i,j-1} + y_{i,j+1}) + 4y_{ij} - h^2 u_{ij}\, y_{ij} &= h^2 g_{ij}, \\
-(p_{i-1,j} + p_{i+1,j} + p_{i,j-1} + p_{i,j+1}) + 4p_{ij} + h^2 y_{ij} - h^2 u_{ij}\, p_{ij} &= h^2 z_{ij}, \\
\nu u_{ij} - y_{ij}\, p_{ij} &= 0.
\end{aligned}
$$

We first set

$$
\begin{aligned}
A_{ij} &= -(y_{i-1,j} + y_{i+1,j} + y_{i,j-1} + y_{i,j+1}) - h^2 g_{ij}, \\
B_{ij} &= -(p_{i-1,j} + p_{i+1,j} + p_{i,j-1} + p_{i,j+1}) - h^2 z_{ij}.
\end{aligned}
\tag{5.149}
$$

Now, the values A_{ij} and B_{ij} are considered constant during the update of the variables at ij. Hence, we have the following system of equations of three variables y_{ij}, u_{ij}, and p_{ij}

$$
\begin{aligned}
A_{ij} + 4y_{ij} - h^2 u_{ij}\, y_{ij} &= 0, \\
B_{ij} + 4p_{ij} + h^2 y_{ij} - h^2 u_{ij}\, p_{ij} &= 0, \\
\nu u_{ij} - y_{ij}\, p_{ij} &= 0.
\end{aligned}
$$

We see that the resulting system of equations is nonlinear, and thus computing the updates for the variables u_{ij}, y_{ij}, and p_{ij} requires us to apply a local Newton step. This approach results in a nonrobust smoothing iteration apparently because we have multiple solutions for u_{ij} that are close. In fact, as we show below, the condition $\nu u_{ij} - y_{ij}(u_{ij})\, p_{ij}(u_{ij}) = 0$ results in a quartic polynomial equation for u_{ij}, and therefore four roots are possible. To determine these solutions we construct the quartic polynomial and solve it exactly by using the Cardano–Tartaglia formula. In this way we can explore among the possible solutions of the optimization step. To construct the quartic polynomial, we can define $y_{ij} = y_{ij}(u_{ij})$ and $p_{ij} = p_{ij}(u_{ij})$ as functions of u_{ij},

$$
\begin{aligned}
y_{ij}(u_{ij}) &= \tfrac{-1}{4-h^2 u_{ij}} A_{ij}, \\
p_{ij}(u_{ij}) &= \tfrac{1}{(4-h^2 u_{ij})^2} (h^2 A_{ij} + h^2 B_{ij} u_{ij} - 4B_{ij}),
\end{aligned}
$$

and equate the reduced gradient to zero, i.e., $\nabla \hat{J}(u) = \nu u - y(u)p(u) = 0$. Hence we have a quartic polynomial equation in u_{ij} given by

$$
\nu h^6 u_{ij}^4 - 12\nu h^4 u_{ij}^3 + 48\nu h^2 u_{ij}^2 - (64\nu + h^2 A_{ij} B_{ij}) u_{ij} - (h^2 A_{ij}^2 - 4A_{ij} B_{ij}) = 0.
$$

The solutions of the quartic polynomial are either four real or two real and two complex. The two complex conjugate solutions can be disregarded. In order to find the minimizer, we choose the minimum real solution of the quartic polynomial which minimizes

$$
\hat{J}_{ij}(u) = \frac{1}{2}(y_{ij}(u) - z_{ij})^2 + \frac{\nu}{2} u_{ij}^2.
$$

With this condition, we get a robust and efficient CSMG smoothing iteration.

We see that in the bilinear case the construction of an appropriate CSMG scheme is more involved. It is therefore interesting to consider the solution of the bilinear control problem using the MGOPT method. To apply this method, we choose a classical optimization scheme as single-grid solver and embed this scheme in the MGOPT algorithm as illustrated in Section 5.4. We consider two gradient-type schemes: the steepest descent method and the NCG scheme discussed in Chapter 4. To evaluate the gradient (5.146) at u, we solve the state and the adjoint equations very accurately. The line search in the coarse-grid correction step uses the Armijo rule [265].

In the following, we report results of numerical experiments obtained with the CSMG scheme and with the MGOPT scheme applied to the minimization of the reduced cost functional.

Results of numerical experiments are obtained with the following setting. Let $\Omega = (0,1)\times(0,1)$, and let $f, z \in L^2(\Omega)$ be given by

$$g(x,y) = 1,$$
$$z(x,y) = \begin{cases} 2 & \text{on } (0.25,0.75)\times(0.25,0.75), \\ 1 & \text{otherwise.} \end{cases}$$

Notice that the target z is not attainable by any control because of its nonzero value at the boundary of the domain.

Numerical results are shown in Tables 5.13 and 5.14 and in Figure 5.11. We can see from Table 5.13 that the CSMG method exhibits almost independence of the number of iterations on ν on the size of the mesh where the problem is being solved. On the other hand the computational effort required by the MGOPT scheme increases as $1/\nu$. In Table 5.13, we also report on a separate column the CPU time for the computation of the roots of the quartic polynomial, as this takes about 70% of the whole computational time. We show in Table 5.14 the computational time for the one-grid optimization scheme using steepest descent and NCG schemes. For solution processes exceeding 20,000 seconds we stopped the calculation. This table shows that MGOPT with NCG is faster than MGOPT with steepest descent.

Table 5.13. *Results of bilinear elliptic optimal control problem using the CSMG method (* time for computing the roots).*

ν	Mesh	Iter.	$\hat{J}$	$\Vert\nabla\hat{J}\Vert_{L_2}$	ρ	Time (sec)	Time (sec)*
	65×65	8	0.863	2.06e-11	0.081	47.6	33.2
1e-2	129×129	9	0.843	1.85e-12	0.083	221.9	155.8
	257×257	9	0.834	1.90e-12	0.084	912.8	641.3
	65×65	10	0.158	1.55e-08	0.287	59.7	41.5
1e-4	129×129	10	0.151	1.46e-08	0.366	243.1	170.1
	257×257	10	0.148	1.45e-08	0.434	997.5	696.7

The presence of constraints on the optimization function u can be easily implemented in the CSMG smoothing by projection of the u-update obtained from the solution of the quartic polynomial equation.

The MGOPT solution of control-constrained bilinear control problems is discussed in [344], where a projected gradient optimization scheme is considered.

Table 5.14. *Results of CPU time (seconds) of bilinear elliptic optimal control problem using the steepest descent (SD) method, MGOPT with SD (MGOPT^1), NCG, and MGOPT with NCG (MGOPT^2) (– longer than* 20,000 *seconds).*

ν	Mesh	SD	MGOPT^1	NCG	MGOPT^2
	65×65	1.2	1.0	0.8	0.8
1e-2	129×129	5.2	4.9	4.0	3.7
	257×257	25.1	19.1	19.0	18.8
	65×65	–	–	17079.0	160.7
1e-4	129×129	–	–	–	805.7
	257×257	–	–	–	4112.4

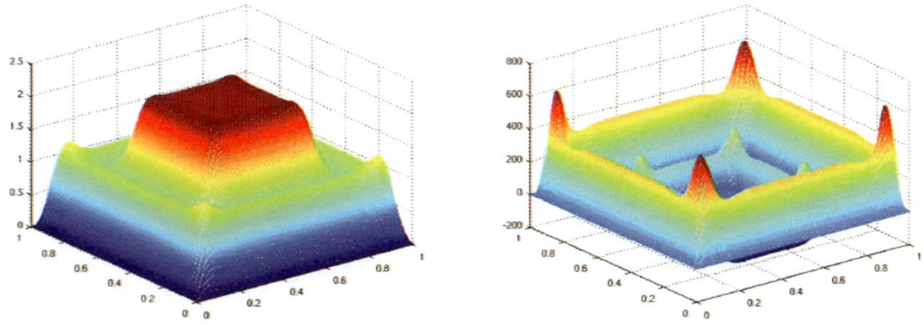

Figure 5.11. *Numerical solutions for the state (left) and control (right) variables of the bilinear elliptic optimal control problem with* $\nu = 10^{-6}$.

For an application, consider a unit square domain $\Omega = (0,1)\times(0,1)$, with $g, z \in L^2(\Omega)$, given by

$$g(x_1,x_2) = \sin(2\pi x_1)\sin(2\pi x_2),$$
$$z(x_1,x_2) = 1+\sin(2\pi x_1)\sin(2\pi x_2).$$

Notice that the target function z is not attainable by any control due to its boundary values. We have box constraints.

The numerical results for computational performance are shown in Table 5.15. In Figure 5.12, the optimal solution for $\nu = 10^{-4}$ is depicted. See [344] for more details.

5.7.3 A CSMG Scheme with FEM Discretization

In this section, we discuss the realization of the CSMG scheme in the case of FEM discretization. Consider the optimality system

$$\begin{aligned} -\nu\Delta y - p &= \nu f \quad \text{in } \Omega, \\ -\Delta p + y &= z \quad \text{in } \Omega, \end{aligned} \tag{5.150}$$

Table 5.15. *Results of control-constrained bilinear elliptic optimal control problem using MGOPT with gradient projection method.*

ν	Mesh	$\|r_y\|_{L^2}$	$\|r_p\|_{L^2}$	$\|u^l - u^l(1)\|_{L^2}$	Time (sec)
	129×129	2.086e-14	1.017e-13	4.581e-06	2.7
10^{-2}	257×257	6.798e-14	3.710e-13	6.044e-06	11.8
	513×513	6.290e-13	3.580e-12	6.277e-06	49.3
	129×129	1.567e-14	9.977e-14	6.057e-06	3.6
10^{-4}	257×257	4.219e-14	2.912e-13	1.841e-06	16.2
	513×513	1.631e-13	9.025e-13	1.364e-06	59.4

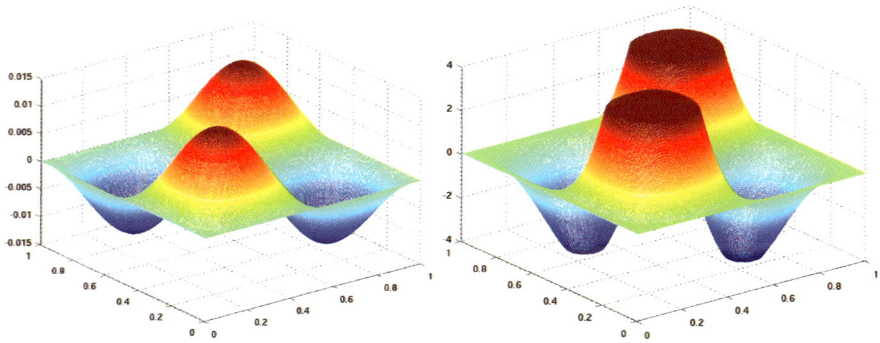

Figure 5.12. *Numerical solutions for the state (left) and control (right) variables of the control-constrained bilinear elliptic optimal control problem using* $\nu = 10^{-4}$.

together with the boundary conditions

$$y = 0 \quad \text{and} \quad p = 0 \quad \text{on} \quad \partial\Omega. \tag{5.151}$$

We convert this problem into its variational formulation as follows: find $(y, p) \in V \times V$ such that

$$\begin{aligned} \int_\Omega (\nu \nabla y \cdot \nabla v_1 - p v_1 - \nu f v_1)\, dx = 0 \quad \forall v_1 \in V, \\ \int_\Omega (\nabla p \cdot \nabla v_2 + y v_2 - z v_2)\, dx = 0 \quad \forall v_2 \in V, \end{aligned} \tag{5.152}$$

where $V \times V$ is the solution space. Next we replace $V \times V$ in (5.152) by a finite-dimensional subspace $V_k \times V_k \subset V \times V$, and the dimension of V_k is n_k. We assume that linearly independent basis functions ϕ_i, $i = 1, 2, \ldots, n_k$, span V_k. Then for $i = 1, 2, \ldots, n_k$, we have

$$\begin{aligned} \int_{\Omega_k} (\nu \nabla_k y_k \cdot \nabla_k \phi_i - p_k \phi_i - \nu f_k \phi_i)\, dx &= 0, \\ \int_{\Omega_k} (\nabla_k p_k \cdot \nabla_k \phi_i + y_k \phi_i - z_k \phi_i)\, dx &= 0. \end{aligned} \tag{5.153}$$

We can write the approximate solutions y_k and p_k in terms of basis functions ϕ_i of V_k as follows

$$y_k(x) = \sum_{j=1}^{n_k} Y_j \phi_j(x) \quad \text{and} \quad p_k(x) = \sum_{j=1}^{n_k} P_j \phi_j(x), \tag{5.154}$$

where Y_j and P_j, $j = 1, \ldots, n_k$, are the unknowns. Thus (5.153) can be rewritten as

$$\begin{pmatrix} \nu Q_k & -M_k \\ M_k & Q_k \end{pmatrix} \begin{pmatrix} Y_k \\ P_k \end{pmatrix} = \begin{pmatrix} \nu G_k^1 \\ G_k^2 \end{pmatrix}, \tag{5.155}$$

where

$$\begin{aligned} Q_{i,j} &= \int_\Omega \nabla\phi_j \cdot \nabla\phi_i \, dx, & M_{i,j} &= \int_\Omega \phi_j \phi_i \, dx, \\ G_i^1 &= \int_\Omega f_h \phi_i \, dx, & G_i^2 &= \int_\Omega z_h \phi_i \, dx, \end{aligned} \tag{5.156}$$

where for simplicity we have omitted the index k. Here Q_k and M_k are called the stiffness and mass matrices, respectively. Let

$$\tilde{A} = \begin{pmatrix} \nu Q & -M \\ M & Q \end{pmatrix}, \quad \tilde{U} = \begin{pmatrix} Y \\ P \end{pmatrix}, \quad \text{and } \tilde{F} = \begin{pmatrix} \nu G^1 \\ G^2 \end{pmatrix};$$

then we have

$$\tilde{A}\tilde{U} = \tilde{F}, \tag{5.157}$$

where $\tilde{A}$ is a $2n_k \times 2n_k$ matrix and vectors $\tilde{U}$ and $\tilde{F}$ are of length $2n_k$. In order to introduce a collective smoothing scheme, we define the following inner product

$$(\tilde{A}_{i+d,\cdot}, \tilde{U})_d = \sum_{\substack{j=1 \\ j \neq \{i, i+n_k\}}}^{2n_k} \tilde{a}_{i+d,j} \tilde{u}_j \, . \tag{5.158}$$

By choosing the ith and $(i+n_k)$th rows of matrix $\tilde{A}$, we get

$$\begin{pmatrix} \tilde{A}_{i,i} & \tilde{A}_{i,i+n_k} \\ \tilde{A}_{i+n_k,i} & \tilde{A}_{i+n_k,i+n_k} \end{pmatrix} \begin{pmatrix} \tilde{U}_i \\ \tilde{U}_{i+n_k} \end{pmatrix} = \begin{pmatrix} \tilde{F}_i - (\tilde{A}_{i,\cdot}, \tilde{U})_0 \\ \tilde{F}_{i+n_k} - (\tilde{A}_{i+n_k,\cdot}, \tilde{U})_{n_k} \end{pmatrix}.$$

Let

$$\tilde{D} = \det \begin{pmatrix} \tilde{A}_{i,i} & \tilde{A}_{i,i+n_k} \\ \tilde{A}_{i+n_k,i} & \tilde{A}_{i+n_k,i+n_k} \end{pmatrix} = \tilde{A}_{i,i}\tilde{A}_{i+n_k,i+n_k} - \tilde{A}_{i,i+n_k}\tilde{A}_{i+n_k,i}.$$

Thus we obtain a collective update given by

$$\begin{pmatrix} \tilde{U}_i \\ \tilde{U}_{i+n_k} \end{pmatrix} = \frac{1}{\tilde{D}} \begin{pmatrix} \tilde{A}_{i+n_k,i+n_k} & -\tilde{A}_{i,i+n_k} \\ -\tilde{A}_{i+n_k,i} & \tilde{A}_{i,i} \end{pmatrix} \begin{pmatrix} \tilde{F}_i - (\tilde{A}_{i,\cdot}, \tilde{U})_0 \\ \tilde{F}_{i+n_k} - (\tilde{A}_{i+n_k,\cdot}, \tilde{U})_{n_k} \end{pmatrix}. \tag{5.159}$$

This collective smoothing solves the system componentwise by treating both variables collectively.

Next, we consider the elliptic optimal control problem with constraints on the control. Recall the state and the adjoint equations in the optimality system,

$$\begin{aligned} -\Delta y - u &= f, \\ -\Delta p + y &= z, \end{aligned} \tag{5.160}$$

together with the homogeneous Dirichlet boundary conditions.

In this case, we have two equations, a variational inequality, and three unknowns. We will use the components of the system matrix $\tilde{A}$ in (5.157) and change the vector $\tilde{U}$ as needed for each equation. Thus, we have

$$\begin{aligned} (Q \quad -M)\begin{pmatrix} Y \\ U \end{pmatrix} &= G^1, \\ (M \quad Q)\begin{pmatrix} Y \\ P \end{pmatrix} &= G^2. \end{aligned} \tag{5.161}$$

Then the following control-constrained optimality system results

$$\begin{aligned} QY - MU &= G^1, \\ QP + MY &= G^2, \\ (\nu U - P, \Phi - U) &\geq 0 \qquad \forall \Phi \in \mathcal{U}_k. \end{aligned} \tag{5.162}$$

A collective smoothing step updates the values Y_i, U_i, and P_i such that the resulting residuals of the state and adjoint equations at that point are zero. We first set

$$C_1 = \sum_{\substack{j=1 \\ j\neq i}}^{n_k} q_{i,j} Y_j, \quad C_2 = \sum_{\substack{j=1 \\ j\neq i}}^{n_k} q_{i,j} P_j, \quad C_3 = \sum_{\substack{j=1 \\ j\neq i}}^{n_k} m_{i,j} Y_j, \text{ and } \quad C_4 = \sum_{\substack{j=1 \\ j\neq i}}^{n_k} m_{i,j} U_j. \tag{5.163}$$

The values C_1, C_2, C_3, and C_4 are considered constant during the updates for the variables at i. Hence, we have the following system of equations of three variables Y_i, U_i, and P_i

$$\begin{aligned} C_1 + q_{i,i} Y_i - C_4 - m_{i,i} U_i - G_i^1 &= 0, \\ C_2 + q_{i,i} P_i + C_3 + m_{i,i} Y_i - G_i^2 &= 0. \end{aligned}$$

Since this is a linear system, we can compute the updates for the variables Y_i and P_i as functions of U_i in the following way:

$$\begin{aligned} Y_i(U_i) &= \frac{1}{q_{i,i}} \left(G_i^1 - C_1 + C_4 + m_{i,i} U_i \right), \\ P_i(U_i) &= \frac{1}{q_{i,i}^2} \left[q_{i,i} \left(G_i^2 - C_2 - C_3 \right) - m_{i,i} \left(G_i^1 - C_1 + C_4 + m_{i,i} U_i \right) \right]. \end{aligned} \tag{5.164}$$

To obtain an update U_i, replace the expression for P_i in the inequality constraint and define the auxiliary variable as

$$\widetilde{U}_i = \frac{1}{\nu q_{i,i}^2 + m_{i,i}^2} \left[q_{i,i} \left(G_i^2 - C_2 - C_3 \right) - m_{i,i} \left(G_i^1 - C_1 + C_4 \right) \right]. \tag{5.165}$$

Then the new value for u_i resulting from the smoothing step is given by

$$U_i = \begin{cases} \overline{u}_i & \text{if} \quad \widetilde{U}_i \geq \overline{u}_i, \\ \widetilde{U}_i & \text{if} \quad \underline{u}_i < \widetilde{U}_i < \overline{u}_i, \\ \underline{u}_i & \text{if} \quad \widetilde{U}_i \leq \underline{u}_i. \end{cases} \tag{5.166}$$

With this new value of U_i, new values for Y_i and P_i are obtained. This completes the description of the collective smoothing step for the finite element control constrained case.

The collective smoothing step defined by (5.164)–(5.166) satisfies the inequality constraint in the optimality system. Consider any grid point wherein $\widetilde{U} \leq \underline{u}$; then, from (5.166), $U = \underline{u}$. Thus, $(\Phi - U) \geq 0$ for any $\Phi \in \mathcal{U}$. On the other hand, we have

$$\begin{aligned} \nu U - P &= \nu U - \frac{1}{q_{i,i}^2}\left[q_{i,i}\left(G_i^2 - C_2 - C_3\right) - m_{i,i}\left(G_i^1 - C_1 + C_4 + m_{i,i}U_i\right)\right] \\ &= \frac{1}{q_{i,i}^2}\left[\left(q_{i,i}^2\nu + m_{i,i}^2\right)U_i - q_{i,i}\left(G_i^2 - C_2 - C_3\right) - m_{i,i}\left(G_i^1 - C_1 + C_4\right)\right] \\ &\geq \frac{1}{q_{i,i}^2}\left[\left(q_{i,i}^2\nu + m_{i,i}^2\right)\widetilde{U} - q_{i,i}\left(G_i^2 - C_2 - C_3\right) - m_{i,i}\left(G_i^1 - C_1 + C_4\right)\right] \\ &= 0. \end{aligned}$$

Therefore, $(\nu U - P, \Phi - U) \geq 0$ for all $\Phi \in \mathcal{U}$. Similarly, we can prove that if $\widetilde{U} \geq \overline{u}$, then the choice $U = \overline{u}$ satisfies the inequality constraint. The case $\underline{u} < \widetilde{U} < \overline{u}$ is obvious.

A typical multigrid method uses a sequence of l nested triangulations of Ω of increasing fineness $\mathcal{T}_1 \subset \mathcal{T}_2 \subset \cdots \subset \mathcal{T}_l$, where $\mathcal{T}_l$ denotes the finest grid. Let $\mathcal{T}_1$ be given and $\mathcal{T}_k, k \geq 2$, be obtained from $\mathcal{T}_{k-1}$ via a regular subdivision; i.e., edge midpoints of $\mathcal{T}_{k-1}$ are connected by new edges to form $\mathcal{T}_k$. Furthermore, h_k denotes the mesh size of $\mathcal{T}_k$, i.e., $h_k := \max_{T \in \mathcal{T}_k} \operatorname{diam} T$ and n_k denotes the number of nodes in $\mathcal{T}_k$, where $n_{k-1} < n_k$. To each triangulation $\mathcal{T}_k, k = 1, \ldots, l$, we define the associated sequence of finite element spaces $V_1 \subset V_2 \subset \cdots \subset V_l$. The mesh hierarchy induces linear systems $A_k u_k = f_k, k = 1, 2, \ldots, l$, on each grid level k. Note that $\mathcal{T}_{k-1} \subset \mathcal{T}_k \Rightarrow V_{k-1} \subset V_k$ and $h_k = \frac{1}{2} h_{k-1}$ since each triangle $T \in \mathcal{T}_{k-1}$ is subdivided into four similar triangles in $\mathcal{T}_k$. In addition to the grid levels, we need transfer operators between coarser and finer grids. Since we have assumed that the finite element discretization has piecewise linear basis functions, we can now define an interpolation operator from a coarse triangulation $\mathcal{T}_{k-1}$ to a fine triangulation $\mathcal{T}_k$ as follows

$$\left(I_{k-1}^k\right)_{ij} = \begin{cases} 1 & \text{for } i = j, \text{ where the nodes } P_i \text{ are both in the triangulations} \\ & \mathcal{T}_k \text{ and } \mathcal{T}_{k-1}\ \forall i, j = 1, 2, \ldots, n_{k-1}, \\ \frac{1}{2} & \text{for } j = i_1 \text{ and } j = i_2, n_{k-1} < i \leq n_k, \\ & \text{where the nodes } P_{i1} \text{ and } P_{i2} \text{ are the boundary nodes of every triangle} \\ & T \in \mathcal{T}_{k-1}, \text{ of where the nodes } P_i \text{ are defined,} \\ 0 & \text{otherwise.} \end{cases}$$

Here, we assume that the nodes are numbered hierarchically: first the nodes of triangulation $\mathcal{T}_1$, then the newly added nodes of $\mathcal{T}_2$, and so on. The restriction operator is defined as the transpose of the interpolation operator as follows

$$I_k^{k-1} = \left(I_{k-1}^k\right)^\top. \tag{5.167}$$

Notice that after the restriction of the residual in the CSMG algorithm, all components which belong to Dirichlet boundary nodes must be set to zero since we know that the residual on the boundary is zero.

Next, we present the results of the numerical experiments using the CSMG scheme with finite elements on different domains. For all computations, we use $\gamma_1 = \gamma_2 = 2$ pre- and postsmoothing steps. We report the CPU time until the norm of the residual, $\|r\|_{L^2}$, satisfies a stopping tolerance of $tol = 10^{-7}$. We also report the observed convergence factor.

First, we consider the following elliptic optimal control problem

$$\begin{aligned} \min J(y,u) &:= \tfrac{1}{2}\|y-z\|_{L^2}^2 + \tfrac{\nu}{2}\|u\|_{L^2}^2, \\ -\Delta y &= f+u \quad \text{in} \quad \Omega, \\ y &= 0 \quad \text{on} \quad \partial\Omega. \end{aligned} \tag{5.168}$$

Let $\Omega = (0,1)\times(0,1)$, and let $f,\ z \in L^2(\Omega)$ be given by

$$\begin{aligned} f &= 0, \\ z &= \sin(2\pi x_1)\sin(\pi x_2). \end{aligned}$$

The target function z is shown in Figure 5.13 together with the chosen domain $\Omega_1 = \Omega$, which is a square with a circle hole with radius $r = 0.25$ and center at $(x_1, x_2) = (0.5, 0.5)$.

Results of numerical experiments for this case are reported in Table 5.16. We can see that choosing different values for the parameter ν, the multigrid method converges within three iterations. That is, the algorithm exhibits an independence of the number of iterations on the parameter ν and also on the size of the mesh. From the CPU times we find that the complexity of the problem is $O(n \log(n))$, where n denotes the number of nodes. These results are in agreement with the estimate given by Theorem 5.19 in the previous section.

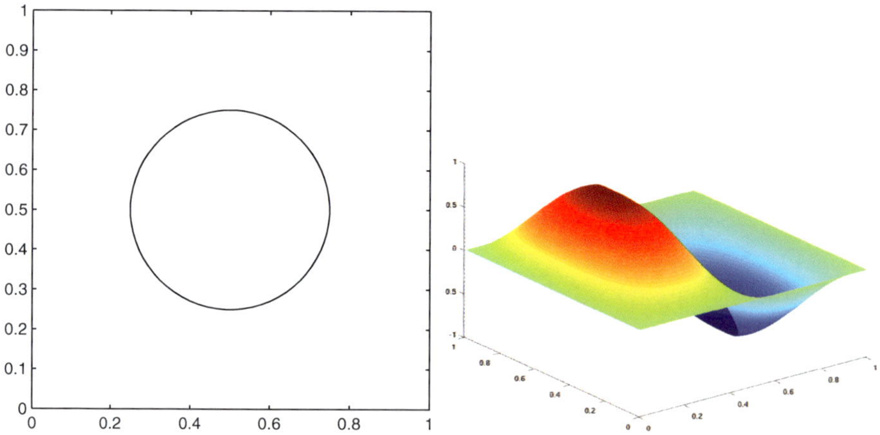

Figure 5.13. *Domain Ω_1: Square on $(0,1)\times(0,1)$ minus a circle with radius $r = 0.25$ and center at $(0.5,0.5)$ (left) and the target function z (right). Reprinted with permission from O. Lass, M. Vallejos, A. Borzì, and C.C. Douglas, Implementation and analysis of multigrid schemes with finite elements for elliptic optimal control problems, Computing,* 84(1-2) (2009), 27–48.

Table 5.16. *Results of the elliptic optimal control problem with unconstrained control on the domain Ω_1 using the CSMG method with finite elements.*

ν	n	Iter.	ρ_y	ρ_p	$\|r_y\|_{L^2}$	$\|r_p\|_{L^2}$	Time (s)
	24960	3	0.081	0.053	3.15e-08	6.81e-07	0.44
10^{-2}	99072	3	0.085	0.058	1.77e-08	4.34e-07	1.93
	394752	3	0.089	0.062	9.73e-09	2.62e-07	8.23
	24960	3	0.039	0.037	3.25e-09	2.84e-07	0.44
10^{-4}	99072	3	0.042	0.039	1.99e-09	1.61e-07	1.99
	394752	3	0.046	0.041	1.19e-09	9.01e-08	8.26
	24960	3	0.043	0.067	3.40e-10	4.22e-07	0.43
10^{-6}	99072	3	0.045	0.069	1.89e-10	2.29e-07	1.94
	394752	3	0.048	0.068	1.04e-10	1.19e-07	8.48

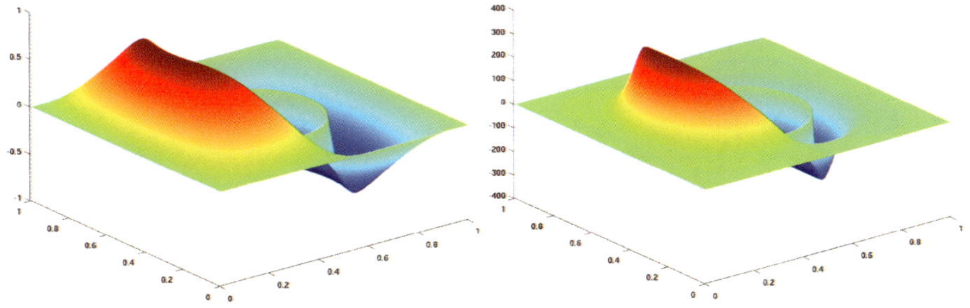

Figure 5.14. *Numerical solutions for the state (left) and control (right) variables of the elliptic optimal control with unconstrained control problem using $\nu = 10^{-6}$. Reprinted with permission from O. Lass, M. Vallejos, A. Borzì, and C.C. Douglas, Implementation and analysis of multigrid schemes with finite elements for elliptic optimal control problems, Computing,* 84(1-2) (2009), 27–48.

The numerical optimal solutions y and u for $\nu = 10^{-6}$ are shown in Figure 5.14.

Next, we discuss the elliptic optimal control problem with constrained control given as follows

$$\begin{aligned} \min J(y,u) &:= \tfrac{1}{2}\|y-z\|_{L^2}^2 + \tfrac{\nu}{2}\|u\|_{L^2}^2, \\ -\Delta y - u &= f \quad \text{in} \quad \Omega, \\ y &= 0 \quad \text{on} \quad \partial\Omega, \end{aligned} \tag{5.169}$$

where $f, z \in L^2(\Omega)$ are the same as in the unconstrained control case. Here, $u \in \mathcal{U} \subset L^2(\Omega)$ since

$$\mathcal{U} = \left\{ u \in L^2(\Omega) | -30 \le u(x) \le 30 \text{ a.e. in } \Omega \right\}. \tag{5.170}$$

First, we consider this problem on the same domain as depicted in Figure 5.13. Results of numerical experiments for this case are given in Table 5.17. We also consider the domain Ω_2 shown in Figure 5.15 defined as $\frac{3}{4}$ of a circle with radius $r = 1$ and center at (0.5,0.5). For this latter case we report results of numerical experiments in Table 5.18.

Table 5.17. *Results of the elliptic optimal control problem with constrained control on the domain Ω_1 using the CSMG method with finite elements.*

ν	n	Iter.	ρ_y	ρ_p	$\|r_y\|_{L^2}$	$\|r_p\|_{L^2}$	Time (s)
	24960	5	0.070	0.075	3.23e-07	3.12e-09	0.55
10^{-4}	99072	5	0.074	0.079	2.03e-07	1.81e-09	2.61
	394752	5	0.077	0.081	1.21e-07	1.01e-09	12.10
	24960	10	0.393	0.391	6.86e-07	3.97e-09	1.09
10^{-6}	99072	9	0.392	0.391	8.97e-07	5.13e-09	4.77
	394752	9	0.391	0.391	4.50e-07	2.55e-09	20.80

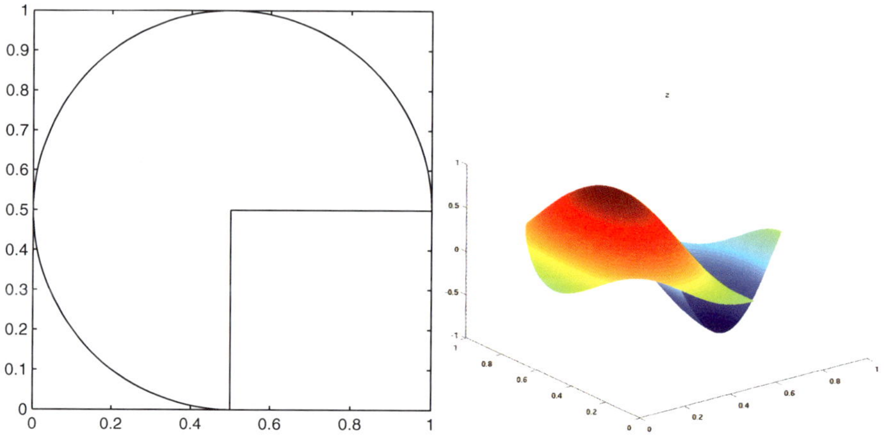

Figure 5.15. *Domain Ω_2: $\frac{3}{4}$ of a circle with radius $r = 1$ and center at* (0.5,0.5) *(left) and the target function z (right). Reprinted with permission from O. Lass, M. Vallejos, A. Borzì, and C.C. Douglas, Implementation and analysis of multigrid schemes with finite elements for elliptic optimal control problems, Computing,* 84(1-2) (2009), 27–48.

Table 5.18. *Results of the elliptic optimal control problem with constrained control on the domain Ω_2 using the CSMG method with finite elements.*

ν	n	Iter.	ρ_y	ρ_p	$\|r_y\|_{L^2}$	$\|r_p\|_{L^2}$	Time (s)
	24865	5	0.083	0.080	3.50e-07	3.03e-09	0.6
10^{-4}	98881	5	0.097	0.084	2.54e-07	1.65e-09	2.7
	394369	5	0.112	0.086	2.24e-07	8.99e-10	11.7
	25361	17	0.762	0.540	9.73e-07	1.67e-09	1.9
10^{-6}	100897	16	0.236	0.364	3.00e-07	1.75e-09	8.6
	402497	15	0.401	0.411	6.17e-07	2.42e-09	35.4

In the constrained control case, we see that by choosing different values for the parameter ν, the multigrid method exhibits a convergence behavior which is almost independent of the mesh size. However, larger values of the convergence factor are obtained by decreasing the value of the optimization weight. This is reasonable since the presence of box constraints may result in steep gradients of the control function. The numerical solutions y and u on Ω_1 and choosing $\nu = 10^{-6}$ are shown in Figure 5.16. The optimal solutions for two different choices of $\nu \in \{10^{-6}, 10^{-4}\}$ computed on Ω_2 are shown in Figure 5.17.

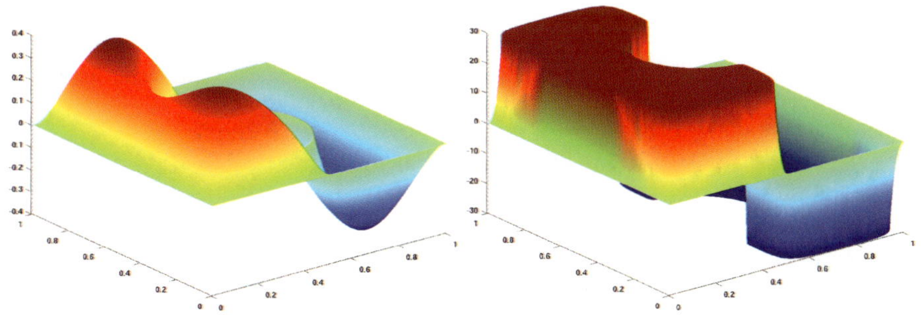

Figure 5.16. *Numerical solutions for the state (left) and control (right) variables of the elliptic optimal control problem with constrained control on the domain* Ω_1 *using* $\nu = 10^{-6}$. *Reprinted with permission from O. Lass, M. Vallejos, A. Borzì, and C.C. Douglas, Implementation and analysis of multigrid schemes with finite elements for elliptic optimal control problems, Computing,* 84(1-2) (2009), 27–48.

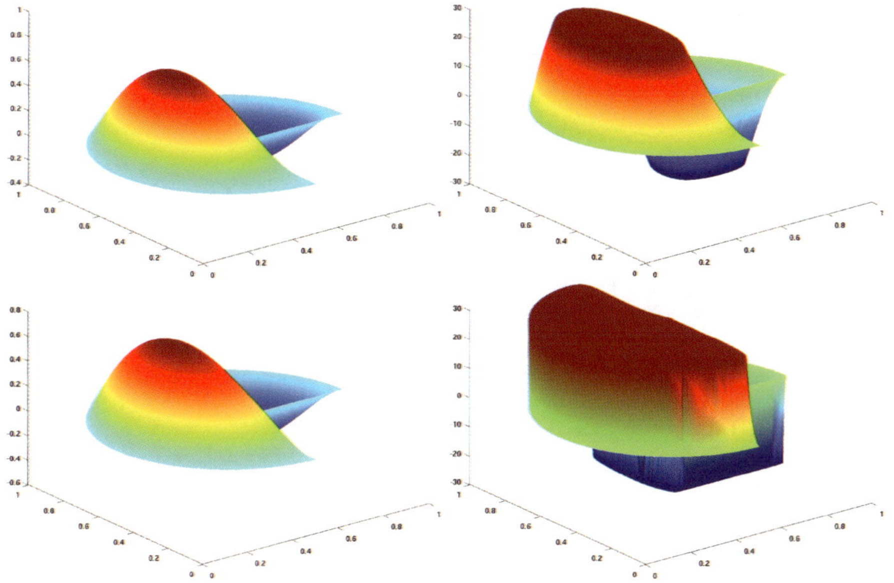

Figure 5.17. *Numerical solutions* y *(top left) and* u *(top right) on the domain* Ω_2 *and parameter* $\nu = 10^{-4}$. *Numerical solutions* y *(bottom left) and* u *(bottom right) on the domain* Ω_2 *and parameter* $\nu = 10^{-6}$. *Reprinted with permission from O. Lass, M. Vallejos, A. Borzì, and C.C. Douglas, Implementation and analysis of multigrid schemes with finite elements for elliptic optimal control problems, Computing,* 84(1-2) (2009), 27–48.

We can see steeper gradients of the control function as we reduce the value of the weight. We remark that quite similar results are obtained for nonconvex domains and in the case of domains with reentrant corners.

5.7.4 CSMG Schemes for Parabolic Control Problems

In this section, we start our discussion on multigrid schemes for parabolic optimal control problems. We describe space-time CSMG multigrid schemes that have been presented in, e.g., [46, 49, 50, 53, 58, 150]. These schemes are based on smoothing methods that provide a robust implementation of the time coupling in the optimality system.

Control of parabolic systems has many applications in biology [61, 180], chemistry [62, 229], and physiology [3, 272]. To solve the related optimal control problems, the solution of the corresponding optimality systems which are governed by reaction-diffusion equations with opposite time orientation is considered. Of particular importance in applications and for benchmark purposes are singular optimal control problems [236]. Especially in these cases, the coupling between state variables and controls must be realized in a robust way in order to guarantee convergence of the algorithms. Robust coupling is also necessary for designing algorithms that permit efficient solution to optimization problems with a computational performance that is independent of the value of the optimization parameters. The results presented in this chapter and in the references mentioned above show that the space-time multigrid methods considered here do meet these requirements. These objectives are achieved by using appropriate smoothing techniques and by solving optimality systems for distributed control or boundary control in one shot in the whole space-time cylinder.

The CSMG approach represents an extension of space-time multigrid schemes for parabolic problems [170, 202, 347] to the case of reaction-diffusion systems with opposite time orientation. For this particular structure, two different smoothing schemes in combination with semicoarsening in space are considered. Other coarsening strategies are possible; see [57, 58]. In the case of tracking along trajectories a pointwise relaxation is presented which can be successfully applied [57, 49, 50] to solve singular parabolic optimal control problems. In the case of terminal observation, block relaxation is the most robust choice. Block smoothing is also advantageous in the case of reaction-diffusion problems with very small diffusion as it occurs in the chemical turbulence modeling and physiology; see [61, 62, 46] and the references given therein.

A disadvantage of any space-time approach is the requirement of storing the dependent variables for all time steps. This is certainly a limitation that arises when open-loop optimal control problems on a large time interval are considered. However, in the limiting case of very long time intervals, this difficulty is overcomed by combining space-time multigrid schemes with receding-horizon techniques [4, 208].

Depending on the application, reaction-diffusion processes can be controlled through source terms or through boundary terms. In the case of distributed control through source terms, the following optimal control problem is formulated

$$\left\{\begin{array}{rcll} & \min J(y,u), & & \\ -\partial_t y + G(y) + \sigma\,\Delta y & = & f+u & \text{in } Q, \\ y & = & y_0 & \text{in } \Omega \times \{t=0\}, \\ y & = & g & \text{on } \Sigma, \end{array}\right. \tag{5.171}$$

where $Q = \Omega \times (0,T)$ and $\Sigma = \partial\Omega \times (0,T)$. To be specific, we take $y_0 \in H_0^1(\Omega)$, $g \in C(\Sigma)$, and $f, u \in L^2(Q)$. In (5.171), the nonlinear term $G(y)$ models the reaction kinetics for the state y and u represents the control function which for the moment we assume to be unconstrained. Here $\sigma > 0$ is the diffusion coefficient.

Alternatively, we consider the following optimal Neumann boundary control problem

$$\begin{cases} \min J(y,u), & & \\ -\partial_t y + G(y) + \sigma \Delta y &= f & \text{in } Q, \\ y &= y_0 & \text{in } \Omega \times \{t = 0\}, \\ -\frac{\partial y}{\partial n} &= u & \text{on } \Sigma, \end{cases} \tag{5.172}$$

where $u \in L^2(\Sigma)$.

Control may be required to track a desired trajectory given by $y_d(\mathbf{x},t) \in L^2(Q)$ or to reach a desired terminal state $y_T(\mathbf{x}) \in L^2(\Omega)$. For this purpose, the following cost functional is considered

$$J(y,u) = \frac{\alpha}{2}||y - y_d||^2_{L^2(Q)} + \frac{\beta}{2}||y(\cdot,T) - y_T||^2_{L^2(\Omega)} + \frac{\nu}{2}||u||^2_{L^2(X)}, \tag{5.173}$$

with $X = Q$ or $X = \Sigma$ corresponding to distributed control or boundary control, respectively. Here, $\nu > 0$ is the weight of the cost of the control and $\alpha \geq 0$, $\beta \geq 0$, $\alpha + \beta > 0$, are optimization parameters. The case $\alpha = 1$, $\beta = 0$ corresponds to tracking without terminal observation, while the case $\alpha = 0$ and $\beta = 1$ means reaching a desired final target configuration no matter which trajectory is followed.

Existence of solutions to the optimal control problems (5.171) and (5.172) can be established under suitable conditions for various forms of the nonlinearity; see, e.g., [49, 148, 207, 216, 236, 262, 339].

The solution to (5.171) is characterized by the following first-order optimality system

$$\begin{aligned} -\partial_t y + G(y) + \sigma \Delta y = f + u \qquad & \text{in } Q, \\ \partial_t p + G'(y)p + \sigma \Delta p + \alpha(y - y_d) = 0 \qquad & \text{in } Q, \\ \nu u - p = 0 \qquad & \text{in } Q, \\ y = g,\ p = 0 \qquad & \text{on } \Sigma. \end{aligned} \tag{5.174}$$

In the case of boundary control (5.172), the optimal solution satisfies the following

$$\begin{aligned} -\partial_t y + G(y) + \sigma \Delta y = f \qquad & \text{in } Q, \\ \partial_t p + G'(y)p + \sigma \Delta p + \alpha(y - y_d) = 0 \qquad & \text{in } Q, \\ \nu u - p = 0 \qquad & \text{on } \Sigma, \\ -\frac{\partial y}{\partial n} = u,\ -\frac{\partial p}{\partial n} = 0 \qquad & \text{on } \Sigma. \end{aligned} \tag{5.175}$$

In both cases we have the initial condition $y(\mathbf{x},0) = y_0(\mathbf{x})$ for the state variable (evolving forward in time). The terminal condition for the adjoint variable (evolving backward in time) is given by

$$p(\mathbf{x},T) = \beta(y(\mathbf{x},T) - y_T(\mathbf{x})). \tag{5.176}$$

Notice that the optimality systems above are unusual in scientific computing because of the parabolic equations with opposite time orientation and the terminal condition.

We assume sufficient regularity of the data, y_d, y_T, such that these functions are properly approximated by their values at grid points. We use the finite difference framework of Chapter 3. We obtain the following

$$\begin{aligned} -\partial_t^+ y_h^m + G(y_h^m) + \sigma \Delta_h y_h^m &= u_h^m, \\ \partial_t^- p_h^m + G'(y_h^m) p_h^m + \sigma \Delta_h p_h^m + \alpha(y_h^m - y_{dh}^m) &= 0, \\ \nu u_h^m - p_h^m &= 0. \end{aligned} \tag{5.177}$$

The discretization of (5.175) is similar, requiring, in addition, specification of the approximation of the Neumann boundary conditions. This is done by considering the optimality system on the boundary and discretizing the boundary derivative using second-order centered finite differences to eliminate the (ghost) variables outside of the domain.

We recall that under suitable conditions the following estimates are obtained

$$||y_h - y||_0 \le c\, h^2, \quad ||p_h - p||_0 \le c\, h^2, \quad \text{and } ||u_h - u||_0 \le c\, h^2,$$

assuming there exist positive constants $c_1 \le c_2$ such that $c_1 h^2 \le \delta t \le c_2 h^2$.

Two CSMG Smoothing Strategies: Linear Case

In this section, we discuss a CSMG scheme to solve reaction-diffusion optimal control problems approximated by finite differences and backward Euler schemes. The space-time CSMG schemes discussed in this section are formulated on the entire space-time cylinder where the parabolic optimality systems are defined. We consider L grid levels indexed by $k = 1, \dots, L$, where $k = L$ refers to the finest grid. The mesh of level k is denoted by $Q_k = Q_{h_k, \delta t_k}$, where $h_k = h_1/2^{k-1}$ and $\delta t_k = \delta t$. This choice corresponds to semicoarsening in space. This choice appears to be the most appropriate for our purpose. However, other choices are possible [49], such as standard coarsening, which results in being robust in the control of transient, i.e., short, time intervals.

In correspondence of semicoarsening in space, no interpolation and restriction in time are needed. However, other choices of intergrid transfer strategies are possible; see [202]. In our implementation, we choose I_k^{k-1} to be the full-weighted restriction operator [340] in space with no averaging in the time direction. The mirrored version of this operator applies also to the boundary points. The prolongation I_{k-1}^k is defined by bilinear interpolation in space.

Now, we discuss the design of two robust collective smoothing schemes for solving (5.174). For simplicity of illustration, we first consider the linear case with $G = 0$ and eliminate the control variable by means of the optimality condition $\nu u_h^m - p_h^m = 0$. We have

$$\begin{aligned} -\left[1 + 4\sigma\gamma\right] y_{i\,j\,m} + \sigma\gamma\left[y_{i+1\,j\,m} + y_{i-1\,j\,m} + y_{i\,j+1\,m} + y_{i\,j-1\,m}\right] + y_{i\,j\,m-1} \\ - \frac{\delta t}{\nu}\, p_{i\,j\,m} = 0, \quad 2 \le m \le N_t + 1, \end{aligned} \tag{5.178}$$

$$\begin{aligned} -\left[1 + 4\sigma\gamma\right] p_{i\,j\,m} + \sigma\gamma\left[p_{i+1\,j\,m} + p_{i-1\,j\,m} + p_{i\,j+1\,m} + p_{i\,j-1\,m}\right] + p_{i\,j\,m+1} \\ + \delta t\, \alpha\,(y_{i\,j\,m} - y_{d\,i\,j\,m}) = 0, \quad 1 \le m \le N_t. \end{aligned} \tag{5.179}$$

Let us define a collective iteration step which is applied at any space-time grid point to update $w_{i\,j\,m} = (y_{i\,j\,m}, p_{i\,j\,m})$. For this purpose consider (5.178) and (5.179) for the two variables $y_{i\,j\,m}$ and $p_{i\,j\,m}$ at the grid point $i\,j\,m$. We can refer to the left-hand sides of (5.178) and (5.179) as the negatives of the residuals $r_y(w_{i\,j\,m})$ and $r_p(w_{i\,j\,m})$, respectively. A step of a collective smoothing iteration at this point consists of a local update given by

$$\begin{pmatrix} y \\ p \end{pmatrix}^{(1)}_{i\,j\,m} = \begin{pmatrix} y \\ p \end{pmatrix}^{(0)}_{i\,j\,m} + \begin{bmatrix} -(1+4\sigma\gamma) & -\delta t/\nu \\ \delta t\,\alpha & -(1+4\sigma\gamma)+\delta t \end{bmatrix}^{(0)\,-1}_{i\,j\,m} \begin{pmatrix} r_y \\ r_p \end{pmatrix}_{i\,j\,m}, \tag{5.180}$$

where r_y and r_p denote the residuals at $i\,j\,m$ prior to the update. While a sweep of this smoothing iteration can be performed in any ordering of i, j, the problem of how to proceed along time direction arises.

To solve this problem the first vector component of (5.180) marching in the forward direction is used to update the state variable and the adjoint variable p is being updated using the second component of (5.180) marching backward in time. In this way a robust iteration is obtained given by the following algorithm; see, e.g., [57, 58, 49].

ALGORITHM 5.10. Time-splitted CGS iteration: linear case (TS-CGS).

1. Set the starting approximation.
2. For $m = 2, \ldots, N_t$ do
3. For $i\,j$ in, e.g., lexicographic order do

$$y^{(1)}_{i\,j\,m} = y^{(0)}_{i\,j\,m} + \frac{[-(1+4\sigma\gamma)]\,r_y(w) + \frac{\delta t}{\nu}\,r_p(w)}{[-(1+4\sigma\gamma)]^2 + \frac{\delta t^2}{\nu}\alpha}\Big|^{(0)}_{i\,j\,m},$$

$$p^{(1)}_{i\,j\,N_t-m+2} = p^{(0)}_{i\,j\,N_t-m+2} + \frac{[-(1+4\sigma\gamma)]\,r_p(w) - \delta t\,\alpha\,r_y(w)}{[-(1+4\sigma\gamma)]^2 + \frac{\delta t^2}{\nu}\alpha}\Big|^{(0)}_{i\,j\,N_t-m+2};$$

4. end.

As in the elliptic case, the TS-CGS scheme applies with few modifications to the case of boundary control [46]. Results of local Fourier analysis, discussed below, show that the TS-CGS scheme has good smoothing properties, independently of the value of ν.

In the regime of small σ (or γ), however, the TS-CGS iteration cannot provide robust smoothing because of lack of strong coupling in the space directions. To overcome this problem, block relaxation of the variables that are strongly connected should be performed. For small σ (or γ) this means solving for the pairs of state and adjoint variables along the time direction for each space coordinate.

To describe this procedure, consider the discrete optimality system (5.178)–(5.179) at any i, j and for all time steps. Thus for each spatial grid point i, j a block-tridiagonal system is obtained, where each block is a 2×2 matrix corresponding to the pair (y, p).

This block-tridiagonal system has the following form

$$M = \begin{bmatrix} A_2 & C_2 & & & \\ B_3 & A_3 & C_3 & & \\ & B_4 & A_4 & C_4 & \\ & & & & \\ & & & & C_{N_t} \\ & & & B_{N_t+1} & A_{N_t+1} \end{bmatrix}. \tag{5.181}$$

Centered at t_m, the entries B_m, A_m, and C_m refer to the variables (y, p) at t_{m-1}, t_m, and t_{m+1}, respectively. The block A_m, $m = 2, \ldots, N_t$, is given by

$$A_m = \begin{bmatrix} -(1+4\sigma\gamma) & -\frac{\delta t}{\nu} \\ \delta t\, \alpha & -(1+4\sigma\gamma) \end{bmatrix}, \tag{5.182}$$

where all functions within the brackets [ ] are evaluated at t_m. Correspondingly, the B_m and C_m blocks are given by

$$B_m = \begin{bmatrix} 1 & 0 \\ 0 & 0 \end{bmatrix} \quad \text{and} \quad C_m = \begin{bmatrix} 0 & 0 \\ 0 & 1 \end{bmatrix}. \tag{5.183}$$

Clearly, for each time step, the variables neighboring the point ij are taken as constant and contribute to the right-hand side of the system.

It remains to discuss the block A_{N_t+1} for $\beta \neq 0$. At $t_m = T$, we have the terminal condition (5.176), which we rewrite as

$$\beta(y_h^m - y_{Th}^m) - p_h^m = 0, \quad m = N_t + 1.$$

Thus, the block A_{N_t+1} is given by

$$A_{N_t+1} = \begin{bmatrix} -(1+4\sigma\gamma) & -\frac{\delta t}{\nu} \\ \beta & -1 \end{bmatrix}. \tag{5.184}$$

For each i, j we have to solve a tridiagonal system $Mw = r$, where $w = (y_h^2, p_h^2, \ldots, y_h^{N_t+1}, p_h^{N_t+1})$ and $r = (r_y(w^2), r_p(w^2), \ldots, r_y(w^{N_t+1}), r_p(w^{N_t+1}))$. In particular we have $r_p(w^{N_t+1}) = p_h^{N_t+1} - \beta(y_h^{N_t+1} - y_{Th}^{N_t+1})$. Block-tridiagonal systems can be solved efficiently with $\mathcal{O}(N_t)$ effort. A block-tridiagonal solver is given in [61]. Summarizing, our collective t-line relaxation is given by the following algorithm [61, 62].

ALGORITHM 5.11. Time-line CGS iteration (TL-CGS).

1. Set the starting approximation.
2. For ij in, e.g., lexicographic order do
$$\begin{pmatrix} y \\ p \end{pmatrix}_{ij}^{(1)} = \begin{pmatrix} y \\ p \end{pmatrix}_{ij}^{(0)} + M^{-1} \begin{pmatrix} r_y \\ r_p \end{pmatrix}_{ij};$$
3. end.

Also in this case r_y and r_p denote the residuals at i, j and for all m prior to the update. Since the solution in time is exact, no time splitting is required.

Two CSMG Smoothing Strategies: Nonlinear Case

In this section, we consider the nonlinear optimality system (5.177) in expanded form. Also in this case, we assume no constraints on the control, and therefore we can eliminate the control variable by means of the optimality condition $\nu u_h^m - p_h^m = 0$. We obtain

$$
\begin{aligned}
&-\left[1+4\sigma\gamma\right] y_{i\,j\,m} + \sigma\gamma\left[y_{i+1\,j\,m} + y_{i-1\,j\,m} + y_{i\,j+1\,m} + y_{i\,j-1\,m}\right] + y_{i\,j\,m-1} \\
&\quad + \delta t\, G_\delta(y_{i\,j\,m}) - \frac{\delta t}{\nu}\, p_{i\,j\,m} = \delta t\, f_{i\,j\,m}, \quad 2 \le m \le N_t + 1,
\end{aligned} \tag{5.185}
$$

$$
\begin{aligned}
&-\left[1+4\sigma\gamma\right] p_{i\,j\,m} + \sigma\gamma\left[p_{i+1\,j\,m} + p_{i-1\,j\,m} + p_{i\,j+1\,m} + p_{i\,j-1\,m}\right] + p_{i\,j\,m+1} \\
&+ \delta t\, G'_\delta(y_{i\,j\,m})\, p_{i\,j\,m} + \delta t\, \alpha\,(y_{i\,j\,m} - y_{d\,i\,j\,m}) = 0, \quad 1 \le m \le N_t.
\end{aligned} \tag{5.186}
$$

In case of terminal observation, at $t_m = T$ we have (5.176) in place of (5.186).

First, we define a pointwise smoothing scheme. Consider a collective Gauss–Seidel (–Newton) step which is applied at each space-time grid point to update $w_{i\,j\,m} = (y_{i\,j\,m}, p_{i\,j\,m})$. For this purpose consider (5.185) and (5.186) for the two variables $y_{i\,j\,m}$ and $p_{i\,j\,m}$ at the grid point $i\,j\,m$. We can refer to the left-hand sides of (5.185) and (5.186) as the negatives of the residuals $r_y(w_{i\,j\,m})$ and $r_p(w_{i\,j\,m})$, respectively. A step of a collective smoothing iteration at this point consists of a local (Newton) update given by

$$
\begin{pmatrix} y \\ p \end{pmatrix}^{(1)}_{i\,j\,m} = \begin{pmatrix} y \\ p \end{pmatrix}^{(0)}_{i\,j\,m} + \begin{bmatrix} -(1+4\sigma\gamma) + \delta t\, G'_\delta & -\delta t/\nu \\ \delta t(\alpha + G''_\delta\, p) & -(1+4\sigma\gamma) + \delta t\, G'_\delta \end{bmatrix}^{(0)\,-1}_{i\,j\,m} \begin{pmatrix} r_y \\ r_p \end{pmatrix}_{i\,j\,m}, \tag{5.187}
$$

where r_y and r_p denote the residuals at $i\,j\,m$ prior to the update. We need to take into account the opposite time orientation of the state equation and of the adjoint equation. For this purpose, to update the state variable we use the first vector component of (5.187) marching in the forward direction and the adjoint variable p is being updated using the second component of (5.187) marching backward in time. In this way a robust iteration is obtained given by the following nonlinear version of the TS-CGS scheme.

ALGORITHM 5.12. Time-splitted CGS iteration: nonlinear case (TS-CGS).

1. Set the starting approximation.
2. For $m = 2, \dots, N_t$ do
3. For $i\,j$ in, e.g., lexicographic order do

$$
y^{(1)}_{i\,j\,m} = y^{(0)}_{i\,j\,m} + \frac{[-(1+4\sigma\gamma) + \delta t G'_\delta]\, r_y(w) + \frac{\delta t}{\nu}\, r_p(w)}{[-(1+4\sigma\gamma) + \delta t G'_\delta]^2 + \frac{\delta t^2}{\nu}(\alpha + G''_\delta p)}\Big|^{(0)}_{i\,j\,m},
$$

$$
p^{(1)}_{i\,j\,N_t-m+2} = p^{(0)}_{i\,j\,N_t-m+2} + \frac{[-(1+4\sigma\gamma) + \delta t G'_\delta]\, r_p(w) - \delta t(\alpha + G''_\delta p)\, r_y(w)}{[-(1+4\sigma\gamma) + \delta t G'_\delta]^2 + \frac{\delta t^2}{\nu}(\alpha + G''_\delta p)}\Big|^{(0)}_{i\,j\,N_t-m+2};
$$

4. end.

As for the linear case, in the regime of small σ (or γ), the TS-CGS iteration cannot provide robust smoothing because the coupling in the space direction becomes weak. To overcome this problem, we develop a time-line block-Newton relaxation of the state and adjoint variables.

To describe this block Gauss–Seidel–Newton procedure, consider the discrete optimality system (5.177) at any i,j and for all time steps. For simplicity, we use the optimality condition to eliminate the control variable. Thus for each spatial grid point i,j a block-tridiagonal system is obtained, where each block is a 2×2 matrix corresponding to the pair (y,p). This block-tridiagonal system has the following form

$$M = \begin{bmatrix} A_2 & C_2 & & & & \\ B_3 & A_3 & C_3 & & & \\ & B_4 & A_4 & C_4 & & \\ & & & & & \\ & & & & C_{N_t} & \\ & & & B_{N_t+1} & A_{N_t+1} & \end{bmatrix}. \tag{5.188}$$

Centered at t_m, the entries B_m, A_m, and C_m refer to the variables (y,p) at t_{m-1}, t_m, and t_{m+1}, respectively. The block A_m, $m=2,\dots,N_t$, is given by

$$A_m = \begin{bmatrix} -(1+4\sigma\gamma)+\delta t G'_\delta & -\frac{\delta t}{\nu} \\ \delta t(\alpha + G''_\delta\, p) & -(1+4\sigma\gamma)+\delta t G'_\delta \end{bmatrix}, \tag{5.189}$$

where all functions within the brackets [ ] are evaluated at t_m. Correspondingly, the B_m and C_m blocks are given by

$$B_m = \begin{bmatrix} 1 & 0 \\ 0 & 0 \end{bmatrix} \quad \text{and} \quad C_m = \begin{bmatrix} 0 & 0 \\ 0 & 1 \end{bmatrix}. \tag{5.190}$$

Clearly, for each time step, the variables neighboring the point ij are taken as constant and contribute to the right-hand side of the system.

It remains to discuss the block A_{N_t+1} for $\beta \neq 0$. At $t_m = T$, we have the terminal condition (5.176), which we rewrite as

$$\beta(y_h^m - y_{Th}^m) - p_h^m = 0, \quad m = N_t + 1.$$

Thus, the block A_{N_t+1} is given by

$$A_{N_t+1} = \begin{bmatrix} -(1+4\sigma\gamma)+\delta t G' & -\frac{\delta t}{\nu} \\ \beta & -1 \end{bmatrix}. \tag{5.191}$$

For each i,j we have to solve a tridiagonal system $Mw=r$, where $w=(y_h^2,p_h^2,\dots,$ $y_h^{N_t+1},p_h^{N_t+1})$ and $r=(r_y(w^2),r_p(w^2),\dots,r_y(w^{N_t+1}),r_p(w^{N_t+1}))$. In particular we have $r_p(w^{N_t+1}) = p_h^{N_t+1} - \beta(y_h^{N_t+1} - y_{Th}^{N_t+1})$. Block-tridiagonal systems can be solved efficiently with $\mathcal{O}(N_t)$ effort. A block-tridiagonal solver is given in [61]. Clearly, the nonlinear version of the TL-CGS relaxation is formally identical to that of the linear case.

TG Fourier Analysis: Linear Parabolic Case

In this section, we illustrate the TG local Fourier analysis of the space-time CSMG scheme with TS-CGS and TL-CGS smoothing applied to a linear parabolic optimal control problem. For the analysis that follows, we consider a linear case with linear reaction $G_\delta(y) = \delta\, y$ and distributed control with the replacement $u = p/\nu$. We denote the parabolic optimality system with $A\,w = f$, where $w = (y, p)$, and assume an infinite grid and one space dimension. On the fine grid, consider the Fourier components $\phi(\boldsymbol{j},\boldsymbol{\theta}) = e^{i\boldsymbol{j}\cdot\boldsymbol{\theta}}$, where i is the imaginary unit, $\boldsymbol{j} = (j_x, j_t) \in \mathbf{Z}\times\mathbf{Z}$, $\boldsymbol{\theta} = (\theta_x,\theta_t) \in [-\pi,\pi)^2$, and $\boldsymbol{j}\cdot\boldsymbol{\theta} = j_x\theta_x + j_t\theta_t$.

In a semicoarsening setting, the frequency domain is spanned by the following two sets of frequencies:

$$\boldsymbol{\theta}^{(0,0)} := (\theta_x,\theta_t) \quad \text{and} \quad \boldsymbol{\theta}^{(1,0)} := (\overline{\theta}_x,\theta_t),$$

where $(\theta_x,\theta_t) \in ([-\pi/2,\pi/2)\times[-\pi,\pi))$ and $\overline{\theta}_x = \theta_x - sign(\theta_x)\pi$. The components $\phi(\cdot,\boldsymbol{\theta}^{\boldsymbol{\alpha}})$ are called harmonics. The first harmonics $\phi(\cdot,\boldsymbol{\theta}^{(0,0)})$ represents LF components in space. The second harmonics $\phi(\cdot,\boldsymbol{\theta}^{(1,0)})$ contains the HF components in space direction. Both have all frequency components in time direction. Using semicoarsening, we have that $\phi(\boldsymbol{j},\boldsymbol{\theta}^{(0,0)}) = \phi(\boldsymbol{j},\boldsymbol{\theta}^{(1,0)})$ on the coarse grid.

The action of the multigrid scheme is to reduce the HF error components by applying the smoothing operator S_k and to reduce the LF error components by coarse-grid correction given by

$$CG_k^{k-1} = [I_k - I_{k-1}^k\,(A_{k-1})^{-1}\,I_k^{k-1}\,A_k].$$

Denote $E_k^\theta = \text{span}[\phi_k(\cdot,\boldsymbol{\theta}^{\boldsymbol{\alpha}}) : \boldsymbol{\alpha} \in \{(0,0),(1,0)\}]$. Under the assumption that all multigrid components are linear and that $(A_{k-1})^{-1}$ exists, we have a representation of the TG operator $\boldsymbol{TG}_k^{k-1}$ on the space $E_k^\theta \times E_k^\theta$ by a 4×4 matrix given by

$$\widehat{TG}_k^{k-1}(\boldsymbol{\theta}) = \hat{S}_k(\boldsymbol{\theta})^{\nu_2}\,\widehat{CG}_k^{k-1}(\boldsymbol{\theta})\,\hat{S}_k(\boldsymbol{\theta})^{\nu_1},$$

where the hat denotes the Fourier symbol [340] of the given operator.

To determine the explicit form of the operator symbols given above, consider the action of the operators on the couple $(\tilde{y},\tilde{p}) \in E_k^\theta \times E_k^\theta$ for a given $(\boldsymbol{j})$, where

$$\tilde{y} = \sum_{\boldsymbol{\theta}}\sum_{p=0,1} Y_{\boldsymbol{\theta}}^{(p,0)}\,\phi_k(\boldsymbol{j},\boldsymbol{\theta}^{(p,0)}) \quad \text{and} \quad \tilde{p} = \sum_{\boldsymbol{\theta}}\sum_{p=0,1} P_{\boldsymbol{\theta}}^{(p,0)}\,\phi_k(\boldsymbol{j},\boldsymbol{\theta}^{(p,0)}), \qquad (5.192)$$

where $\sum_{\boldsymbol{\theta}}$ denotes formal summation in $\boldsymbol{\theta} = (\theta_x,\theta_t) \in ([-\pi/2,\pi/2)\times[-\pi,\pi))$ and $\tilde{W}_{\boldsymbol{\theta}}^{\boldsymbol{\alpha}} = (\tilde{Y}_{\boldsymbol{\theta}}^{\boldsymbol{\alpha}}, \tilde{P}_{\boldsymbol{\theta}}^{\boldsymbol{\alpha}})$ are the corresponding Fourier coefficients.

In the Fourier space the action of one smoothing step can be expressed by $\tilde{W}_{\boldsymbol{\theta}}^{(new)} = \hat{S}(\boldsymbol{\theta})\,\tilde{W}_{\boldsymbol{\theta}}^{(old)}$, where $\hat{S}(\boldsymbol{\theta})$ is the Fourier symbol [340] of the smoothing iteration. This operator applies to the two equations (5.185)–(5.186) acting on LF and HF components. It has the following form

$$\hat{S}_k(\boldsymbol{\theta}) = \text{diag}\{\hat{s}(\boldsymbol{\theta}^{(0,0)}), \hat{s}(\boldsymbol{\theta}^{(1,0)})\},$$

where $\hat{s}(\boldsymbol{\theta})$ is the 2×2 Fourier symbol of the smoothing scheme for a generic $\boldsymbol{\theta}$.

A way to characterize the smoothing property of the operator S_k is to assume an ideal coarse-grid correction which annihilates the LF error components and leaves the HF error components unchanged. That is, one defines the projection operator $\widehat{Q}_k^{k-1}$ on $E_k^\theta \times E_k^\theta$ by

$$\widehat{Q}_k^{k-1}(\boldsymbol{\theta}) = \begin{bmatrix} Q_k^{k-1}(\boldsymbol{\theta}) & 0 \\ 0 & Q_k^{k-1}(\boldsymbol{\theta}) \end{bmatrix}, \quad \text{where } Q_k^{k-1}(\boldsymbol{\theta}) = \begin{cases} \operatorname{diag}\{0,0\} & \text{if } \boldsymbol{\theta} = \boldsymbol{\theta}^{(0,0)}, \\ \operatorname{diag}\{1,1\} & \text{if } \boldsymbol{\theta} = \boldsymbol{\theta}^{(1,0)}. \end{cases}$$

In this framework the smoothing property of S_k is defined as follows

$$\mu = \max\{r(\widehat{Q}_k^{k-1}(\boldsymbol{\theta})\, \hat{S}_k(\boldsymbol{\theta})) : \boldsymbol{\theta} \in ([-\pi/2, \pi/2) \times [-\pi, \pi))\}, \tag{5.193}$$

where r is the spectral radius. Compare with (5.105) and with (5.9).

Now consider applying the TS-CGS step. We obtain

$$\begin{pmatrix} -(1+2\sigma\gamma - \delta t\delta) + \sigma\gamma e^{-i\theta_x} & -\frac{\delta t}{\nu} \\ \alpha\delta t & -(1+2\sigma\gamma - \delta t\delta) + \sigma\gamma e^{-i\theta_x} \end{pmatrix} \begin{pmatrix} \tilde{Y}_{\boldsymbol{\theta}}^{(new)} \\ \tilde{P}_{\boldsymbol{\theta}}^{(new)} \end{pmatrix}$$
$$= \begin{pmatrix} -(e^{-i\theta_t} + \sigma\gamma e^{i\theta_x}) & 0 \\ 0 & -(e^{i\theta_t} + \sigma\gamma e^{i\theta_x}) \end{pmatrix} \begin{pmatrix} \tilde{Y}_{\boldsymbol{\theta}}^{(old)} \\ \tilde{P}_{\boldsymbol{\theta}}^{(old)} \end{pmatrix}.$$

Hence

$$\hat{s}(\boldsymbol{\theta}) = \begin{pmatrix} -(1+2\sigma\gamma - \delta t\delta) + \sigma\gamma e^{-i\theta_x} & -\frac{\delta t}{\nu} \\ \alpha\delta t & -(1+2\sigma\gamma - \delta t\delta) + \sigma\gamma e^{-i\theta_x} \end{pmatrix}^{-1} \times \begin{pmatrix} -(e^{-i\theta_t} + \sigma\gamma e^{i\theta_x}) & 0 \\ 0 & -(e^{i\theta_t} + \sigma\gamma e^{i\theta_x}) \end{pmatrix}. \tag{5.194}$$

Next consider the case of TL-CGS relaxation. The Fourier symbol of the smoothing operator is given by the following 2×2 matrix

$$\hat{s}(\boldsymbol{\theta}) = -(A + B\, e^{-i\theta_t} + C\, e^{i\theta_t} + \tilde{I}\, e^{-i\theta_x})^{-1}(\tilde{I}\, e^{i\theta_x}),$$

where

$$A = \begin{bmatrix} -(1+2\sigma\gamma - \delta t\delta) & -\frac{\delta t}{\nu} \\ \alpha\delta t & -(1+2\sigma\gamma - \delta t\delta) \end{bmatrix}, \; B_m = \begin{bmatrix} 1 & 0 \\ 0 & 0 \end{bmatrix}, \text{ and } C_m = \begin{bmatrix} 0 & 0 \\ 0 & 1 \end{bmatrix},$$

and $\tilde{I} = \sigma\gamma\, I$, I being the 2×2 identity matrix.

To investigate the action of the coarse-grid correction, we consider a full-weighting restriction operator whose symbol is given by

$$\hat{I}_k^{k-1}(\boldsymbol{\theta}) = \frac{1}{2} \begin{bmatrix} (1+\cos(\theta_x)) & 0 & (1-\cos(\theta_x)) & 0 \\ 0 & (1+\cos(\theta_x)) & 0 & (1-\cos(\theta_x)) \end{bmatrix}. \tag{5.195}$$

For the linear prolongation operator we have $\hat{I}_{k-1}^k(\boldsymbol{\theta}) = \hat{I}_k^{k-1}(\boldsymbol{\theta})^T$. The symbol of the fine grid operator is

$$\widehat{A}_k(\boldsymbol{\theta}) = \begin{bmatrix} a_y(\boldsymbol{\theta}^{(0,0)}) & -\delta t/\nu & 0 & 0 \\ \alpha\delta t & a_p(\boldsymbol{\theta}^{(0,0)}) & 0 & 0 \\ 0 & 0 & a_y(\boldsymbol{\theta}^{(1,0)}) & -\delta t/\nu \\ 0 & 0 & \alpha\delta t & a_p(\boldsymbol{\theta}^{(1,0)}) \end{bmatrix}, \tag{5.196}$$

where

$a_y(\boldsymbol{\theta}) = 2\sigma\gamma\cos(\theta_x) - e^{-i\theta_t} - 2\sigma\gamma + \delta t\delta - 1$ and $a_p(\boldsymbol{\theta}) = 2\sigma\gamma\cos(\theta_x) - e^{i\theta_t} - 2\sigma\gamma + \delta t\delta - 1$.

The symbol of the coarse-grid operator follows

$$\widehat{A}_{k-1}(\boldsymbol{\theta}) = \begin{bmatrix} \sigma\gamma\cos(2\theta_x)/2 - e^{-i\theta_t} - \sigma\gamma/2 + \delta t\delta - 1 & -\delta t/\nu \\ \alpha\delta t & \sigma\gamma\cos(2\theta_x)/2 - e^{i\theta_t} - \sigma\gamma/2 + \delta t\delta - 1 \end{bmatrix}.$$

Notice that on the coarser grid δt remains unchanged while $\gamma \to \gamma/4$ by coarsening.

Based on the representation on TG_k^{k-1} by a 4×4 matrix $\widehat{TG}_k^{k-1}(\boldsymbol{\theta})$ we can calculate the TG convergence factor given by

$$\rho(TG_k^{k-1}) = \sup\left\{ r\left(\widehat{TG}_k^{k-1}(\boldsymbol{\theta})\right) : \boldsymbol{\theta} \in ([-\pi/2, \pi/2) \times [-\pi, \pi)) \right\},$$

which requires determination of the spectral radius of a 4×4 matrix. It results that μ and ρ are almost independent of the value of the weight ν and of the mesh parameter γ for a large range of choices of values of these parameters. For $\sigma = 0$ no spatial coupling is present and the TL-CGS scheme becomes an exact solver; i.e., $\mu = 0$ results. On the other hand, for moderate values of ν and corresponding to smaller values of σ the convergence factor of the TS-CGS multigrid scheme worsens. In Table 5.19, local Fourier analysis quantitative estimates of the convergence factor of TL-CGS and TS-CGS multigrid schemes are given. We see that both schemes result in convergence factors that are typical of multigrid schemes for Poisson problems.

Table 5.19. *The convergence factor ρ for TL-CGS and TS-CGS multigrid schemes ($\nu_1 = \nu_2 = 1$); $\delta t = 1/64$, $\delta = 0$, $\sigma = 1$, $\alpha = 1$.*

	TL-CGS			TS-CGS		
$\gamma \setminus \nu$	10^{-8}	10^{-6}	10^{-4}	10^{-8}	10^{-6}	10^{-4}
16	0.01	0.12	0.12	0.01	0.12	0.14
32	0.04	0.13	0.13	0.04	0.13	0.14
48	0.08	0.13	0.13	0.08	0.13	0.13
64	0.11	0.13	0.13	0.11	0.13	0.13

The CSMG Scheme for Parabolic Control-Constrained Problems and Higher-Order Time Discretization

Parabolic control problems result in very large-sized algebraic systems so that techniques are needed in order to reduce the computational time required to determine the optimal control function. A possible strategy is to construct reduced models via, e.g., proper orthogonal decomposition [227] or consider adaptivity [190, 253]. In this section, we consider the idea of reducing the size of the algebraic problems by considering uniform high-order discretization of the optimality system characterizing the optimal solution. In fact, the use of higher-order schemes allows us to attain the required accuracy with much coarser meshes, thus reducing considerably the size of the algebraic problems to be solved while keeping the computational cost at a minimum.

In addition to higher-order time discretization, in the following we discuss the presence of constraints on the control and present generalization of the TS-CGS and TL-CGS smoothing to accommodate these constraints.

We consider the parabolic distributed control problem (5.171) and we require that $u \in U_{ad}$, where $U_{ad} \subset L^2(Q)$ represents the following closed and convex set of admissible controls

$$U_{ad} := \{u \in L^2(Q) : \underline{u}(\mathbf{x},t) \leq u(\mathbf{x},t) \leq \overline{u}(\mathbf{x},t) \text{ a.e. in } Q\}, \tag{5.197}$$

where $\underline{u}$ and $\overline{u}$ are elements of $L^\infty(Q)$. We take $g = 0$ and assume an initial condition $y_0 \in H_0^1(\Omega)$.

Recall the following first-order optimality system

$$\begin{aligned} -\partial_t y + G(y) + \sigma\Delta y = f + u \quad & \text{in } Q, \\ \partial_t p + G'(y)p + \sigma\Delta p + \alpha(y - y_d) = 0 \quad & \text{in } Q, \\ y = 0,\ p = 0 \quad & \text{on } \Sigma, \\ (\nu u - p, v - u) \geq 0 \quad & \forall v \in U_{ad}, \end{aligned} \tag{5.198}$$

with initial condition $y(x,0) = y_0(x)$ for the state variable (evolving forward in time). The terminal condition for the adjoint variable (evolving backward in time) is given by

$$p(x,T) = \beta(y(x,T) - y_T(x)). \tag{5.199}$$

The optimality system corresponding to $\nu = 0$ is given by (5.198) with the (inequality) optimality condition replaced with

$$p = \min\{0, p + u - \underline{u}\} + \max\{0, p + u - \overline{u}\} \qquad \text{in } Q.$$

Following [59, 147] and considering the linear case $G = 0$, one can prove that the solution to (5.198) with $\nu = 0$ exists and is unique.

Now, we consider the second-order backward differentiation (BDF2) formula together with the Crank–Nicolson (CN) method in order to obtain a second-order time discretization scheme; see Chapter 3.

We have the following discrete optimality system

$$\begin{aligned} -\partial_{BD}^{+} y_h^m + G(y_h^m) + \sigma\Delta_h y_h^m &= f_h^m + u_h^m, \\ \partial_{BD}^{-} p_h^m + G'(y_h^m)p_h^m + \sigma\Delta_h p_h^m + \alpha(y_h^m - y_{dh}^m) &= 0, \\ (\nu u_h^m - p_h^m, v_h^m - u_h^m) &\geq 0 \quad \forall v \in U_{ad}^h, \end{aligned} \tag{5.200}$$

where we assume sufficient regularity of the data, y_d, y_T, and f, such that these functions are properly approximated by their values at grid points.

In order to develop the CSMG smoothing scheme, let us write the discretized optimality system (5.200), in expanded form, for a space-time grid point $(i\,j\,m)$. We have

$$-\left(\tfrac{3}{2}+4\sigma\gamma\right)y_{ijm}+\sigma\gamma\left[y_{i+1jm}+y_{i-1jm}+y_{ij+1m}+y_{ij-1m}\right]+2y_{ijm-1}$$
$$-\tfrac{1}{2}y_{ijm-2}+\delta t\,G(y_{ijm})-\delta t\,u_{ijm}=\delta t\,f_{ijm},\quad 3\le m\le N_t+1, \tag{5.201a}$$

$$-\left(\tfrac{3}{2}+4\sigma\gamma\right)p_{ijm}+\sigma\gamma\left[p_{i+1jm}+p_{i-1jm}+p_{ij+1m}+p_{ij-1m}\right]+2p_{ijm+1}$$
$$-\tfrac{1}{2}p_{ijm+2}+\delta t\,G'(y_{ijm})p_{ijm}+\alpha\delta t(y_{ijm}-y_{dijm})=0,\quad 1\le m\le N_t-1, \tag{5.201b}$$

$$\nu(u_{ijm}-p_{ijm})\cdot(v_{ijm}-u_{ijm})\ge 0\quad \forall v_h\in U^h_{ad}. \tag{5.201c}$$

In the case of terminal observation, at $t_m=T$ $(m=N_t+1)$, we have (5.199) in place of (5.201b).

Further, since we use the CN method to calculate the required first steps to initialize the BDF2 method, we also write the expanded form of (3.26) and (3.27) to analyze the cases $m=2$, corresponding to the instant δt, and $m=N_t$, corresponding to the instant $T-\delta t$. Therefore, for the case $m=2$ we approximate the state variable with the following CN discretization

$$\begin{aligned}
&-(1+2\sigma\gamma)y_{ijm}+\tfrac{\sigma\gamma}{2}\left[y_{i+1jm}+y_{i-1jm}+y_{ij+1m}+y_{ij-1m}\right]\\
&\quad+(1-2\sigma\gamma)y_{ijm-1}+\tfrac{\delta t}{2}G(y_{ijm})-\tfrac{\delta t}{2}u_{ijm}\\
&\quad+\tfrac{\sigma\gamma}{2}\left[y_{i+1jm-1}+y_{i-1jm-1}+y_{ij+1m-1}+y_{ij-1m-1}\right]\\
&\quad+\tfrac{\delta t}{2}G(y_{ijm-1})-\tfrac{\delta t}{2}u_{ijm-1}=\tfrac{\delta t}{2}(f_{ijm-1}+f_{ijm}),\quad m=2,
\end{aligned} \tag{5.202}$$

and we approximate the corresponding adjoint variable p_{ij2} with (5.201b). Moreover, for the case $m=N_t$ we approximate the corresponding state variable y_{ijN_t} with (5.201a), and we calculate the adjoint variable with the following CN discretization

$$\begin{aligned}
&-(1+2\sigma\gamma)p_{ijm}+\tfrac{\sigma\gamma}{2}\left[p_{i+1jm}+p_{i-1jm}+p_{ij+1m}+p_{ij-1m}\right]\\
&\quad+(1-2\sigma\gamma)p_{ijm+1}+\tfrac{\delta t}{2}G'(y_{ijm})p_{ijm}+\tfrac{\alpha\delta t}{2}(y_{ijm}-y_{dijm})\\
&\quad+\tfrac{\sigma\gamma}{2}\left[p_{i+1jm+1}+p_{i-1jm+1}+p_{ij+1m+1}+p_{ij-1m+1}\right]\\
&\quad+\tfrac{\delta t}{2}G'(y_{ijm+1})p_{ijm+1}+\tfrac{\alpha\delta t}{2}(y_{ijm+1}-y_{dijm+1})=0,\ m=N_t.
\end{aligned} \tag{5.203}$$

In this case, the approximation of the control is obtained equally as in (5.201c).

Now, we can proceed with the construction of a pointwise smoothing scheme following the same lines of development of the TS-CGS scheme in case of backward Euler discretization. Consider the optimality system (5.201) at the space-time grid points $(i\,j\,m)$ for $m=3,\dots,N_t-1$. A similar discussion follows for $m=2$ and $m=N_t$, where the BDF2 and the CN discretization both appear in the optimality system. We start by defining the following quantities

$$\begin{aligned}
a &= \left(\tfrac{3}{2}+4\sigma\gamma\right),\\
S_{ijm} &= \sigma\gamma\left[y_{i+1\,jm}+y_{i-1\,jm}+y_{i\,j+1m}+y_{i\,j-1m}\right]+2y_{ijm-1}-\tfrac{1}{2}y_{ijm-2}-\delta t\,f_{i\,jm},\\
R_{ijm} &= \sigma\gamma\left[p_{i+1\,jm}+p_{i-1\,jm}+p_{i\,j+1m}+p_{i\,j-1m}\right]+2p_{ijm+1}-\tfrac{1}{2}p_{ijm+2}-\delta t\,\alpha\,y_{di\,jm}.
\end{aligned}$$

With this notation, the optimality system (5.201) at (i, j, m) can be written as follows

$$-a\, y_{ijm} + S_{ijm} + \delta t G(y_{ijm}) - \delta t u_{ijm} = 0, \tag{5.204a}$$

$$-a\, p_{ijm} + R_{ijm} + \delta t G'(y_{ijm}) p_{ijm} + \alpha \delta t\, y_{ijm} = 0, \tag{5.204b}$$

$$(\nu u_{ijm} - p_{ijm})(v_{ijm} - u_{ijm}) \geq 0 \quad \forall v_h \in U_{ad,h}. \tag{5.204c}$$

This is a nonlinear problem that includes an inequality constraint. To solve this problem, we generalize the nonlinear version of the TS-CGS scheme discussed in the case of backward discretization.

Consider (5.204a) and (5.204b). The Jacobian of these two equations is given by

$$J_{ijm} := \begin{pmatrix} -a + \delta t G' & 0 \\ \delta t(\alpha + G'' p_{ijm}) & -a + \delta t G' \end{pmatrix}$$

and its inverse is

$$J_{ijm}^{-1} = \frac{1}{(-a + \delta t G')^2} \begin{pmatrix} -a + \delta t G' & 0 \\ -\delta t(\alpha + G'' p_{ijm}) & -a + \delta t G' \end{pmatrix}. \tag{5.205}$$

Notice that, in the case of nonmonotone nonlinearity (e.g., singular control problems), we should choose δt sufficiently small to guarantee that $(-a + \delta t G')^2 \neq 0$.

Now, for a given u_{ijm}, a classical local Newton update for the auxiliary state and adjoint variables $\widehat{y}_{ijm}$ and $\widehat{p}_{ijm}$ is given by

$$\begin{pmatrix} \widehat{y} \\ \widehat{p} \end{pmatrix}_{ijm} = \begin{pmatrix} y \\ p \end{pmatrix}_{ijm} + J_{ijm}^{-1} \begin{pmatrix} r_y \\ r_p \end{pmatrix}_{ijm}, \tag{5.206}$$

where $(r_y)_{ijm}$ and $(r_p)_{ijm}$ denote the residuals of (5.204a) and (5.204b), respectively. In the case of the BDF2 discretization, these residuals are given by

$$\begin{aligned} (r_y)_{ijm} &= a\, y_{ijm} - S_{ijm} - \delta t G(y_{ijm}) + \delta t u_{ijm} && \text{for } m = 3, \ldots, N_t + 1, \\ (r_p)_{ijm} &= a\, p_{ijm} - R_{ijm} - \delta t G'(y_{ijm}) p_{ijm} - \alpha \delta t\, y_{ijm} && \text{for } m = 1, \ldots, N_t - 1. \end{aligned}$$

Since $(r_y)_{ijm}$ depends explicitly on u_{ijm}, we can write $\widehat{p}_{ijm}$ as a function of u_{ijm} as follows

$$\begin{aligned} \widehat{p}_{ijm}(u_{ijm}) = p_{ijm} &+ \frac{-\delta t(\alpha + G'' p_{ijm})[a\, y_{ijm} - S_{ijm} - \delta t G(y_{ijm}]}{(-a + \delta t G')^2} \\ &+ \frac{a\, p_{ijm} - R_{ijm} - \delta t G'(y_{ijm}) p_{ijm} - \alpha \delta t\, y_{ijm}}{(-a + \delta t G')} \\ &- \frac{\delta t^2 (\alpha + G'' p) u_{ijm}}{(-a + \delta t G')^2}. \end{aligned} \tag{5.207}$$

We use $\widehat{p}_{ijm}$ in order to obtain the update for the control variable u_{ijm}. Let us recall that the gradient of the objective functional is given as $\nabla \hat{J}(u) = \nu u - p$. Therefore, from

$\nu\widetilde{u}_{ijm} - \widehat{p}_{ijm} = 0$, we obtain the auxiliary variable

$$\widetilde{u}_{ijm} = \left(\nu + \frac{\delta t^2(\alpha + G''p)u_{ijm}}{(-a+\delta t G')^2}\right)^{-1}$$
$$\times \left[p_{ijm} + \frac{-\delta t(\alpha + G'' p_{ijm})[a\, y_{ijm} - S_{ijm} - \delta t G(y_{ijm})]}{(-a+\delta t G')^2} + \frac{a\, p_{ijm} - R_{ijm} - \delta t G'(y_{ijm}) p_{ijm} - \alpha \delta t\, y_{ijm}}{(-a+\delta t G')} \right]. \tag{5.208}$$

Then, the new value for u_{ijm} resulting from the smoothing step is obtained by projection as follows

$$u_{ijm} = \begin{cases} \overline{u}_{ijm} & \text{if} \quad \tilde{u}_{ijm} > \overline{u}_{ijm}, \\ \widetilde{u}_{ijm} & \text{if} \quad \underline{u}_{ijm} \le \widetilde{u}_{ijm} \le \overline{u}_{ijm}, \\ \underline{u}_{ijm} & \text{if} \quad \tilde{u}_{ijm} < \underline{u}_{ijm}. \end{cases} \tag{5.209}$$

With u_{ijm} given, we can use (5.206) to obtain new values for y_{ijm} and p_{ijm}.

An iteration step of this smoothing scheme can be performed in any ordering for the spatial variables, and in order to take into account the opposite time orientation of the state and the adjoint equations we update the state variable y using the first vector component of (5.206) marching in the forward direction, and the adjoint variable p is being updated using the second component of (5.206) marching backward in time. In this way a robust iteration is obtained. Let us recall that the calculation of the initialization steps y_{ij2} and p_{ijN_t} are carried out in the same way, but using the combination of the CN scheme with the BDF2 method, as described above.

ALGORITHM 5.13. Projected time-splitted CGS iteration (P-TS-CGS).

1. Set the starting approximation: calculate y_{ij2}, p_{ij2}, u_{ij2}, y_{ijN_t}, p_{ijN_t}, and u_{ijN_t}.

2. For ij in, e.g., lexicographic order do

3. For $m = 3,\dots,N_t - 1$: compute $(r_y)_{ijm}$, $\tilde{u}_{ijm}$ and projection u_{ijm}. Then, the state update is given by
$$y^{(1)}_{i\,j\,m} = y^{(0)}_{i\,j\,m} + \frac{(r_y)_{ijm}}{(-a+\delta t G')}.$$

4. For $m' = N_t - 1,\dots,3$ (backward): compute $(r_y)_{ijm'}$, $(r_p)_{ijm'}$, $\tilde{u}_{ijm'}$ and projection $u_{ijm'}$. Then, the adjoint update is given by
$$p^{(1)}_{i\,j\,m'} = p^{(0)}_{i\,j\,m'} + \frac{(-a+\delta t G')(r_p)_{ijm'} - \delta t(\alpha + G''p)(r_y)_{ijm'}}{(-a+\delta t G')^2}.$$

5. end.

In the control-unconstrained case, the iteration above applies without projection and it becomes equivalent to the time-splitted CGS iteration.

In order to construct a robust block time-line smoothing scheme, we proceed in a way similar to that followed for the pointwise approach in order to obtain an approximation for the controls u_{ijm}. Thus, we consider the residual equations $(r_y)_{ijm} = 0$ and $(r_p)_{ijm} = 0$ for all $m = 1, \ldots, N_t + 1$ at any fixed (i, j). The solution of this problem provides the mapping $u_{ijm} \rightarrow y_{ijm}$ and $u_{ijm} \rightarrow p_{ijm}$, and by requiring us to satisfy the (unconstrained) optimality condition, we obtain $\tilde{u}_{ijm}$ followed by projection, thus obtaining a new approximation for the control and consequently the updates for the state and adjoint variables.

Let us recall that the residuals $(r_y)_{ijm}$ are defined as follows

$$
\begin{aligned}
(r_y)_{ijm} &= y_{ijm} - \psi_{ijm} \quad \text{for } m = 1, \\
(r_y)_{ijm} &= (1+2\sigma\gamma)\, y_{ijm} - \tfrac{\sigma\gamma}{2}\left[y_{i+1jm} + y_{i-1jm} + y_{ij+1m} + y_{ij-1m}\right] - (1-2\sigma\gamma)\, y_{ijm-1} \\
&\quad - \tfrac{\delta t}{2} G(y_{ijm}) + \tfrac{\delta t}{2} u_{ijm} - \tfrac{\sigma\gamma}{2}\left[y_{i+1jm-1} + y_{i-1jm-1} + y_{ij+1m-1} + y_{ij-1m-1}\right] \\
&\quad - \tfrac{\delta t}{2} G(y_{ijm-1}) + \tfrac{\delta t}{2} u_{ijm-1} + \tfrac{\delta t}{2}(f_{ijm-1} + f_{ijm}) \quad \text{for } m = 2, \\
(r_y)_{ijm} &= \left(\tfrac{3}{2} + 4\sigma\gamma\right) y_{ijm} - \sigma\gamma\left[y_{i+1jm} + y_{i-1jm} + y_{ij+1m} + y_{ij-1m}\right] - 2y_{ijm-1} \\
&\quad + \tfrac{1}{2} y_{ijm-2} - \delta t\, G(y_{ijm}) + \delta t\, u_{ijm} + \delta t\, f_{ijm} \quad \text{for } 3 \leq m \leq N_t + 1.
\end{aligned}
$$

Note that the definition of $(r_y)_{ij1}$ is introduced considering a function ψ_h^m associated with the initial condition y_0. The introduction of this residual helps us to completely describe the system of equations in the time interval $[0, T]$, making it possible to obtain a desirable structure for the matrices involved in the solution of the equation.

Furthermore, the residuals $(r_p)_{ijm}$ are given by

$$
\begin{aligned}
(r_p)_{ijm} &= \left(\tfrac{3}{2} + 4\sigma\gamma\right) p_{ijm} - \sigma\gamma\left[p_{i+1jm} + p_{i-1jm} + p_{ij+1m} + p_{ij-1m}\right] - 2p_{ijm+1} \\
&\quad + \tfrac{1}{2} p_{ijm+2} - \delta t\, G'(y_{ijm}) p_{ijm} - \alpha\delta t(y_{ijm} - y_{dijm}) = 0 \quad \text{for } 1 \leq m \leq N_t + 1, \\
(r_p)_{ijm} &= (1+2\sigma\gamma)\, p_{ijm} - \tfrac{\sigma\gamma}{2}\left[p_{i+1jm} + p_{i-1jm} + p_{ij+1m} + p_{ij-1m}\right] \\
&\quad - (1-2\sigma\gamma)\, p_{ijm+1} - \tfrac{\delta t}{2} G'(y_{ijm}) p_{ijm} - \tfrac{\alpha\delta t}{2}(y_{ijm} - y_{dijm}) \\
&\quad - \tfrac{\sigma\gamma}{2}\left[p_{i+1jm+1} + p_{i-1jm+1} + p_{ij+1m+1} + p_{ij-1m+1}\right] \\
&\quad - \tfrac{\delta t}{2} G'(y_{ijm+1}) p_{ijm+1} - \tfrac{\alpha\delta t}{2}(y_{ijm+1} - y_{dijm+1}) = 0 \quad \text{for } m = N_t, \\
(r_p)_{ijm} &= \beta(y_{ijm} - y_{Tijm}) - p_{ijm} \quad \text{for } m = N_t + 1.
\end{aligned}
$$

Note that $(r_p)_{ijN_t+1}$ corresponds to the terminal condition (5.199).

To describe the block Gauss–Seidel procedure, consider the residual equations at any fixed i, j and for all time steps. For each spatial grid point i, j, the pair of state and adjoint equations corresponding to (y, p) at a given m corresponds to five 2×2 blocks for the pairs $(y, p)_{m-2}$, $(y, p)_{m-1}$, $(y, p)_m$, $(y, p)_{m+1}$, $(y, p)_{m+2}$. Considering all time steps, we obtain a block-pentadiagonal system. This system has the following form

$$M = \begin{bmatrix} A_1 & D_1 & E_1 & & & & & \\ C_2 & A_2 & D_2 & E_2 & & & & \\ B_3 & C_3 & A_3 & D_3 & E_3 & & & \\ & & \ddots & \ddots & \ddots & \ddots & \ddots & \\ & & & B_{N_t-1} & C_{N_t-1} & A_{N_t-1} & D_{N_t-1} & E_{N_t-1} \\ & & & & B_{N_t} & C_{N_t} & A_{N_t} & D_{N_t} \\ & & & & & B_{N_t+1} & C_{N_t+1} & A_{N_t+1} \end{bmatrix}. \tag{5.210}$$

Centered at t_m, the entries B_m C_m, A_m, D_m, and E_m refer to the variables (y,p) at t_{m-2}, t_{m-1}, t_m, t_{m+1}, and t_{m+2}, respectively. For $m = 3,\ldots,N_t-1$, the block A_m is given by

$$A_m = \begin{bmatrix} -(\frac{3}{2}+4\sigma\gamma)+\delta t G' & -\chi_{ijm}\frac{\delta t}{\nu} \\ \delta t(\alpha+G'' p) & -(\frac{3}{2}+4\sigma\gamma)+\delta t G' \end{bmatrix}, \tag{5.211}$$

where all functions within the brackets [] are evaluated at t_m, and the indicator function χ_{ijm} is defined, for $m = 1,\ldots,N_t+1$, by

$$\chi_{ijm} := \begin{cases} 1 & \text{if } \underline{u}_{ijm} \le \tilde{u}_{ijm} \le \overline{u}_{ijm}, \\ 0 & \text{otherwise,} \end{cases} \tag{5.212}$$

where $\tilde{u}_{ijm}$ is given by (5.208). Thanks to the introduction of this indicator term, we guarantee a correct updating of the state and adjoint variables, mainly in the grid points, where $\underline{u}_{ijm} \le \tilde{u}_{ijm} \le \overline{u}_{ijm}$.

The B_m, C_m, D_m, and E_m blocks are given by

$$B_m = \begin{bmatrix} -\frac{1}{2} & 0 \\ 0 & 0 \end{bmatrix}, C_m = \begin{bmatrix} 2 & 0 \\ 0 & 0 \end{bmatrix}, D_m = \begin{bmatrix} 0 & 0 \\ 0 & 2 \end{bmatrix}, \text{ and } E_m = \begin{bmatrix} 0 & 0 \\ 0 & -\frac{1}{2} \end{bmatrix}. \tag{5.213}$$

Clearly, for each time step, the variables neighboring the point ij are taken as constant and contribute to the right-hand side of the system.

Next, according to the proposed approach, the blocks A_2, C_2, as well as A_{N_t} and D_{N_t}, are defined in a different way because a CN method is used at $t_2 = \delta t$ and at $t_{N_t} = T - \delta t$. We then obtain that

$$\begin{aligned} A_2 &= \begin{bmatrix} -(1+2\sigma\gamma)+\frac{\delta t}{2}G' & -\chi_{ij2}\frac{\delta t}{2\nu} \\ \delta t(\alpha+G'' p) & -(\frac{3}{2}+4\sigma\gamma)+\delta t G' \end{bmatrix}, \\ C_2 &= \begin{bmatrix} (1-2\sigma\gamma)+\frac{\delta t}{2}G' & -\chi_{ij1}\frac{\delta t}{2\nu} \\ 0 & 0 \end{bmatrix}, \end{aligned} \tag{5.214}$$

where the functions within the brackets [] in A_2 are evaluated at $t_2 = \delta t$, while the functions within the brackets in C_2 are evaluated at $t_1 = 0$. The matrices D_2 and E_2 are defined in the same way as in (5.213).

Further, we have that

$$A_{N_t} = \begin{bmatrix} -(\frac{3}{2}+4\sigma\gamma)+\delta t G' & -\chi_{ijN_t}\frac{\delta t}{\nu} \\ \frac{\delta t}{2}(\alpha+G''p) & -(1+2\sigma\gamma)+\frac{\delta t}{2}G' \end{bmatrix}, \tag{5.215}$$

$$D_{N_t} = \begin{bmatrix} 0 & 0 \\ \frac{\delta t}{2}(\alpha+G''p) & (1-2\sigma\gamma)+\frac{\delta t}{2}G' \end{bmatrix},$$

where the functions within the brackets [] in A_{N_t} are evaluated at $t_{N_t} = T - \delta t$, while the functions within the brackets in D_{N_t} are evaluated at $t_{N_t+1} = T$. Matrices B_{N_t} and C_{N_t} are also defined in the same way as in (5.213).

It remains to discuss the blocks A_1 and A_{N_t+1} for $\beta \neq 0$. As stated before, at $t_m = t_0$, we impose the following condition

$$y_{ijm} - \psi_{ijm} = 0, \quad m = 1, \tag{5.216}$$

where ψ_h^m is a given function associated with the initial condition. Thus, the block A_1 is given by

$$A_1 = \begin{bmatrix} 1 & 0 \\ \delta t(\alpha+G''p) & -(\frac{3}{2}+4\sigma\gamma)+\delta t G' \end{bmatrix}. \tag{5.217}$$

The other blocks D_1 and E_1 are defined equally as in (5.213). Clearly, thanks to the introduction of this superfluous condition (5.216), we obtain the structure of a pentadiagonal block matrix for system matrix M.

Further, at $t_m = T$, we have the terminal condition (5.199), which we rewrite as

$$\beta(y_{ijm} - y_{T,ijm}) - p_{ijm} = 0, \quad m = N_t + 1.$$

Thus, the block A_{N_t+1} is given by

$$A_{N_t+1} = \begin{bmatrix} -(1+4\sigma\gamma)+\delta t G' & -\chi_{ijN_t+1}\frac{\delta t}{\nu} \\ \beta & -1 \end{bmatrix}, \tag{5.218}$$

and the blocks B_{N_t+1} and E_{N_t+1} are as in (5.213).

Thus, for each i, j we have to solve a pentadiagonal system $Mw = r$, where $w = (y_h^2, p_h^2, \ldots, y_h^{N_t+1}, p_h^{N_t+1})$ and $r = (r_y(w^2), r_p(w^2), \ldots, r_y(w^{N_t+1}), r_p(w^{N_t+1}))$. Summarizing, our collective t-line relaxation is given by the following algorithm.

ALGORITHM 5.14. Projected time-line CGS iteration (P-TL-CGS).

1. Set the starting approximation.
2. For ij in, e.g., lexicographic order do: calculate u_{ij} by using (5.208) and (5.209), and construct $(r_y)_{ij}$ and $(r_p)_{ij}$. Then

$$\begin{pmatrix} y \\ p \end{pmatrix}_{ij}^{(1)} = \begin{pmatrix} y \\ p \end{pmatrix}_{ij}^{(0)} + M^{-1}\begin{pmatrix} r_y \\ r_p \end{pmatrix}_{ij};$$

3. end.

Also in this case we consider that the residuals r_y and r_p are constructed at i,j and for all m prior to the update. Since the solution in time is exact, no time splitting is required.

In the case of the control-unconstrained problem, the algorithm described above does not need the projection step to calculate the control. We use the optimality condition $\nu u_h^m - p_h^m = 0$ in order to eliminate the control in the construction of $(r_y)_{ijm}$, and, consequently, the indicator term χ_{ijm} is no longer needed in the sense that we can set $\chi_{ijm} = 1$ for all grid points i,j,m and the TL-CGS smoother results.

In Algorithm 5.14, the problem of how to solve the block-pentadiagonal system $Mw = r$ arises. Efficient solvers for these kinds of systems can be constructed generalizing the Thomas algorithm, which is a well-known method to solve block-tridiagonal systems (see [353]). For this purpose, in [150] a solver is presented that solves pentadiagonal systems of order N_t with $\mathcal{O}(4(N_t+1))$ effort; see also [25].

In the control-unconstrained case, results in [150] show that the CSMG scheme with TS-CGS and with TL-CGS smoothing, and BDF2 discretization, provides space-time second-order accurate solutions with optimal computational complexity. In the following, we report results obtained in [150] for a control-constrained optimal control problem. The multigrid algorithm with the P-TS-CGS smoothing scheme is tested with an exact solution, which is constructed as follows. We choose $\underline{u}(\mathbf{x},t) = -1/2$ and $\overline{u}(\mathbf{x},t) = 1/2$ and the following

$$
\begin{aligned}
y(x_1,x_2,t) &= (1-t)\sin(\pi x_1)\sin(\pi x_2), \\
p(x_1,x_2,t) &= \nu(1-t)\sin(2\pi x_1)\sin(2\pi x_2), \\
u(x_1,x_2,t) &= \max\{-0.5, \min\{0.5, p(x_1,x_2,t)/\nu\}\}.
\end{aligned}
\tag{5.219}
$$

The corresponding data is given by

$$
\begin{aligned}
f &= -u - \partial_t y + \Delta y, \\
y_d &= y + \partial_t p + \Delta p.
\end{aligned}
\tag{5.220}
$$

Results of experiments are reported in Table 5.20. In this case, second-order accuracy is no longer achieved due to a loss of regularity of the control function when the constraints become active. In this situation, more advanced techniques [193, 289] are required to recover high-order accuracy.

Table 5.20. *Accuracy results for a constrained-control problem:* $\sigma = 1$, $\alpha = 1$, *and* $\beta = 0$.

ν	$N_x \times N_y \times N_t$	γ	$\|y - y_h\|$	$\|p - p_h\|$	$\|u - u_h\|$
	$64 \times 64 \times 64$	64	$1.88 \cdot 10^{-4}$	$4.49 \cdot 10^{-6}$	$3.56 \cdot 10^{-3}$
10^{-3}	$128 \times 128 \times 128$	128	$5.07 \cdot 10^{-5}$	$1.10 \cdot 10^{-6}$	$8.74 \cdot 10^{-4}$
	$256 \times 256 \times 256$	256	$1.32 \cdot 10^{-5}$	$2.68 \cdot 10^{-7}$	$2.13 \cdot 10^{-4}$
	$64 \times 64 \times 64$	64	$3.31 \cdot 10^{-5}$	$1.81 \cdot 10^{-7}$	$2.52 \cdot 10^{-2}$
10^{-7}	$128 \times 128 \times 128$	128	$7.60 \cdot 10^{-6}$	$3.03 \cdot 10^{-8}$	$8.67 \cdot 10^{-3}$
	$256 \times 256 \times 256$	256	$1.85 \cdot 10^{-6}$	$5.93 \cdot 10^{-9}$	$2.94 \cdot 10^{-3}$

In Table 5.21, we report results concerning the computational performance of the CSMG approach with constrained-control problems. We obtain convergence rates that show robustness and typical multigrid efficiency that improves on finer meshes. This is due to the fact that on fine meshes the active sets are better resolved. On the other hand,

Table 5.21. *Numerical results for the constrained tracking problem with TS-CGS and TL-CGS smoothing schemes. Parameters:* $\sigma = 1$, $\alpha = 1$, *and* $\beta = 0$. *Initial condition for state equation:* $y_0 = y(\mathbf{x}, 0)$.

	ν	γ	ρ	$\|r_y\|$	$\|r_p\|$	$\|y_h - R_h y_d\|$
TS-CGS	10^{-2}	32	0.034	$1.32 \cdot 10^{-9}$	$6.56 \cdot 10^{-11}$	$2.26 \cdot 10^{-1}$
		64	0.030	$6.32 \cdot 10^{-9}$	$3.41 \cdot 10^{-10}$	$2.29 \cdot 10^{-1}$
		128	0.029	$6.17 \cdot 10^{-8}$	$4.09 \cdot 10^{-9}$	$2.30 \cdot 10^{-1}$
	10^{-4}	32	0.081	$3.67 \cdot 10^{-9}$	$5.86 \cdot 10^{-12}$	$2.30 \cdot 10^{-3}$
		64	0.029	$4.43 \cdot 10^{-10}$	$5.03 \cdot 10^{-12}$	$2.29 \cdot 10^{-3}$
		128	0.028	$2.87 \cdot 10^{-9}$	$6.21 \cdot 10^{-11}$	$2.30 \cdot 10^{-3}$
	10^{-6}	32	0.548	$7.15 \cdot 10^{-6}$	$6.27 \cdot 10^{-9}$	$1.75 \cdot 10^{-4}$
		64	0.293	$1.95 \cdot 10^{-8}$	$1.94 \cdot 10^{-11}$	$4.35 \cdot 10^{-5}$
		128	0.091	$1.76 \cdot 10^{-9}$	$4.99 \cdot 10^{-13}$	$2.52 \cdot 10^{-5}$
TL-CGS	10^{-2}	32	0.034	$1.34 \cdot 10^{-9}$	$6.57 \cdot 10^{-11}$	$2.26 \cdot 10^{-1}$
		64	0.030	$6.30 \cdot 10^{-9}$	$3.41 \cdot 10^{-10}$	$2.29 \cdot 10^{-1}$
		128	0.029	$6.17 \cdot 10^{-8}$	$4.09 \cdot 10^{-9}$	$2.30 \cdot 10^{-1}$
	10^{-4}	32	0.082	$3.67 \cdot 10^{-9}$	$5.66 \cdot 10^{-12}$	$2.29 \cdot 10^{-3}$
		64	0.029	$6.82 \cdot 10^{-10}$	$5.05 \cdot 10^{-12}$	$2.29 \cdot 10^{-3}$
		128	0.028	$3.42 \cdot 10^{-9}$	$6.21 \cdot 10^{-11}$	$2.30 \cdot 10^{-3}$
	10^{-6}	32	0.503	$4.18 \cdot 10^{-3}$	$2.20 \cdot 10^{-5}$	$1.30 \cdot 10^{-4}$
		64	0.294	$1.56 \cdot 10^{-8}$	$1.32 \cdot 10^{-11}$	$3.86 \cdot 10^{-5}$
		128	0.149	$2.18 \cdot 10^{-9}$	$1.74 \cdot 10^{-12}$	$2.48 \cdot 10^{-5}$

smaller values of ν result in larger active sets and steeper gradient of the control function arise, making the problem more difficult to solve and thus explaining the worsening of the convergence factor.

In Figure 5.18, the state y and the control u for two instants are depicted. We observe that for small time periods, the control is active, but as the time period increases, the control becomes nonactive, since $p \to 0$ as $t \to T = 1$.

We complete this section reporting results of experiments for the limit case of $\nu = 0$. This discussion is possible due to the robustness of the CSMG approach where the smoothing scheme remains well defined choosing a zero weight of the control. This appears to be a unique feature of this solution strategy. In particular, with this scheme it is possible to investigate bang-bang control problems that arise, e.g., taking $\nu = 0$ and nonattainable target functions. This is a less investigated subject due to the difficulty of computing bang-bang solutions; see [48, 59, 147, 370].

We take $f = 0$ and $y_0(x) = 0$, $\underline{u}(\mathbf{x},t) = -1$ and $\overline{u}(\mathbf{x},t) = 1$, and consider the following target trajectory

$$y_d(x_1, x_2, t) = \sin(2\pi t)\sin(3\pi x_1)\sin(3\pi x_2).$$

With this target function and $\nu = 0$ we obtain a control which is everywhere active; that is, we have bang-bang control. In Figure 5.19, the optimal control and the corresponding state for $\nu = 0$ are depicted for two different instants of time.

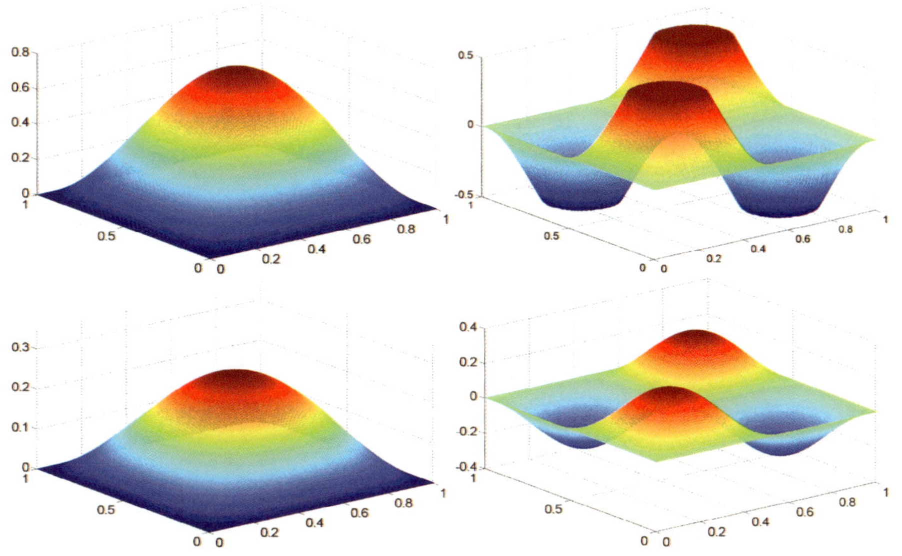

Figure 5.18. *Control constrained tracking problem: state y (left column) and u (right column) for* $t = T/4 = 0.25$ *and* $t = 3T/4 = 0.75$. *Parameters:* $\alpha = 1$, $\beta = 0$, $\nu = 10^{-7}$, $\gamma = 64$, *and* $\sigma = 1$. *Reprinted with permission from S. Gonzalez Andrade and A. Borzì, Multigrid second-order accurate solution of parabolic control-constrained problems, Computational Optimization and Applications, to appear.*

Local Fourier Analysis for BDF2 Discretization

We can use the local Fourier analysis framework to analyze the TS-CGS smoothing scheme applied to (5.201a)–(5.201b) resulting from the BDF2 discretization and without constraints on the control. To analyze the update procedure at (i, m), we consider that y_{im-1} and y_{im-2} as well as p_{im+1} and p_{im+2} have been updated in the previous iteration step and consequently they have the superindex (1). We obtain that

$$\begin{pmatrix} -(\frac{3}{2}+2\sigma\gamma)+\sigma\gamma e^{-i\theta_x} & -\frac{\delta t}{\nu} \\ \alpha\delta t & -(\frac{3}{2}+2\sigma\gamma)+\sigma\gamma e^{-i\theta_x} \end{pmatrix} \begin{pmatrix} \tilde{Y}^{(1)}_{\boldsymbol{\theta}} \\ \tilde{P}^{(1)}_{\boldsymbol{\theta}} \end{pmatrix}$$

$$+ \begin{pmatrix} 2e^{-i\theta_t} - \frac{e^{-2i\theta_t}}{2} & 0 \\ 0 & 2e^{i\theta_t} - \frac{e^{2i\theta_t}}{2} \end{pmatrix} \begin{pmatrix} \tilde{Y}^{(1)}_{\boldsymbol{\theta}} \\ \tilde{P}^{(1)}_{\boldsymbol{\theta}} \end{pmatrix} = \begin{pmatrix} -\sigma\gamma e^{i\theta_x} & 0 \\ 0 & -\sigma\gamma e^{i\theta_x} \end{pmatrix} \begin{pmatrix} \tilde{Y}^{(0)}_{\boldsymbol{\theta}} \\ \tilde{P}^{(0)}_{\boldsymbol{\theta}} \end{pmatrix}.$$

Hence, we obtain the Fourier symbol of the TS-CGS scheme as follows

$$\hat{s}(\boldsymbol{\theta}) = \begin{pmatrix} -(\frac{3}{2}+2\sigma\gamma)+\sigma\gamma e^{-i\theta_x}+2e^{-i\theta_t} - \frac{e^{-2i\theta_t}}{2} & -\frac{\delta t}{\nu} \\ \alpha\delta t & -(\frac{3}{2}+2\sigma\gamma)+\sigma\gamma e^{-i\theta_x}+2e^{i\theta_t} - \frac{e^{2i\theta_t}}{2} \end{pmatrix}^{-1} \times \begin{pmatrix} -\sigma\gamma e^{i\theta_x} & 0 \\ 0 & -\sigma\gamma e^{i\theta_x} \end{pmatrix}.$$

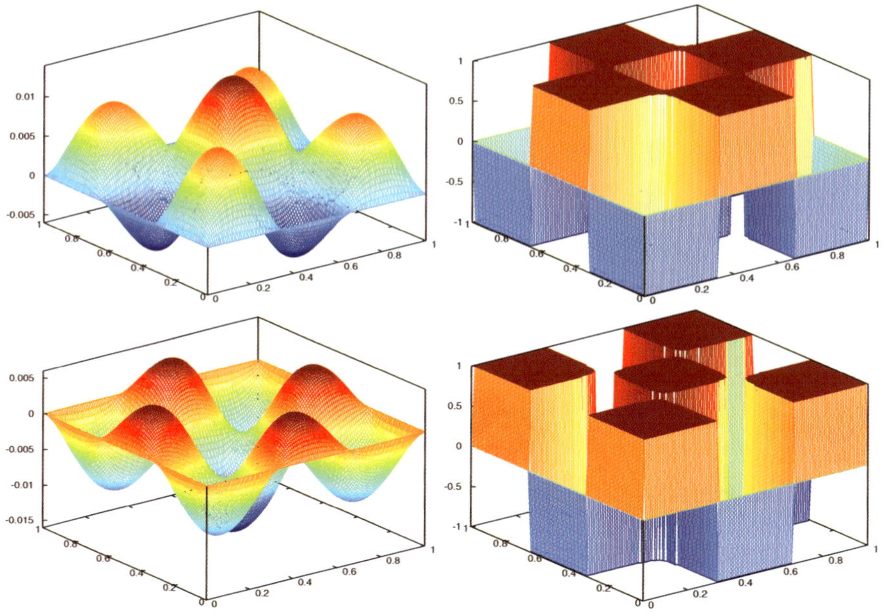

Figure 5.19. *Numerical bang-bang control solutions with* $\nu = 0$ *at* $t = T/4$ *(top) and* $t = 3T/4$ *(bottom). The state (left) and the control (right);* $128 \times 128 \times 128$ *mesh. Reprinted with permission from S. Gonzalez Andrade and A. Borzì, Multigrid second-order accurate solution of parabolic control-constrained problems, Computational Optimization and Applications, to appear.*

The numerical evaluation of the smoothing factor $\mu(S_k)$ of the TS-CGS scheme demonstrates that the TS-CGS scheme for the BDF2 discretization is mesh independent and optimization-parameter independent for a very large range of meshes and of values of the weight of the cost of the control.

Next, consider the case of TL-CGS relaxation. In this case, the Fourier symbol of the smoothing operator is given by the following 2×2 matrix

$$\hat{s}(\boldsymbol{\theta}) = -(A + B\, e^{-2i\theta_t} + C\, e^{-i\theta_t} + D\, e^{i\theta_t} + E\, e^{2i\theta_t} + \tilde{I}\, e^{-i\theta_x})^{-1}(\tilde{I}\, e^{i\theta_x}),$$

where

$$A = \begin{bmatrix} -(\frac{3}{2} + 2\sigma\gamma) & -\frac{\delta t}{\nu} \\ \alpha\delta t & -(\frac{3}{2} + 2\sigma\gamma) \end{bmatrix}, \ B_m = \begin{bmatrix} -\frac{1}{2} & 0 \\ 0 & 0 \end{bmatrix},$$

$$C_m = \begin{bmatrix} 2 & 0 \\ 0 & 0 \end{bmatrix}, \ D_m = \begin{bmatrix} 0 & 0 \\ 0 & 2 \end{bmatrix}, \ E_m = \begin{bmatrix} 0 & 0 \\ 0 & -\frac{1}{2} \end{bmatrix}, \text{ and } \tilde{I} = \begin{bmatrix} \sigma\gamma & 0 \\ 0 & \sigma\gamma \end{bmatrix}.$$

Since the system is solved for all m at once, we have that y_{im-1}, y_{im-2}, p_{im+1}, and p_{im+2} have superindex (0). We obtain that the TL-CGS scheme has smoothing properties similar to those of TS-CGS scheme. This fact appears also from results of numerical experiments.

Next, we investigate the action of the coarse-grid correction to obtain TG convergence estimates. We consider the full-weighting restriction operator (5.195) and the symbol of the fine-grid operator is as in (5.196), where

$$\begin{aligned} a_y(\boldsymbol{\theta}) &= 2\sigma\gamma\cos(\theta_x) + 2e^{-i\theta_t} - \tfrac{1}{2}e^{-2i\theta_t} - 2\sigma\gamma - \tfrac{3}{2}, \\ a_p(\boldsymbol{\theta}) &= 2\sigma\gamma\cos(\theta_x) + 2e^{i\theta_t} - \tfrac{1}{2}e^{2i\theta_t} - 2\sigma\gamma - \tfrac{3}{2}. \end{aligned}$$

The symbol of the coarse-grid operator follows

$$\widehat{A}_{k-1}(\boldsymbol{\theta}) = \begin{bmatrix} \frac{\sigma\gamma\cos(\theta_x)}{2} + 2e^{-i\theta_t} - \frac{1}{2}e^{-2i\theta_t} - \frac{\sigma\gamma+3}{2} & -\delta t/\nu \\ \alpha\delta t & \frac{\sigma\gamma\cos(\theta_x)}{2} + 2e^{i\theta_t} - \frac{1}{2}e^{2i\theta_t} - \frac{\sigma\gamma+3}{2} \end{bmatrix}.$$

Notice that on the coarser grid δt remains unchanged, since we do not apply coarsening in the time direction, while $\gamma \to \gamma/4$ by coarsening.

Based on the representation on TG_k^{k-1} by a 4×4 matrix $\widehat{TG_k^{k-1}}(\boldsymbol{\theta})$ we can calculate the convergence factor given by $\rho(TG_k^{k-1})$ that requires us to determine the spectral radius of a 4×4 matrix.

We obtain that the convergence factor ρ is almost independent of the value of the weight ν and of the discretization parameter γ for both choices of the smoothing scheme. Notice that our result predicts convergence factors that improve for smaller values of the optimization parameter. This is a unique feature of the CSMG multigrid approach. The estimates obtained with TG local Fourier analysis are sharp, and in order to facilitate comparison with values of convergence factors obtained with numerical experiments, we report in Tables 5.22 and 5.23 the local Fourier analysis quantitative estimates of the smoothing factor and the convergence factor of TL-CGS- and TS-CGS-multigrid schemes.

Table 5.22. *Smoothing factor $\mu(S_k)$ and convergence factor $\rho(TG_k^{k-1})$ for the TS-CGS multigrid scheme ($\nu_1 = \nu_2 = 1$). Parameters: $\delta t = 1/64$, $\sigma = 1$, $\alpha = 1$, and $\beta = 0$.*

TS-CGS	γ \ ν	10^{-8}	10^{-6}	10^{-4}	10^{-2}
$\mu(S_k)$	32	0.2289	0.4843	0.4516	0.4493
	48	0.3317	0.4737	0.4502	0.4486
	64	0.4056	0.4677	0.4494	0.4483
$\rho(TG_k^{k-1})$	32	0.0427	0.1317	0.1361	0.1347
	48	0.0822	0.1352	0.1354	0.1344
	64	0.1147	0.1368	0.1350	0.1342

5.7.5 Projected Collective Smoothing Schemes and the Semismooth Newton Method

The CSMG collective smoothing schemes are formulated based on the idea of obtaining at the grid-point (grid-block) level the mappings $y = y(u)$ and $p = p(u)$ and using these mappings in the optimality condition to compute the unconstrained control. In the presence of constraints the projection of this control on the given bounds gives the required solution. We show that the control obtained using this procedure can be equally obtained applying a local or block semismooth Newton (SSN) methods [191, 281, 343].

Table 5.23. *Smoothing factor $\mu(S_k)$ and convergence factor $\rho(TG_k^{k-1})$ for the TL-CGS multigrid schemes ($\nu_1 = \nu 2 = 1$). Parameters: $\delta t = 1/64$, $\sigma = 1$, $\alpha = 1$, and $\beta = 0$.*

TL-CGS	γ \ ν	10^{-8}	10^{-6}	10^{-4}	10^{-2}
$\mu(S_k)$	32	0.2289	0.4843	0.4516	0.4493
	48	0.3317	0.4737	0.4502	0.4486
	64	0.4056	0.4677	0.4494	0.4483
$\rho(TG_k^{k-1})$	32	0.0427	0.1300	0.1282	0.1266
	48	0.0822	0.1330	0.1303	0.1290
	64	0.1147	0.1345	0.1313	0.1301

Now, we analyze the application of a local SSN method applied to (5.204) to show that the resulting iterative scheme is equivalent to the P-TS-CGS scheme. Recall that (5.204c) is equivalent to the following (see [245])

$$u(\mathbf{x},t) = \max\left\{\underline{u}(\mathbf{x},t), \min\left\{\overline{u}(\mathbf{x},t), \frac{1}{\nu}p(\mathbf{x},t)\right\}\right\} \text{ a.e. in } Q \text{ and for } \nu > 0. \tag{5.221}$$

We consider (5.221) at a grid point $(i\,j\,m)$ and, for the ease of illustration, we study the system (5.204) with $G(y) = 0$. Further, we denote (5.204) as the following operator equation

$$\Phi(y_{ijm}, p_{ijm}, u_{ijm}) := \begin{bmatrix} -a\, y_{ijm} + S_{ijm} - \delta t u_{ijm} \\ -a\, p_{ijm} + R_{ijm} + \alpha\delta t\, y_{ijm} \\ u_{ijm} - \max\left\{\underline{u}_{ijm}, \min\left\{\overline{u}_{ijm}, \frac{1}{\nu}p_{ijm}\right\}\right\} \end{bmatrix} = 0. \tag{5.222}$$

We can state that both the max and the min functions involved in (5.222) are semismooth. Indeed, it is well known (see Lemma 3.1 in [191]) that the mappings $y \to \max(0, y)$ and $y \to \min(0, y)$, from $\mathbb{R}^n$ to $\mathbb{R}^n$, $n \in \mathbb{N}$, are Newton differentiable with Newton derivatives given by the diagonal matrices

$$(\Gamma_{max})_{ii} := \begin{cases} 1 & \text{if } y_i \geq 0, \\ 0 & \text{if } y_i < 0, \end{cases} \quad \text{and} \quad (\Gamma_{min})_{ii} := \begin{cases} 1 & \text{if } y_i \leq 0, \\ 0 & \text{if } y_i > 0, \end{cases} \quad i = 1, \ldots, n, \tag{5.223}$$

respectively. As a result, from Theorem 4.6 in [322] it is implied that the real function $\max\{\underline{u}_{ijm}, \min\{\overline{u}_{ijm}, \frac{1}{\nu}p_{ijm}\}\}$ is Newton differentiable, with respect to p_{ijm}, and its Newton derivative is given by

$$\Gamma_p := \chi_{\mathcal{A}_+}\, \chi_{\mathcal{A}_-}\, \frac{1}{\nu},$$

where $\chi_{\mathcal{A}_+}$ and $\chi_{\mathcal{A}_-}$ are defined by

$$\chi_{\mathcal{A}_+} := \begin{cases} 1 & \text{if } \min\{\overline{u}_{ijm}, \frac{1}{\nu}p_{ijm}\} \geq \underline{u}_{ijm}, \\ 0 & \text{if } \min\{\overline{u}_{ijm}, \frac{1}{\nu}p_{ijm},\} < \underline{u}_{ijm}, \end{cases} \quad \text{and} \quad \chi_{\mathcal{A}_-} := \begin{cases} 1 & \text{if } \frac{1}{\nu}p_{ijm} \leq \overline{u}_{ijm}, \\ 0 & \text{if } \frac{1}{\nu}p_{ijm} > \overline{u}_{ijm}, \end{cases} \tag{5.224}$$

respectively. Consequently, we obtain the SSN step applied to the operator equation (5.222) as follows

$$\begin{bmatrix} -a & 0 & -\delta t \\ \delta t\alpha & -a & 0 \\ 0 & -\chi_{\mathcal{A}_+}\chi_{\mathcal{A}_-}\frac{1}{\nu} & 1 \end{bmatrix}_{ijm} \begin{pmatrix} \delta_y \\ \delta_p \\ \delta_u \end{pmatrix}_{ijm} = \begin{pmatrix} r_y \\ r_p \\ r_u \end{pmatrix}_{ijm}. \tag{5.225}$$

Since (5.225) is an uncoupled system of equations for the residual $(\delta_u)_{ijm}$, we have

$$u_{ijm}^{(1)} = \max\left\{\underline{u}_{ijm}, \min\left\{\overline{u}_{ijm}, \frac{1}{\nu}p_{ijm}^{(0)}\right\}\right\} + \chi_{\mathcal{A}_+}\chi_{\mathcal{A}_-}\frac{1}{\nu}(\delta_p)_{ijm}, \tag{5.226}$$

which yields that

$$\begin{aligned} \begin{pmatrix} y \\ p \end{pmatrix}_{ijm}^{(1)} &= \begin{pmatrix} y \\ p \end{pmatrix}_{ijm}^{(0)} \\ &+ \begin{bmatrix} -a & -\delta t\,\chi_{\mathcal{A}_+}\chi_{\mathcal{A}_-}\frac{1}{\nu} \\ \alpha\,\delta t & -a \end{bmatrix}_{ijm}^{(0)\,-1} \begin{pmatrix} a\,y - S + \delta t U \\ a\,p - R - \alpha\delta t\,y \end{pmatrix}_{ijm}^{(0)}, \end{aligned} \tag{5.227}$$

where $U_{ijm}^{(0)} := \max\{\underline{u}_{ijm}, \min\{\overline{u}_{ijm}, \frac{1}{\nu}p_{ijm}^{(0)}\}\}$.

Now, we show that one iteration step given by (5.226)–(5.227) is equivalent to one iteration of the algorithm P-TS-CGS in the sense that the two methods compute the same update for the control, the state, and the adjoint variables. Indeed, the local SSN iteration must be performed in the forward time direction to calculate the updates for y_{ijm} and u_{ijm} and in the backward time direction to calculate the updates for p_{ijm}.

Consider the three possible occurrences given by (5.209):

(i) $\frac{1}{\nu}p_{ijm} > \overline{u}_{ijm}$. Here, we have that $\chi_{\mathcal{A}_-} = 0$, and we obtain that $U_{ijm} := \overline{u}_{ijm}$. Therefore, from (5.226), we obtain that $u_{ijm}^{(1)} = \overline{u}_{ijm}$ and, from (5.227), the following updates for y_{ijm} and p_{ijm}:

$$\begin{aligned} y_{ijm}^{(1)} &= y_{ijm}^{(0)} + \frac{(r_y)_{ijm}}{-a}, \\ p_{ijm}^{(1)} &= p_{ijm}^{(0)} + \frac{-\alpha\,\delta t\,(r_y)_{ijm} - a\,(r_p)_{ijm}}{a^2} \end{aligned} \tag{5.228}$$

with $(r_y)_{ijm} := a\,y_{ijm}^{(0)} - S_{ijm} + \delta t\overline{u}_{ijm}$ and $(r_p)_{ijm} := a\,p_{ijm}^{(0)} - R_{ijm} - \alpha\delta t\,y_{ijm}$.

(ii) $\frac{1}{\nu}p_{ijm} < \underline{u}_{ijm}$. In this case, we have that $\chi_{\mathcal{A}_+} = 0$, since $\min\{\overline{u}_{ijm}, \frac{1}{\nu}p_{ijm}\} < \underline{u}_{ijm}$. Hence, we have that $U_{ijm} = \underline{u}_{ijm}$ and (5.226) implies that $u_{ijm}^{(1)} = \underline{u}_{ijm}$. Further, (5.227) gives the following updates for y_{ijm} and p_{ijm}:

$$\begin{aligned} y_{ijm}^{(1)} &= y_{ijm}^{(0)} + \frac{(r_y)_{ijm}}{-a}, \\ p_{ijm}^{(1)} &= p_{ijm}^{(0)} + \frac{-\alpha\,\delta t\,(r_y)_{ijm} - a\,(r_p)_{ijm}}{a^2} \end{aligned} \tag{5.229}$$

with $(r_y)_{ijm} := a\,y_{ijm}^{(0)} - S_{ijm} + \delta t\underline{u}_{ijm}$ and $(r_p)_{ijm} := a\,p_{ijm}^{(0)} - R_{ijm} - \alpha\delta t\,y_{ijm}$.

(iii) $\underline{u}_{ijm} \leq \frac{1}{\nu} p_{ijm} \leq \overline{u}_{ijm}$. In this case, we have that $\chi_{\mathcal{A}_+} = \chi_{\mathcal{A}_-} = 1$. Thus, $U_{ijm} = \frac{1}{\nu} p^{(0)}_{ijm}$ and (5.226) yields that

$$\begin{aligned} u^{(1)}_{ijm} &= \tfrac{1}{\nu} p^{(0)}_{ijm} + \tfrac{1}{\nu}(\delta_p)_{ijm} \\ &= \tfrac{1}{\nu} p^{(0)}_{ijm} + \tfrac{1}{\nu}(p^{(1)}_{ijm} - p^{(0)}_{ijm}) = \tfrac{1}{\nu} p^{(1)}_{ijm}. \end{aligned} \tag{5.230}$$

Moreover, by solving the system (5.227) we obtain the following updates for y_{ijm} and p_{ijm}

$$y^{(1)}_{ijm} = \frac{\nu a\, S_{ijm} - \nu\, \delta t\, R_{ijm}}{\nu a^2 + \alpha\, \delta t^2} \quad \text{and} \quad p^{(1)}_{ijm} = \frac{\nu a\, R_{ijm} - \nu \alpha \delta t\, S_{ijm}}{\nu\, a^2 + \alpha \delta t^2}. \tag{5.231}$$

The equivalence between the SSN iteration and the P-TS-CGS iteration is clear in the cases (i) and (ii). Furthermore, in the case (iii), since $\widetilde{u}_{ijm}$ is given by

$$\widetilde{u}_{ijm} = \frac{a\, R_{ijm} + \delta t \alpha\, S_{ijm}}{a^2\, \delta t + \alpha\, \delta t^2},$$

we obtain the same expressions (5.231), by plugging $\widetilde{u}_{ijm}$ in $(r_y)_{ijm}$. Therefore, the equivalence between the P-TS-CGS and the SSN iteration (5.226)–(5.227) is totally established.

We complete this section analyzing the equivalence between the P-TL-CGS approach and an SSN scheme. We require that the controls u_{ijm} be available to construct the residuals $(r_y)_{ijm}$ and $(r_p)_{ijm}$, prior to the calculation of the generalized Newton step, when solving the residual equations. We consider that the controls u_{ijm} are given, for $i = 1, \ldots, N_x$, $j = 1, \ldots, N_y$, and $m = 1, \ldots, N_t + 1$, by

$$u_{ijm} := \max\left\{\underline{u}_{ijm}, \min\left\{\overline{u}_{ijm}, \frac{1}{\nu} p_{ijm}\right\}\right\}. \tag{5.232}$$

Further, for ease of illustration, we consider the linear case $G(y) = 0$. Therefore, due to the fact that function (5.232) is Newton differentiable, we calculate an SSN step to solve the residual equations as follows

$$\begin{pmatrix} \delta_y \\ \delta_p \end{pmatrix}_{ij} = -J_E^{-1} \begin{pmatrix} r_y \\ r_p \end{pmatrix}_{ij} \quad \text{for } i = 1, \ldots, N_x, \text{ and } j = 1, \ldots, N_y, \tag{5.233}$$

where J_E stands for the generalized Jacobian of E_{ij}. We immediately observe that this Jacobian can be written as the following block pentadiagonal matrix

$$-J_E = \begin{bmatrix} A_1 & D_1 & E_1 & & & & & \\ \widetilde{C}_2 & \widetilde{A}_2 & \widetilde{D}_2 & E_2 & & & & \\ B_3 & C_3 & \widetilde{A}_3 & D_3 & E_3 & & & \\ & & \ddots & \ddots & \ddots & \ddots & \ddots & \\ & & & B_{N_t-1} & C_{N_t-1} & \widetilde{A}_{N_t-1} & \widetilde{D}_{N_t-1} & E_{N_t-1} \\ & & & & B_{N_t} & C_{N_t} & \widetilde{A}_{N_t} & D_{N_t} \\ & & & & & B_{N_t+1} & C_{N_t+1} & \widetilde{A}_{N_t+1} \end{bmatrix}, \tag{5.234}$$

where the diagonal block entries are given by

$$\widetilde{A}_m = \begin{bmatrix} -(\frac{3}{2}+4\sigma\gamma) & -\delta t\, \chi_{\mathcal{A}_+}\chi_{\mathcal{A}_-}\frac{1}{\nu} \\ \delta t\, \alpha & -(\frac{3}{2}+4\sigma\gamma) \end{bmatrix} \quad \text{for } m = 3,\dots,N_t-1.$$

Next, due to the combination of CN and BDF2 schemes for $m = 2$ and $m = N_t$, we have that

$$\widetilde{A}_2 = \begin{bmatrix} -(1+2\sigma\gamma) & -\delta t\, \chi_{\mathcal{A}_+}\chi_{\mathcal{A}_-}\frac{1}{2\nu} \\ \delta t\, \alpha & -(\frac{3}{2}+4\sigma\gamma) \end{bmatrix} \text{ and } \widetilde{A}_{N_t} = \begin{bmatrix} -(\frac{3}{2}+4\sigma\gamma) & -\delta t\, \chi_{\mathcal{A}_+}\chi_{\mathcal{A}_-}\frac{1}{\nu} \\ \frac{\delta t}{2}\alpha & -(1+2\sigma\gamma) \end{bmatrix}.$$

The terminal block is given by

$$\widetilde{A}_{N_t+1} = \begin{bmatrix} -(\frac{3}{2}+4\sigma\gamma) & -\delta t\, \chi_{\mathcal{A}_+}\chi_{\mathcal{A}_-}\frac{1}{\nu} \\ \beta & 1 \end{bmatrix},$$

and the matrix $\widetilde{C}_2$ takes the following form

$$\widetilde{C}_2 = \begin{bmatrix} -(1-2\sigma\gamma) & -\delta t\, \chi_{\mathcal{A}_+}\chi_{\mathcal{A}_-}\frac{1}{2\nu} \\ 0 & 0 \end{bmatrix}.$$

As before, the functions within the brackets [ ] are evaluated at the corresponding t_m. Further, all the other constituent block matrices of $-J_E$ are defined in the same way as the corresponding constituent blocks of the matrix M. Furthermore, notice that the function $\chi_{\mathcal{A}_+}\chi_{\mathcal{A}_-}$ can be rewritten as

$$\chi_{ijm} := \begin{cases} 1 & \text{if } \underline{u}_{ijm} \le \frac{1}{\nu} p_{ijm} \le \overline{u}_{ijm}, \\ 0 & \text{otherwise,} \end{cases}$$

which is, by construction, equivalent to the indicator function (5.212). Therefore, the matrix M and the generalized Jacobian $-J_E$ are equivalent. Thanks to this argumentation, we conclude that the algorithm P-TL-CGS is equivalent to the SSN strategy given by the generalized Newton step (5.233).

5.7.6 Multigrid Receding-Horizon Approach

It is possible to combine multigrid schemes with receding-horizon techniques [208] to develop an efficient optimal control algorithm for tracking a desired trajectory over very long time intervals. In the following, we sketch the implementation of the multigrid receding-horizon scheme. For an application of this scheme in physiology see [46].

Consider the optimal control problem of tracking y_d for $t \ge 0$. Define time windows of size Δt. In each time window, an optimal control problem with tracking ($\alpha = 1$) and terminal observation ($\beta = 1$) is solved with desired trajectory given by $y_d(x,t)$, $t \in$

$(n\Delta t,(n+1)\Delta t)$, and terminal state given by $y_T(x) = y_d(x,(n+1)\Delta t)$. The resulting optimal state at $n\Delta t$ defines the initial condition for the next optimal control problem defined in $(n\Delta t,(n+1)\Delta t)$. The following algorithm results.

ALGORITHM 5.15. Multigrid receding-horizon scheme (MG-RH).

1. Set $y(\mathbf{x},0) = y_0(\mathbf{x})$ and $n = 0$.
2. Set $y_T(\mathbf{x}) = y_d(\mathbf{x},(n+1)\Delta t)$.
3. Apply the CSMG scheme to solve the optimal control problem in $(n\Delta t,(n+1)\Delta t)$.
4. Update $n := n+1$, set $y_0(\mathbf{x}) = y(\mathbf{x},n\Delta t)$, and goto 2.

To show the ability of the CSMG multigrid receding-horizon approach to track over long time intervals, we consider the following desired trajectory

$$y_d(x_1,x_2,t) := t\ \sin(2\pi t)(x_1 - x_1^2)(x_2 - x_2^2)$$

over the time interval $(0,T)$ with $T = 5$, using 10 time windows of size $\Delta t = 0.5$. We test the receding-horizon algorithm considering a control-unconstrained and a control-constrained problem. In both cases, we take $\sigma = 0.01$ and we use the CSMG algorithm with the P-TL-CGS smoothing schemes because this smoothing scheme is efficient and robust with small values of the diffusion coefficient.

For the control-unconstrained receding-horizon case, we take $\alpha = 1$, $\beta = 1$, and $\nu = 10^{-6}$. We obtain that in each time window the optimal control problem is solved to the required tolerance by 3 STMG-$V(2,2)$-cycles on a grid with $\gamma = 64$. In Figure 5.20(left) the time evolution of the state variable compared to the desired trajectory is depicted and shows accurate tracking ($\| y_h - R_h\, y_d \| \approx 10^{-6}$). Also in Figure 5.20(right), the control function is depicted that appears to be smooth across time windows.

For the control-constrained receding-horizon case, we study the tracking of the given trajectory, considering the following constraints for the control $\underline{u}(\mathbf{x},t) = -1$ and $\overline{u}(\mathbf{x},t) = 1$. Our experience shows that fast accurate tracking is obtained taking $\alpha = 1$, $\beta = 0.1$, and $\nu = 10^{-4}$. In this case, the optimal control problem is solved to the required tolerance,

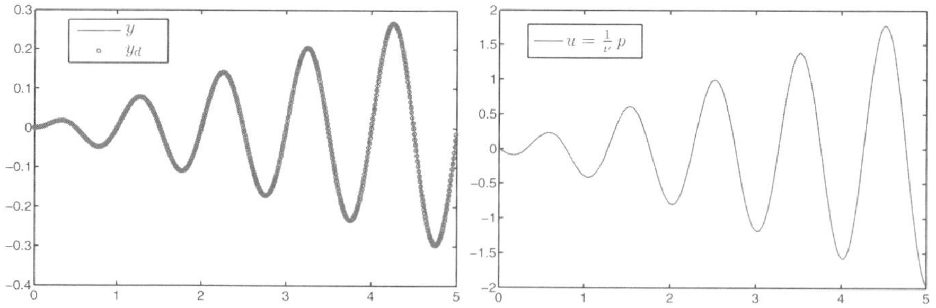

Figure 5.20. *Receding-horizon solution for the control-unconstrained tracking problem. Left: time evolution of the state y (solid line) and the desired trajectory y_d (dots) at $(x_1,x_2) = (0.5,0.5)$. Right: optimal control $u = \frac{1}{\nu}\, p$ at $(x_1,x_2) = (0.5,0.5)$.*

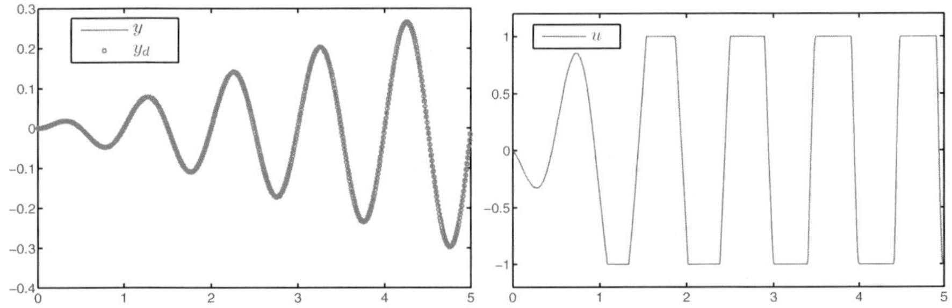

Figure 5.21. *Receding-horizon solution for the control-constrained tracking problem. Left: time evolution of the state y (solid line) and the desired trajectory y_d (dots) at $(x_1,x_2) = (0.5,0.5)$. Right: optimal control u at $(x_1,x_2) = (0.5,0.5)$. Reprinted with permission from S. Gonzalez Andrade and A. Borzì, Multigrid second-order accurate solution of parabolic control-constrained problems, Computational Optimization and Applications, to appear.*

on a grid with $\gamma = 64$, by 5 STMG-$W(2,2)$-cycles, in average in each time window. In Figure 5.21(left), the time evolution of the state variable compared to the desired trajectory is depicted that shows accurate tracking ($\|y_h - R_h\, y_d\| \approx 10^{-4}$). In Figure 5.21(right), the control function is depicted. We observe that initially the control constraints appear to be nonactive but as the time increases, the constraints become active.

For a challenging application of the MG-RH scheme to control the reaction-diffusion model of cardiac arrhythmia [3, 272], see [46].

5.7.7 A CSMG Scheme for Fredholm Control Problems

We discuss the CSMG solution of an optimal control problem governed by a Fredholm integral equation of the second kind. Specifically, we consider the Fredholm integral equations of the second kind with linear distributed control mechanism introduced in Chapter 3. We have

$$\min J(y,u) := \frac{1}{2}\|y - z\|^2_{L^2(\Omega)} + \frac{\nu}{2}\|u\|^2_{L^2(\Omega)}, \tag{5.235}$$

$$y = f(y) + u + g \qquad \text{in } \Omega, \tag{5.236}$$

where $g, z \in L^2(\Omega)$ are given and $\nu > 0$. The term $f(y)$ is supposed to be symmetric and is given by

$$f(y)(x) = \int_\Omega K(x,t)\, y(t)\, dt. \tag{5.237}$$

Existence and uniqueness of solution for (5.235)–(5.237) and related discretization issues are discussed in Section 3.4.

Without constraints on the control u, we have the scalar equation $\nu\, u - p = 0$, which we use to eliminate $u = p/\nu$ and obtain the following equivalent system

$$\left\{ \begin{array}{lcl} y - f(y) - p/\nu & = & g, \\ p - f(p) + y & = & z. \end{array} \right. \tag{5.238}$$

This system corresponds to two coupled integral equations. As discussed in Chapter 3, we use the Nyström method [14, 176] to discretize this problem. We obtain the following discrete optimality system

$$\begin{aligned} y_i - h \sum_{j=-N}^{N} w_{ij}\, y_j - p_i/\nu &= g_i, \\ p_i - h \sum_{j=-N}^{N} w_{ij}\, p_j + y_i &= z_i. \end{aligned} \tag{5.239}$$

Now, we define an iterative procedure on Ω_h that belongs to the class of CGS schemes. It results from a sequential update of the optimization variables at each grid point. The update of the variables (y_i, p_i) at grid point x_i is obtained by solving exactly the discrete optimality system with respect to (y_i, p_i) and considering the remaining variables as constant. This iterative method is given by the following algorithm. Let an initial approximation $(y^{(0)}, p^{(0)})$ be given. Here, tol is the required tolerance on the L^2-norm of the residual of the constraint equation.

ALGORITHM 5.16. CGS iteration scheme.

1. For $m = 0, 1, 2, \ldots, \ell$ do
2. If $\|g_h - y_h^m + f_h(y_h^m) - p_h^m/\nu\|_{L_h^2} < tol$ then stop.
3. For $i = -N, -N+1, \ldots, 0, 1, \ldots, N-1, N$ (lexicographic order) do

$$\begin{pmatrix} y_i \\ p_i \end{pmatrix}^{(m+1)} = \begin{pmatrix} 1 - h w_{ii} & -1/\nu \\ 1 & 1 - h w_{ii} \end{pmatrix}^{-1} \times \left[\begin{pmatrix} g_i \\ z_i \end{pmatrix} + h \sum_{j<i} w_{ij} \begin{pmatrix} y_j \\ p_j \end{pmatrix}^{(m+1)} + h \sum_{j>i} w_{ij} \begin{pmatrix} y_j \\ p_j \end{pmatrix}^{(m)} \right]. \tag{5.240}$$

4. End.

Notice that the summation in (5.240) can be implemented in a fast procedure with recursive subtraction and addition of the quantity being updated. In [6], this smoothing procedure is implemented in a CSMG scheme obtaining a fast and robust multigrid solver for Fredholm integral control problems. This result is in agreement with estimates of local Fourier analysis presented in [6].

Next, we report results of numerical experiments to validate the convergence performance of the CGS and CSMG schemes. We consider an application corresponding to a kernel that represents the covariance function of an Ornstein–Uhlenbeck stochastic process at the equilibrium that arises in statistical communication theory [213, 279]. We have $K(x,t) = -e^{-\alpha|x-t|}/2$, where $\alpha > 0$ represents the characteristic correlation time of the process. Here, $y(x)$ represents a signal and $u(x)$ a control for the signal. In this case, the norm of the kernel $K(x,t)$ as defined in (3.37) is approximately equal to 0.142, so that we can state existence and uniqueness of solution for a given control.

In addition, we take $g(x) = 2/\pi, u(x) = \sin(\pi x)$ and a target function which is discontinuous as typical in the modeling of signals. We choose

$$z(x) = \lfloor 5[x(x-1)/\pi + (1+1/\pi^2)\sin(\pi x)] \rfloor /5,$$

where $\lfloor\ \rfloor$ is the floor function.

In Table 5.24, results obtained with the CGS iteration are reported. We see that tracking improves as ν decreases. We obtain robust tracking despite z being discontinuous. Next, we consider the same setting and apply the CSMG multigrid scheme with the smoother given by Algorithm 5.16. Results with the multigrid scheme are reported in Table 5.25. We obtain that both the CGS and the multigrid scheme converge efficiently to the solution and the observed convergence rate is weakly dependent on the mesh size. As ν becomes smaller, convergence rates improve, thus showing robustness. We see that multigrid convergence rates are two orders of magnitude better than the CGS convergence rates. Moreover, the observed rates are in good agreement with the corresponding estimates by local Fourier analysis. In Figure 5.22, we depict the computed optimal state and control solutions.

Table 5.24. *Results with the CGS scheme; $\alpha = 1$.*

ν	N	$\|y-z\|$	$\|\|res\|\|_2$	ρ	N_{iter}
1e-03	512	6.66e-04	9.03e-015	7.02e-04	6
1e-03	256	6.66e-04	8.89e-015	7.01e-04	6
1e-03	128	6.66e-04	8.60e-015	6.99e-04	6
1e-04	512	6.67e-05	1.32e-013	4.10e-05	5
1e-04	256	6.67e-05	1.30e-013	4.09e-05	5
1e-04	128	6.66e-05	1.26e-013	4.06e-05	5
1e-05	512	6.67e-06	8.33e-014	4.10e-06	4
1e-05	256	6.67e-06	8.27e-014	4.09e-06	4
1e-05	128	6.66e-06	8.12e-014	4.06e-06	4
1e-06	512	6.67e-07	8.56e-016	4.21e-07	4
1e-06	256	6.67e-07	8.32e-016	4.12e-07	4
1e-06	128	6.66e-07	8.08e-016	4.04e-07	4

Local Fourier Analysis of a CSMG Solver of Integral Control Problems

In this section, we discuss the local Fourier analysis of the CSMG scheme for solving Fredholm control problems. We consider the case of two levels with a fine-grid level with mesh size $h = h_k$ and the coarse-grid problem is constructed on the grid with mesh size $H = h_{k-1}$. Recall the optimality system

$$y_h - f_h(y_h) - p_h/\nu = g_h, \tag{5.241}$$

$$p_h - f_h^T(p_h) + y_h = z_h. \tag{5.242}$$

Let $(e_y(j), e_p(j)) = \sum_\theta W_\theta\, \phi(j,\theta)$ denote the errors for the state and the adjoint variables on the grid points $x_j = jh$. Here, $W_\theta = (Y_\theta, P_\theta)$ are the corresponding Fourier coefficients. The action of one smoothing step on the errors can be expressed by $W_\theta^{(1)} = \hat{S}(\theta)\, W_\theta^{(0)}$.

Table 5.25. *Results with the CSMG scheme; $\alpha = 1$ and $m_1 = 1, m_2 = 1$ pre- and postsmoothing sweeps.*

ν	N	$\|y-z\|$	$\|\|res\|\|_2$	ρ	N_{cycle}
1e-03	512	6.66e-04	3.74e-016	1.07e-05	3
1e-03	256	6.66e-04	3.65e-016	1.04e-05	3
1e-03	128	6.66e-04	3.48e-016	1.00e-05	3
1e-04	512	6.67e-05	3.52e-013	1.04e-06	2
1e-04	256	6.67e-05	3.51e-013	1.03e-06	2
1e-04	128	6.66e-05	3.48e-013	1.03e-06	2
1e-05	512	6.67e-06	3.52e-015	1.03e-07	2
1e-05	256	6.67e-06	3.52e-015	1.03e-07	2
1e-05	128	6.66e-06	3.48e-015	1.02e-07	2
1e-06	512	6.67e-07	1.62e-016	4.76e-08	2
1e-06	256	6.67e-07	1.27e-016	3.75e-08	2
1e-06	128	6.66e-07	1.08e-016	3.18e-08	2

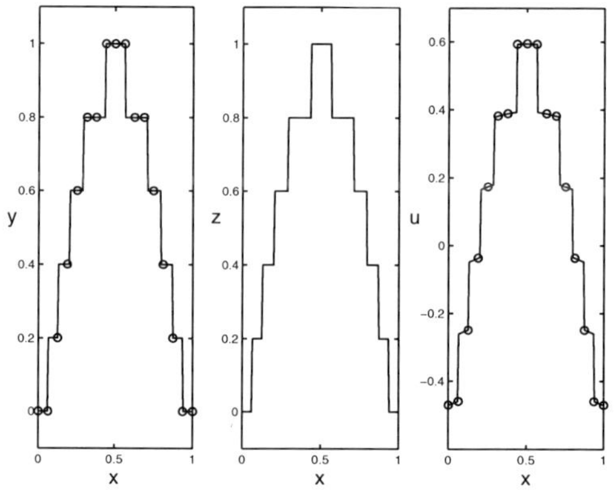

Figure 5.22. *Optimal solution y_N (left), target z, and control u_N (right) for $\alpha = 1$, $\nu = 10^{-3}$, $N = 8$, with the Nyström formula. Circles are the interpolation points. Reprinted with permission from M. Annunziato and A. Borzì, Fast solvers of Fredholm optimal control problems, Numer. Math. Theory Methods Appl.,* 3(4) (2010), 431–448.

Now, consider applying the CGS step for solving our distributed control problem. We assume that the kernel is symmetric, i.e., $w_{i-j} = w_{|i-j|}$, and that it is decaying sufficiently fast such that we can truncate the sum approximating the integral, $\sum_{k=-\ell}^{\ell} \phi(k) \approx \sum_{k=-N}^{N} \phi(k)$. Substituting $(e_y(j), e_p(j))$ in (5.239) and applying the CGS Algorithm 5.16, we obtain

$$\begin{pmatrix} (1 - h\sum_{k=-\ell}^{0} w_{|k|} e^{i\theta k}) & -\frac{1}{\nu} \\ 1 & (1 - h\sum_{k=-\ell}^{0} w_{|k|} e^{i\theta k}) \end{pmatrix} \begin{pmatrix} Y_\theta^{(1)} \\ P_\theta^{(1)} \end{pmatrix} = \begin{pmatrix} h\sum_{k=1}^{\ell} w_{|k|} e^{i\theta k} & 0 \\ 0 & h\sum_{k=1}^{\ell} w_{|k|} e^{i\theta k} \end{pmatrix} \begin{pmatrix} Y_\theta^{(0)} \\ P_\theta^{(0)} \end{pmatrix}.$$

Hence

$$\hat{S}(\theta) = \begin{pmatrix} (1 - h\sum_{k=-\ell}^{0} w_{|k|} e^{i\theta k}) & -\frac{1}{\nu} \\ 1 & (1 - h\sum_{k=-\ell}^{0} w_{|k|} e^{i\theta k}) \end{pmatrix}^{-1} \times \begin{pmatrix} h\sum_{k=1}^{\ell} w_{|k|} e^{i\theta k} & 0 \\ 0 & h\sum_{k=1}^{\ell} w_{|k|} e^{i\theta k} \end{pmatrix}. \tag{5.243}$$

With local Fourier analysis, we can estimate the smoothing factor μ. It appears that μ is almost independent of the value of the discretization parameter h and increases by increasing the value of the weight ν. Also in this case, we can perform a TG Fourier analysis to estimate the convergence factor of the multigrid iteration.

Consider the optimality system (5.241)–(5.242). We have that the symbol of the fine-grid operator is as follows

$$\widehat{A}_h(\theta) = \begin{bmatrix} a_y^h(\theta) & -1/\nu & 0 & 0 \\ 1 & a_p^h(\theta) & 0 & 0 \\ 0 & 0 & a_y^h(\bar{\theta}) & -1/\nu \\ 0 & 0 & 1 & a_p^h(\bar{\theta}) \end{bmatrix},$$

where

$$a_y^h(\theta) = 1 - h\sum_{k=-\ell}^{\ell} w_{|k|} e^{i\theta k}, \qquad a_p^h(\theta) = a_y^h(\theta).$$

Similarly, for the frequency represented on the coarse grid, the symbol of the coarse-grid operator is as follows

$$\widehat{A}_H(\theta) = \begin{bmatrix} a_y^H(2\theta) & -1/\nu \\ 1 & a_p^H(2\theta) \end{bmatrix}.$$

In Table 5.26, we report theoretical estimates of $\rho(TG_h^H)$ resulting from the TG convergence analysis. Comparison with results of numerical experiments show that these estimates are sharp; see Table 5.25 for results of numerical experiments. We can see that the local Fourier analysis predicts mesh-independent smoothing factors and convergence factors and these factors improve as ν becomes smaller.

Table 5.26. *Estimates for $\rho(TG_h^H)$ for the case of $m_1 = m_2 = 1$ smoothing steps; $w_{|i-j|} = -\frac{1}{2}\exp(-|i-j|h)$.*

ν	$N = 64$	$N = 128$	$N = 256$	$N = 512$
1.0e-01	3.31e-03	3.35e-03	3.36e-03	3.37e-03
1.0e-02	3.74e-04	3.77e-04	3.79e-04	3.80e-04
1.0e-03	3.78e-05	3.82e-05	3.84e-05	3.85e-05
1.0e-04	3.79e-06	3.83e-06	3.85e-06	3.86e-06
1.0e-05	3.79e-07	3.83e-07	3.85e-07	3.86e-07
1.0e-06	3.80e-08	3.83e-08	3.85e-08	3.86e-08

5.7.8 Optimization Properties of the CSMG Scheme

The CSMG strategy aims at solving optimality systems that represent first-order necessary conditions for PDE-based optimization problems. We notice that intermediate steps of the CSMG solution process may well be unfeasible or result in increasing values of the objective to be minimized. These phenomena can be observed in the case of nonconvex and nonlinear optimization problems. In fact, in the construction of the CSMG scheme it is not required to provide a minimizing sequence, but to solve the optimality system. Nevertheless, as discussed in [60], the CSMG scheme can be defined in such a way to provide a minimizing sequence as in the case of the MGOPT method.

For this discussion consider the following optimization problem

$$\begin{cases} \min J(y,u) := h(y) + \nu\, g(u), \\ \qquad\qquad c(y,u) = 0 \quad \text{in } \Omega. \end{cases} \tag{5.244}$$

Here g and h are required to be continuously differentiable, bounded from below, and such that $g(u) \to \infty$ as $||u|| \to \infty$. Allowing g and h to be locally nonconvex and e to be possibly nonlinear, (5.244) may have multiple extremals including minima, maxima, and saddle points. Local minima satisfy the first-order necessary conditions given by the optimality system

$$\begin{aligned} c(y,u) &= 0, \\ c_y(y,u)^* p &= -h'(y), \\ \nu\, g'(u) + c_u^* p &= 0. \end{aligned} \tag{5.245}$$

To guarantee a CSMG cycle that is minimizing the objective, we should define the smoothing process based on the gradient of the reduced cost functional and show that the coarse-grid correction step provides a descent update.

In order to update the control function in the smoothing process, we could use the following descent scheme (as in [10])

$$u_h^{new} = u_h - \beta(\nu\, g'(u_h) + c_u^*\, p_h(u_h)), \tag{5.246}$$

where optimal choice of the scaling factor $\beta > 0$ may be done using linesearch methods. Alternatively, one could use a smoothing step based on subspace correction methods [334]; see also [60]. However, notice that local Newton updates are not a suitable approach in the case of nonconvex problems because the Newton scheme may well converge to a maximum instead. In this case, in [60] the search for possible negative eigenvalues of the reduced Hessian considered at the coarsest grid of the multigrid process is discussed. If negative eigenvalues are detected, a globalization step in the direction of negative curvature is performed to escape undesired maxima or saddle points.

Now, let us discuss the coarse-grid correction procedure. Consider the case where $c(y,u) = -\Delta y - u$ and $h(y) = ||y - z||^2_{L^2(\Omega)}/2$, and assume that g' is Lipschitz continuous with Lipschitz constant $\gamma > 0$ and satisfies the following monotonicity requirement

$$(g'(u) - g'(v), u - v) \geq \delta\, \|u - v\|^2_{L^2(\Omega)}$$

for some $\delta > 0$. We require that this property hold after discretization and we show that the CSMG coarse-grid correction provides a descent direction in the sense that

$$(\nu\, g'(u_h) - p_h, I_H^h(u_H - \hat{I}_h^H u_h))_h < 0,$$

unless $u_H = \hat{I}_h^H u_h$, occurring at convergence.

Starting from an initial approximation and after a few presmoothing steps the resulting triple (y_h, u_h, p_h) satisfies the optimality system up to residuals (d_h^1, d_h^2, d_h^3), that is,

$$\begin{aligned} -\Delta_h y_h - u_h &= d_h^1, \\ -\Delta_h p_h + y_h - z_h &= d_h^2, \\ \nu g'(u_h) - p_h &= d_h^3. \end{aligned} \tag{5.247}$$

For the coarse-grid process, we take $\hat{I}_h^h = I_h^H$, where I_h^H is the full-weighting restriction operator. For I_H^h we choose bilinear interpolation, i.e., $(I_h^H u_h, v_H)_H = (u_h, I_H^h v_H)_h$. We define $z_H = I_h^H z_h$. With this setting, we obtain the following coarse-grid equations

$$\begin{aligned} -\Delta_H y_H - u_H &= I_h^H \Delta_h y_h - \Delta_H I_h^H y_h, \\ -\Delta_H p_H + y_H - z_H &= I_h^H \Delta_h p_h - \Delta_H I_h^H p_h, \\ \nu g'(u_H) - p_H &= 0. \end{aligned} \tag{5.248}$$

As usual in TG convergence analysis, we assume that this coarse system of equations is solved exactly. From the first equation of (5.248), and using the corresponding equation in (5.247), we obtain

$$u_H - I_h^H u_h = -\Delta_H (y_H - I_h^H y_h) + I_h^H d_h^1. \tag{5.249}$$

Combining the fine and coarse adjoint equations we have

$$p_H - I_h^H p_h = \Delta_H^{-1} (y_H - I_h^H y_h) + \Delta_H^{-1} I_h^H d_h^2. \tag{5.250}$$

Let us assume that

$$(g'(v_H) - I_h^H g'(v_h), v_H - I_h^H v_h)_H \geq \delta' \, \|v_H - I_h^H v_h\|_H^2 \tag{5.251}$$

for some $\delta' > 0$ independent of v_h and v_H. Note that (5.251) is satisfied, for example, if g' is linear or if I_h^H is strict injection and g is strictly convex.

With these preparations we are ready to show that the update step of the CSMG coarse-grid correction follows a descent direction

$$\begin{aligned}
&(\nu g'(u_h) - p_h, I_H^h (u_H - I_h^H u_h))_h = (I_h^H (\nu g'(u_h) - p_h), u_H - I_h^H u_h)_H \\
&= (\nu I_h^H g'(u_h) - I_h^H p_h, u_H - I_h^H u_h)_H \\
&= (\nu I_h^H g'(u_h) - p_H + \Delta_H^{-1} (y_H - I_h^H y_h) + \Delta_H^{-1} I_h^H d_h^2, u_H - I_h^H u_h)_H \\
&= -\nu (g'(u_H) - I_h^H g'(u_h), u_H - I_h^H u_h)_H \\
&\quad + (\Delta_H^{-1} (y_H - I_h^H y_h) + \Delta_H^{-1} I_h^H d_h^2, -\Delta_H (y_H - I_h^H y_h) + I_h^H d_h^1)_H \\
&= -\nu (g'(u_H) - I_h^H g'(u_h), u_H - I_h^H u_h)_H - (y_H - I_h^H y_h, y_H - I_h^H y_h)_H \\
&\quad + (\Delta_H^{-1} (y_H - I_h^H y_h), I_h^H d_h^1)_H - (\Delta_H^{-1} I_h^H d_h^2, \Delta_H (y_H - I_h^H y_h))_H \\
&\quad + (\Delta_H^{-1} I_h^H d_h^2, I_h^H d_h^1)_H \\
&\leq -\nu (g'(u_H) - I_h^H g'(u_h), u_H - I_h^H u_h)_H \\
&\quad + \frac{1}{2} (\|\Delta_H^{-1} I_h^H d_h^1\|_H^2 + \|\Delta_H^{-1} I_h^H d_h^2\|_H^2 + \|I_h^H d_h^1\|_H^2 + \|I_h^H d_h^2\|_H^2) \\
&\leq -\nu \delta' \|u_H - I_h^H u_h\|_H^2 \\
&\quad + \frac{1}{2} (\|\Delta_H^{-1} I_h^H d_h^1\|_H^2 + \|\Delta_H^{-1} I_h^H d_h^2\|_H^2 + \|I_h^H d_h^1\|_H^2 + \|I_h^H d_h^2\|_H^2).
\end{aligned}$$

Therefore

$$(\nu\, g'(u_h) - p_h, I_H^h(u_H - I_h^H u_h))_h < 0$$

if (5.251) holds and the residuals d_h^1 and d_h^2 are sufficiently small.

Finally we show that the coarse-grid correction step does not produce overshooting in the sense that $(\hat{J}'(u_h)_h, \hat{J}'(u_h^{new})_h)_h \geq 0$. We consider the case where $g'(u) = u$. We have the following

$$\begin{aligned}
&(\hat{J}'(u_h)_h, \hat{J}'(u_h^{new})_h)_h \\
&= (\nu\, u_h - p_h, \nu\,(u_h + I_H^h(u_H - I_h^H u_h)) - (p_h + I_H^h(p_H - I_h^H p_h))_h \\
&= \|\nu\, u_h - p_h\|_h^2 + (\nu\, u_h - p_h, I_H^h[\nu\,(u_H - I_h^H u_h) - (p_H - I_h^H p_h)])_h \\
&= \|\nu\, u_h - p_h\|_h^2 - \|I_h^H(\nu\, u_h - p_h)\|_H^2 \geq 0,
\end{aligned}$$

where we use $\|I_h^H\| \leq 1$.

Chapter 6

PDE Optimization with Uncertainty

6.1 Introduction

The importance of PDE-constrained optimization is derived from the demand in applications where PDEs represent the constitutive models and the environment of the implementation of the results of optimization strategies is not exactly known. Often, critical modeling parameters are only known to lie in a certain region and the practical implementation of the design and control solutions is subject to a priori unknown perturbations. We illustrate recent approaches to compute solutions to PDE-constrained optimization problems which are robust with respect to the stochasticity of the application framework. This emerging field of research in PDE optimization is very important, on the one hand because it addresses the need to solve real application problems, and on the other hand because it poses new analytical and computational challenges to the PDE optimization community. While we do not provide much background material on some techniques concerning the representation and treatment of uncertainty and rely on references, we focus on the combination of the solution methodologies presented in previous chapters with these techniques.

We know that the efficient computation of the optimal solutions can be achieved only if structures in the underlying problem are exploited. Besides the structures naturally arising in PDE-constrained optimization problems, which are exploited here in the form of multigrid methods and one-shot methods, a new structure comes into focus because of the stochasticity involved. Although the exact perturbations involved in practical applications are in general unknown, usually a good estimate of the probability distribution of the perturbations is available. This knowledge is to be exploited in the treatment of uncertainty in PDE-constrained optimization.

In the following sections, we discuss the treatment of uncertainty in the case where the coefficients of the PDE models are subject to known random field perturbations. Afterwards, we consider semi-infinite robust design for aerodynamic optimization under geometric uncertainty. In these problems, the uncertainty is related to the PDE models which are deterministic and to the data which is assumed to be known. On the other hand, we shall also discuss cases where the model is stochastic and the PDE optimization framework enters in the form of a Fokker–Planck system to control the probability density function distribution of the state of the model. Further, we shall discuss a Bayesian approach to quantify uncertainty in PDE optimization problems where the data is only known through some characterizing statistical quantities.

We start our discussion considering the following general problem formulation

$$\min\ J(y,u), \tag{6.1}$$
$$c(y,u) = 0, \tag{6.2}$$

where J is a real-valued differentiable objective function and c denotes the PDE constraint. For a fixed u, we require the existence of a unique $y = y(u)$ that satisfies the PDE constraint.

In this introductory section, we discuss for ease of presentation a finite-dimensional stochastic perturbation σ "polluting" part of the data of the PDE-constrained optimization problem. This stochastic random variable σ is assumed to be a mapping

$$\sigma : \mathcal{O} \to \mathbb{R}^n,$$

where the triple $(\mathcal{O}, \mathcal{A}, \mathcal{P})$ denotes a probability space, $\mathcal{O}$ is the space of elementary events, and $\mathcal{A}$ is the sigma-algebra of subsets of $\mathcal{O}$ measured by the probability measure $\mathcal{P}$ on $\mathcal{O}$; see, e.g., [237]. Thus, we define the expected (mean) value of the stochastic variable σ as

$$\sigma_o := \mathbb{E}(\sigma) = \int_{\mathcal{O}} \sigma(\omega) d\mathcal{P}(\omega).$$

Now, we focus on the following three reformulations of problem (6.1)–(6.2) subject to uncertainty. We have

$$u_0 := \arg\min_u \{J(y,u) \,|\, c(y,u,\sigma_0) = 0\}, \tag{6.3}$$
$$\bar{u} := \mathbb{E}(\arg\min_u \{J(y,u) \,|\, c(y,u,\sigma) = 0\}), \tag{6.4}$$
$$\hat{u} := \arg\min_u \mathbb{E}(\{J(y,u) \,|\, c(y,u,\sigma) = 0\}). \tag{6.5}$$

The optimal design/control u_0 corresponds to the standard approach by just performing a deterministic optimization, where the stochastic variable is substituted by its expected value. We denote with $\bar{u}$ the average optimization function obtained as the expectation of the optimization functions corresponding to all possible random configurations. With $\hat{u}$, we denote the design/control function that optimizes the expectation of the reduced objective.

Because of the arising nonlinearities in all three problem formulations, we observe in general that $u_0 \neq \bar{u} \neq \hat{u} \neq u_0$. The problem formulation (6.4) is highly modular and can be readily coupled with existing solver technologies for PDE-constrained optimization. This formulation will be investigated in the next section. The problem formulation (6.5) will be discussed in Section 6.3 for the problem of robust aerodynamic shape optimization, where also additional scalar-valued state constraints are added. In Section 6.4, we discuss the solution of (6.5) by model reduction using proper orthogonal decomposition. In this discussion, distributed uncertainty plays an important role and the Karhunen–Loève (KL) expansion is used extensively. Solution approaches for finitely many scalar uncertainties, as discussed above for ease of presentation, can be found in, e.g., [313], where problem formulation (6.5) is used. In Section 6.5, we discuss a strategy to control probability density functions of stochastic processes using the Fokker–Planck equation. The last section of this chapter is concluded with a discussion on recent techniques for Bayesian uncertainty quantification related to inverse problems governed by PDE models.

6.2 PDE Control Problems with Uncertain Coefficients

In this section, we consider nonlinear parabolic control problems with uncertain coefficients. Uncertainty may appear in nonlinear reaction terms or in the diffusion coefficient. We consider the case of distributed controls as in [44, 67]. We have the following reaction-diffusion model

$$\begin{cases} -\partial_t y + G_\delta(y) + \sigma\,\Delta y = f & \text{in } Q = \Omega \times (0,T), \\ y = y_0 & \text{in } \Omega \times \{t=0\}, \\ y = 0 & \text{on } \Sigma = \partial\Omega \times (0,T), \end{cases} \tag{6.6}$$

where $\Omega \subset \mathbb{R}^d$, $d \geq 1$, and T is the time horizon where the reaction-diffusion process is considered. The nonlinear term $G_\delta(y)$ models the reaction kinetics for the state y. Here δ is the reaction parameter and σ is the diffusion coefficient, both described by random fields.

We have $\sigma = \sigma(x,\omega)$, where $x \in \Omega$ and $\omega \in \mathcal{O}$. The values of the stochastic function $\sigma(x,\omega)$ are usually spatially correlated in a way characterized by a covariance structure. Clearly, we cannot model numerically the resulting infinite-dimensional coefficient space, and a suitable finite-dimensional approximation must be introduced. For this purpose, a common approach is to use the KL expansion [237] of the random field $\sigma(x,\omega)$ that is based on a spectral decomposition of the covariance kernel of the stochastic process. We assume that the mean and the covariance functions of $\sigma(x,\omega)$ are known, respectively, as

$$\sigma_0(x) = \mathbb{E}(\sigma)(x) := \int_{\mathcal{O}} \sigma(x,\omega)d\mathcal{P}(\omega) \tag{6.7}$$

and

$$C_\sigma(x,x') = \int_{\mathcal{O}} (\sigma(x,\omega) - \sigma_0(x))(\sigma(x',\omega) - \sigma_0(x'))d\mathcal{P}(\omega). \tag{6.8}$$

We see that C_σ defines the kernel of a compact, positive, and self-adjoint operator. Denote with λ_j the real positive eigenvalues and with $z_j(x)$ the corresponding orthonormal eigenfunctions of C_σ as follows. We have

$$\int_\Omega C_\sigma(x,x')z_j(x')dx' = \lambda_j\, z_j(x), \qquad x \in \Omega,$$

where we assume to order the eigenvalues in decreasing order. Having the eigenpairs, we can define the following uncorrelated random variables

$$Y_j(\omega) = \frac{1}{\sqrt{\lambda_j}} \int_\Omega (\sigma(x,\omega) - \sigma_0(x))\, z_j(x)dx, \qquad j = 1,2,\ldots,$$

with zero mean and unit variance, i.e., $\int_{\mathcal{O}} Y_i(\omega)Y_j(\omega)d\mathcal{P}(\omega) = \delta_{ij}$. We assume that the images $\Gamma_j = Y_j(\mathcal{O})$, $j = 1,2,\ldots$, are bounded intervals in $\mathbb{R}$ and consider the case where the random variables Y_j, with density probability ρ_j, are independent.

Now, the truncated KL expansion of $\sigma(x,\omega)$ is given by

$$\sigma_N(x,\omega) = \sigma_0(x) + \sum_{j=1}^{N} \sqrt{\lambda_j}\, z_j(x)\, Y_j(\omega), \tag{6.9}$$

where N denotes the number of terms in the truncation. Following [363], we define $\Gamma = \Pi_{j=1}^{N} \Gamma_j$, and the joint probability density function with $\rho = \Pi_{j=1}^{N} \rho_j$. We see that $\sigma_N(x,\omega)$ may provide a suitable approximation to $\sigma(x,\omega)$ assuming that the eigenvalues decay sufficiently fast and N is sufficiently large. For correlation lengths comparable to the size of the domain, a small value of N can be sufficient to obtain an accurate representation of the random field.

In particular, consider random fields characterized by squared exponential covariance

$$C_\sigma(x,x') = s^2 \exp\left(-\frac{|x-x'|^2}{\ell^2} \right), \qquad x,x' \in \Omega,$$

with mean σ_0 and variance s^2. The degree of variability of this random field can be characterized by the ratio s/σ_0, while the frequency of variation of this field is related to the ratio L/ℓ, where L is the characteristic length of the physical domain and ℓ represents the physical correlation length. We have that the eigenvalues have exponential decay as $\lambda \sim c_1 \ell \exp(-c_2 \ell^2)$ for some positive constants c_1 and c_2; see, e.g., [204, 263].

We consider random fields of the type given above that can be approximated by (6.9) with good accuracy for moderate values of N. Therefore, assuming finite-dimensional random fields, we can write the explicit dependence of σ on the random variables $Y = [Y_1,\dots,Y_N]$. We have

$$\sigma(x,\omega) \approx \sigma(x,Y_1,\dots,Y_N).$$

Similarly, we can assume that the reaction dynamics of (6.6) can be modeled by a KL-truncated temporal random field as follows

$$\delta(t,\omega) \approx \delta(t,W_1,\dots,W_M) = \delta_0 + \sum_{j=1}^{M} \sqrt{\mu_j}\, v_j(t)\, W_j(\omega),$$

where the $W = [W_1,\dots,W_M]$ are uncorrelated random variables with zero mean and unit variance.

With the setting above and deterministic initial and boundary conditions, we have that the solution to (6.6) can be described in terms of spatial and time coordinates and the set of random variables $[Y_1,\dots,Y_N; W_1,\dots,W_M]$ as follows

$$y(x,t,\omega) \approx y(x,t,Y_1,\dots,Y_N,W_1,\dots,W_M).$$

Now, we formulate a parabolic optimal control problem with random coefficients; the discussion on the elliptic case follows along the same lines of reasoning; see [44]. We have

$$\min_{y,u} J(y,u) := \frac{\alpha}{2}\|y-y_d\|^2_{L^2(Q)} + \frac{\beta}{2}\|y(\cdot,T)-y_T\|^2_{L^2(\Omega)} + \frac{\nu}{2}\|u\|^2_{L^2(Q)}, \tag{6.10}$$

$$-\partial_t y + G_{\delta(W(\omega))}(y) + \sigma(Y(\omega))\,\Delta y = f + u \quad \text{in } Q \times \mathcal{O}, \tag{6.11}$$

where we take homogeneous boundary conditions and deterministic initial condition $y_0 \in H_0^1(\Omega)$. The objective (6.10) models the requirement to track a desired deterministic trajectory $y_d \in L^2(Q)$ and to reach a desired deterministic terminal state $y_T \in L^2(\Omega)$. Here, $\nu > 0$ is the weight of the cost of the control and $\alpha \ge 0$, $\beta \ge 0$, $\alpha + \beta > 0$, are optimization parameters. For any given event ω, we assume that the resulting random field and the nonlinearity are such that the optimal solution has the required regularity; see [268, 316] for a related discussion. To simplify notation, whenever possible in the following we omit the dependence of the variables from ω.

The optimal control problem (6.10)–(6.11) is stochastic in the sense that for any event ω, it provides a different control. Each single event corresponds to a deterministic optimal control problem with $\delta = \delta(t,\omega)$ and $\sigma = \sigma(x,\omega)$, and therefore, corresponding to each event, the discussion on the existence and uniqueness of optimal solutions and their numerical determination proceeds as in the deterministic cases discussed in Chapters 3 and 5.

We have that the solution to (6.10)–(6.11) is characterized by the following first-order optimality system

$$\begin{aligned} -\partial_t y + G_\delta(y) + \sigma \Delta y &= f + u && \text{in } Q, \\ \partial_t p + G_\delta'(y) p + \sigma \Delta p + \alpha(y - y_d) &= 0 && \text{in } Q, \\ \nu u - p &= 0 && \text{in } Q, \\ y = 0,\ p &= 0 && \text{on } \Sigma, \end{aligned} \tag{6.12}$$

with initial condition $y(x,0) = y_0(x)$ for the state variable and terminal condition for the adjoint variable given by $p(x,T) = \beta(y(x,T) - y_T(x))$.

Further, we assume that our setting is such that a unique mapping $y = y(u)$, and we recall the reduced cost functional $\hat{J}(u) = J(y(u),u)$. We have $\nabla \hat{J}(u) = \nu u - p(u)$.

By solving optimal control problems with random coefficients, we explore the space of controls depending on the configuration of the parameters. By definition, this space represents the solution of the stochastic control problem. However, a possible task is to deliver a unique control for the governing random PDE model. This control should be at least suboptimal and provide good tracking features for all configurations of the coefficients and thus define a robust deterministic control. For this reason, we consider the following problem

$$\min_{u \in L^2(Q)} \int_{\mathcal{O}} \hat{J}(u;\omega)\, d\mathcal{P}(\omega), \tag{6.13}$$

where $\hat{J}(u;\omega) = \hat{J}(y(u;\omega),u)$. This formulation corresponds to problem (6.5) in the introduction. One can see that this formulation requires us to solve an optimization problem with infinitely many terms in the objective (the integral in $\mathcal{O}$) and infinitely many PDE constraints that characterize $y(u;\omega)$. Notice that we do not have a theory for (6.13) and we can only argue that a local optimal u exists. However, after discretization of the stochastic space, an optimization problem with a very large but finite number of PDE constraints must be solved. In this case, existence of optimal solutions can be proved with standard arguments.

Next, in an attempt to avoid solving (6.13), we explore alternative formulations, in particular formulations corresponding to (6.4). On the other hand, in Section 6.4 we discuss a proper orthogonal decomposition framework to determine robust controls with (6.13).

Following the discussion in [313, 280], given a functional F depending on a set of random input variables (σ, δ), with mean $(\mathbb{E}(\sigma), \mathbb{E}(\delta))$ and standard deviations (s_σ, s_δ), we have the following second-order Taylor expansion:

$$\mathbb{E}(F(\sigma,\delta)) = F(\mathbb{E}(\sigma),\mathbb{E}(\delta)) + \frac{1}{2}\frac{\partial^2 F}{\partial \sigma^2} s_\sigma^2 + \frac{1}{2}\frac{\partial^2 F}{\partial \delta^2} s_\delta^2 + \frac{1}{2}\frac{\partial^2 F}{\partial \sigma \partial \delta} s_\sigma s_\delta + O(s^4),$$

where the second derivatives are evaluated at $(\sigma_0, \delta_0) = (\mathbb{E}(\sigma), \mathbb{E}(\delta))$. From this formula we conclude that a first-order approximation to (6.13) corresponds to the optimal u_0, with the state equation having σ_0 and δ_0 as diffusion and reaction coefficients, respectively. Further notice that the state equation is nonlinear and that therefore the control obtained using the mean diffusion and reaction coefficients is not the same as the mean of the controls $\bar{u} = \mathbb{E}(u)$ obtained by averaging over all controls corresponding to all (σ, δ) configurations. Therefore, while the Taylor series expansion above shows that $u_0 = u_{\mathbb{E}(\sigma),\mathbb{E}(\delta)}$ is a first-order approximate solution to (6.13), it appears reasonable to ask whether or not $\bar{u}$, as defined in (6.4), can also be an approximate minimizer. This is not clear, since, in general, we expect

$$J(y(u_0), u_0) \neq J(y(\bar{u}), \bar{u}),$$

as already noted in the introduction. In Section 6.2.3, we compare these two approximation strategies in the effort of designing a robust optimal control.

6.2.1 Discretization of the Probabilistic Space

Recent research on stochastic PDE problems has focused on deterministic PDE systems with coefficients modeled by random fields [352] that can be accurately represented by a truncated KL expansion, since in this case a finite-dimensional approximation of the original infinite-dimensional stochastic space is possible. Based on the approximation of stochastic fields with the KL expansion or the polynomial chaos expansion [142], it has been possible to investigate the solution of PDE problems with random coefficients [17, 132, 263, 290, 363].

In the framework given by the generalized polynomial chaos expansion [142, 364], the dependent variables are considered as random processes and are written in the form $y(x,t,\omega) = \sum_{i=1}^{K} y_i(x,t)\psi_i(\xi(\omega))$, where $y_i(x,t)$ are deterministic functions and the ψ_i are the Askey polynomials which are chosen depending on the type of stochastic process $\xi(\omega)$ as input. A challenging aspect of the polynomial chaos approach is the solution of very large-sized coupled systems; see, e.g., [290].

On the other hand, a stochastic-grid collocation approach [17, 263, 363] allows us to avoid the solution of PDE problems coupled in the probability space. However, in the collocation case, a uniform Cartesian-mesh discretization of the probability parameter space may result in a curse of dimensionality, and in order to solve this problem, a Smolyak sparse-grid scheme [98, 222] has been introduced to model high-dimensional stochastic-coefficient spaces; see, e.g., [44, 67, 263]. Alternatively, Monte Carlo and quasi-Monte Carlo methods have been used to explore the probability space of PDEs with random coefficients; see, e.g., [18, 151].

In the following, we discuss the solution of (6.10)–(6.11) using a collocation approach. On the stochastic $(N+M)$-dimensional grid, the coordinates of the grid points represent the coefficients of the truncated KL expansion, while the physical space-time domain is approximated using classical discretization schemes. On the stochastic grid, we need to solve (6.10)–(6.11) for each grid point, and all statistical quantities related to these solutions are obtained by integration on this grid resulting from the Cartesian product of $\mathcal{N} = N + M$ one-dimensional grids, with $(m_1, \ldots, m_N, m_{N+1}, \ldots, m_{N+M})$ interpolation nodes in the respective dimensions. Let $m_j = 2^{j-1} + 1$ be the number of nodes for a quadrature scheme of order $j > 1$ and $m_1 = 1$. Following [263], we consider Chebyshev nodes given by

$$z_i^j = -\cos\frac{\pi(i-1)}{m_j - 1}, \quad i = 1,\ldots,m_j.$$

In addition to having an explicit formula, the Chebyshev nodes are chosen for their low Lebesgue number and because they form a nested set of nodes [132, 222, 263, 363]. In fact, if we write the set of Chebyshev nodes of order j as $\{z^j\} = \{z_1^j,\ldots,z_{m_j}^j\}$, then the nesting property gives that $\{z^j\} \subset \{z^{j+1}\}$.

Let $\{w^j\}$ be the set of the quadrature weights in one dimension which correspond to the nodes $\{z^j\}$; then the weights in multiple dimensions on a tensor product grid are written as a tensor product of one-dimensional weights $w = w^{j_1} \otimes \cdots \otimes w^{j_n}$. Thus, in particular, the mean value of a function $f : \mathbb{R}^{\mathcal{N}} \to \mathbb{R}$ is given by

$$\mathbb{E}(f) = \sum_{i_1=1}^{m_1} \cdots \sum_{i_{\mathcal{N}}=1}^{m_{\mathcal{N}}} f(z_{i_1}^{j_1},\ldots,z_{i_{\mathcal{N}}}^{j_{\mathcal{N}}})(w^{j_1} \otimes \cdots \otimes w^{j_{\mathcal{N}}}),$$

and the variance is $Var(f) = \mathbb{E}([f - \mathbb{E}(f)]^2)$.

We see that the formulae above require $m_1 \times \cdots \times m_{\mathcal{N}}$ function evaluations, and this effort grows exponentially with the number of dimensions. To circumvent this limitation, a Smolyak scheme can be used [132, 135, 263, 363] that constructs multivariate interpolation as a linear combination of tensor-product formulas on a minimal number of nodes of the multidimensional space. A full $\mathcal{N}$-dimensional grid of order $\mathcal{J}$, with $j_1 = j_2 = \cdots = j_{\mathcal{N}} = \mathcal{J}$ in each dimension, is formed using a tensor product of the constituent one-dimensional grids and has total order $j_1 + j_2 + \cdots + j_{\mathcal{N}} = \mathcal{N}\mathcal{J}$. On the other hand, the sparse grid of order $\mathcal{J}$ is composed of a strict subset of full grids, where $\mathcal{J}$ is the order of the largest allowed grid and the orders of the constituent grids add up to a total order given by $\mathcal{J} + \mathcal{N} - 1$. To better explain this point, consider $\mathcal{N} = 2$ and a sparse grid of order $\mathcal{J} = 4$. This grid results in $\{z^1 \otimes z^4\} \bigcup \{z^2 \otimes z^3\} \bigcup \{z^3 \otimes z^2\} \bigcup \{z^4 \otimes z^1\}$; see, e.g., [44, 67] for details. With χ we denote the total number of sparse-grid collocation points.

To define the weights for sparse-grid integration, we need to define the difference weights of order j, which are $\eta^j = w^j - w^{j-1}$. The quadrature weights on a full grid with order $\mathbf{j} = (j_1,\ldots,j_{\mathcal{N}})$ would be the tensor product $w = w^{j_1} \otimes \cdots \otimes w^{j_{\mathcal{N}}}$. On the sparse grid of dimension $\mathcal{N}$ and total order $\mathcal{J}$, the weights are given by

$$w^{\mathcal{J}} = \sum_{\ell=1}^{\mathcal{N}+\mathcal{J}-1} \sum_{|\mathbf{j}|=\ell} (\eta^{j_1} \otimes \cdots \otimes \eta^{j_{\mathcal{N}}}).$$

For more details on sparse grids, see [98, 222].

Following [17, 263], it is possible to estimate the global error affecting the solution of the optimal control problem with random coefficients. Let e_h be the bound of the solution error of the deterministic problem, denote with $e_{\mathcal{N}}$ the error due to truncation of the KL expansion, and let e_S denote the sparse-grid interpolation error; then $e_h + e_{\mathcal{N}} + e_S$ provides an estimate of the global error.

6.2.2 Sparse-Grid CSMG Methods

The advantage of the stochastic collocation approach on sparse grids with respect to, e.g., Monte Carlo simulation, is to greatly reduce the number of solver calls. Nevertheless, a large number of optimality system solves is still required, which demands efficient solution strategies that are robust with respect to the choice of values of the optimization parameters and to the changes of the coefficients' configurations due to randomness. We have shown that these requirements are met by the CSMG strategy. Therefore in the following, we present results of experiments with optimal control problems with random coefficients where each random configuration corresponds to a sparse-grid point and the resulting deterministic optimality system is discretized in the physical space and time.

Results of numerical experiments show that the combination of sparse-grid and CSMG techniques results in a solution process with optimal computational complexity with respect to the sizes of the physical and probability grids.

6.2.3 Experiments with a Parabolic Control Problem

We consider (6.11) with a nonlinear reaction term given by $G_\delta(y) = \delta e^y$ that is used to model explosive combustion phenomena [57]. We assume that the diffusivity and reactivity are random fields that are sufficiently well approximated by a truncated KL expansion. In the case of the diffusion coefficient, we ensure positiveness assuming, as in [263], a KL expansion in the form (6.9) for $\log(\sigma(\omega) - \sigma_0)$. Therefore, we consider a random diffusivity coefficient given by

$$\sigma(x_1, x_2, \omega) = \sigma_0 + \exp\big\{[Y_1(\omega)\cos(\pi x_2) + Y_3(\omega)\sin(\pi x_2)]e^{-1/8} \\ + [Y_2(\omega)\cos(\pi x_1) + Y_4(\omega)\sin(\pi x_1)]e^{-1/8}\big\},$$

where $\sigma_0 = 1/100$ and $Y_j \in [-1,1]$, $j = 1,2,3,4$. This field is characterized by a squared exponential covariance typical of Gaussian processes. Two realizations of $\sigma(x_1, x_2, \omega)$ are depicted in Figure 6.1.

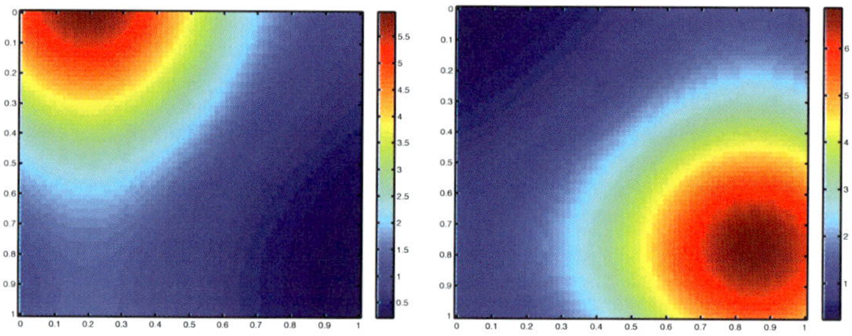

Figure 6.1. *Two realizations of* $\sigma(x_1, x_2, \omega)$. *Left:* $Y_1 = 0.8$, $Y_2 = 1.0$, $Y_3 = -0.2$, $Y_4 = 0.7$. *Right:* $Y_1 = -0.8$, $Y_2 = -1.0$, $Y_3 = 0.7$, $Y_4 = 0.5$. *This figure first appeared in A. Borzì and G. von Winckel, Multigrid methods and sparse-grid collocation techniques for parabolic optimal control problems with random coefficients, SIAM J. Sci. Comput.*, 31(3) (2009), 2172–2192.

For the reaction term, we consider the following model of uncertainty:

$$\delta(t, \omega) = \delta_0 + W(\omega)\sin(2\pi t/T),$$

where $\delta_0 = 1$ and $W \in [-1,1]$. For simplicity, we consider there are no background thermal sources so that $f = 0$. The desired target trajectory is given by the following oscillating function with increasing amplitude

$$y_d(x,t) = (1+t)(x_1 - x_1^2)(x_2 - x_2^2)\cos(4\pi t).$$

We take $y_T(x) = y_d(x,T)$.

We use one pre- and postsmoothing, and the domain $\Omega = (0,1)^2$ is discretized by a hierarchy of meshes and $h = 1/4$ is the coarsest space-mesh size. We take $T = 1$ and $\delta t = T/N_t$. Two different grids $N_x \times N_y \times N_t$ are considered, $32 \times 32 \times 32$ and $64 \times 64 \times 64$, which result in $\gamma = 32$ and $\gamma = 64$, respectively. For the discretization of the stochastic space we use a five-dimensional sparse grid and present results with orders $\mathcal{J} = 2$ and $\mathcal{J} = 3$. These correspond to a total number of sparse-grid points of $\chi = 61$ and $\chi = 241$, respectively.

To describe the results of the experiments, we report the observed multigrid convergence factors ρ_{obs} and CPU times. The tracking ability of the sparse-grid multigrid algorithm is given in terms of the mean values of the norms of the tracking error $\mathbb{E}(\|y - y_d\|)$ and of the terminal observation error $\mathbb{E}(|y - y_T|)$. Other important statistical observable are the variance Var of the tracking error and its skewness $Skew$.

The results are reported in Table 6.1 in the case of tracking of trajectories. They illustrate efficiency and robustness of the proposed multigrid solvers. Usual multigrid convergence speeds are obtained which appear to be independent of the mesh size and of ν. Notice that the CPU times scale linearly as a factor of 2^3 as the space-time mesh is refined. Moreover, as we move from an order-two to an order-three stochastic sparse grid we also obtain optimal complexity in CPU time. The results in Table 6.1 also demonstrate the tracking ability of the optimal control formulation. As the value of ν decreases, smaller values of $\mathbb{E}(\|y - y_d\|)$ are obtained as desired in an optimal control framework.

Table 6.1. *Results for $\alpha = 1$, $\beta = 0$, with the CSMG multigrid scheme. Denote $\|\Delta y\| = \|y - y_d\|$.*

$\mathcal{J}=2,\ \chi=61$						
ν	$N_x \times N_y \times N_t$	CPU(s)	$\mathbb{E}(\|\Delta y\|)$	$Var(\|\Delta y\|)$	$Skew(\|\Delta y\|)$	$\mathbb{E}(\rho_{obs})$
10^{-6}	$32 \times 32 \times 32$	87.89	$1.15\,10^{-3}$	$5.84\,10^{-7}$	$5.79\,10^{-10}$	0.08
10^{-8}	$32 \times 32 \times 32$	69.75	$3.48\,10^{-5}$	$6.45\,10^{-10}$	$2.59\,10^{-14}$	0.08
10^{-6}	$64 \times 64 \times 64$	722.46	$1.18\,10^{-3}$	$5.94\,10^{-6}$	$5.93\,10^{-10}$	0.12
10^{-8}	$64 \times 64 \times 64$	795.68	$4.75\,10^{-5}$	$7.08\,10^{-10}$	$2.80\,10^{-14}$	0.12
$\mathcal{J}=3,\ \chi=241$						
ν	$N_x \times N_y \times N_t$	CPU(s)	$\mathbb{E}(\|\Delta y\|)$	$Var(\|\Delta y\|)$	$Skew(\|\Delta y\|)$	$\mathbb{E}(\rho_{obs})$
10^{-6}	$32 \times 32 \times 32$	353.51	$1.15\,10^{-3}$	$6.27\,10^{-7}$	$1.75\,10^{-9}$	0.12
10^{-8}	$32 \times 32 \times 32$	297.82	$3.49\,10^{-5}$	$4.83\,10^{-10}$	$2.02\,10^{-14}$	0.08
10^{-6}	$64 \times 64 \times 64$	2867.01	$1.18\,10^{-3}$	$6.37\,10^{-7}$	$1.80\,10^{-9}$	0.13
10^{-8}	$64 \times 64 \times 64$	3152.42	$4.73\,10^{-5}$	$5.60\,10^{-10}$	$2.14\,10^{-14}$	0.10

Once a (finite) set of controls approximating the solution of the stochastic PDE optimal control problem is obtained, the question arises of how to construct a unique control that provides good tracking features for all configurations of the coefficients. As a possible control function, we consider the function $\bar{u} = \mathbb{E}(u)$ defined as the mean function of the optimal controls corresponding to each point of the discretized configuration space. Alternatively, we may consider the control function obtained using average parameter values. We denote the corresponding state functions as $y_{\bar{u}}$ and y_{σ_0,δ_0}, respectively. In Table 6.2 we present a comparison of the performance of the two controls. We see that the proposed controls have similar tracking performance, although the approach with $\bar{u}$ performs slightly better.

Table 6.2. *Results with* $\nu = 10^{-6}$; $N_x \times N_y \times N_t = 64 \times 64 \times 64$, $\chi = 241$.

		$\mathbb{E}(\|y-y_d\|)$	$Var(\|y-y_d\|)$	$Skew(\|y-y_d\|)$
$\alpha=1$	$y_{\bar{u}}$	$1.33\,10^{-1}$	$1.20\,10^{-2}$	$3.26\,10^{-4}$
$\beta=0$	y_{σ_0,δ_0}	$1.43\,10^{-1}$	$1.40\,10^{-2}$	$4.44\,10^{-2}$
$\alpha=0$	$y_{\bar{u}}$	$2.86\,10^{-1}$	$5.79\,10^{-4}$	$3.99\,10^{-5}$
$\beta=1$	y_{σ_0,δ_0}	$5.25\,10^{-1}$	$2.46\,10^{-6}$	$-1.80\,10^{-9}$

Similar results are obtained with the sparse-grid multigrid scheme applied to elliptic control problems with random coefficients [44].

6.3 Aerodynamic Design under Geometric Uncertainty

Uncertainty arises in flight conditions like the angle of incidence, the velocity (Mach number) of the plane, the density of the air, or the Reynolds number. The formulation of robust optimization problems as well as their numerical treatment is studied in [313]. Here we focus on the aircraft geometry itself as an uncertainty source [65, 300]. The real shape may vary from the planned shape due to manufacturing tolerances, temporary factors like icing, or material fatigue. Since there are so many factors having effects on the shape, this uncertainty has to be considered in the optimization problem in order to produce a design which is robust to small perturbations of the shape itself. In the literature, there are only a few papers on this topic investigating the influence of variations of the profile; see, e.g., [156, 238]. The general approach that we discuss is again to apply a KL expansion to approximate the infinite-dimensional probability space of geometrical uncertainty. To better overcome the curse of dimensionality, an adaptively refined sparse grid is used in order to compute statistics of the solution.

6.3.1 Modeling Geometric Uncertainty

The geometrical uncertainty depends on the geometry itself, so it is modeled as a Gaussian random field $\sigma : \Gamma \times \mathcal{O} \to \mathbb{R}$, defined on a probability space $(\mathcal{O}, \mathcal{A}, \mathcal{P})$ and on the shape of the airfoil Γ. In each point x of the shape Γ, the uncertainty is described by a normally distributed random variable $\sigma(x,\cdot)$. Additionally, the second order statistics, the mean value, and the covariance function are given to fully describe the random field. We have

$$\begin{aligned} \sigma_0(x) &= 0, \\ C_\sigma(x,y) &= b^2 \exp\left(-\frac{\operatorname{arc}(x,y)^2}{l^2}\right) \end{aligned} \tag{6.14}$$

for all $x, y \in \Gamma$. The function arc $\Gamma \times \Gamma \to \mathbb{R}^+$ measures the shortest distance between two points on the curved geometry Γ. The parameter l determines how quickly the covariance falls off and b controls the magnitude of the bumps. A squared exponential covariance function is chosen, since the resulting perturbed geometry is smooth due to the smoothness of the random field.

Then, a perturbed geometry is given as

$$v(x,\omega) = x + \sigma(x,\omega)\,\vec{n}(x), \tag{6.15}$$

where $\vec{n}$ is the unit vector in x normal to the profile Γ. The approximation of the random field σ is again performed by applying the KL expansion as in Section 6.2.

6.3.2 Semi-infinite Robust Design

The usual single setpoint aerodynamic shape optimization problem can be described in the following rather abstract form of an optimization problem with equality and inequality constraints

$$\min J(y,p), \tag{6.16}$$
$$c(y,p), = 0, \tag{6.17}$$
$$h(y,p) \geq 0. \tag{6.18}$$

Now, we think of (6.17) as the discretized outer flow equation around, e.g., an airfoil described by the geometry parameters $p \in \mathbb{R}^{n_p}$. Rather than more general investigations on aerodynamic shape optimization as in [123], we assume that a finite parametrization of the shape to be optimized is given. The vector y is the state vector (velocities, pressure,...) of the flow model (6.17), and we assume that (6.17) can be solved uniquely for y for all reasonable geometries p. The objective in (6.16) typically is the drag to be minimized. The restriction (6.18) typically denotes lift or pitching moment requirements. To make the discussion here simpler, we assume a scalar-valued restriction, i.e., $h(y,p) \in \mathbb{R}$. The generalization of the discussions below to more than one restriction is straightforward. Here, we treat the angle of attack as a fixed parameter which is not adjusted to reach the required lift (cf., e.g., [233, 205, 270]). The general deterministic problem formulation (6.16)–(6.18) is influenced by stochastic perturbations. In the literature, min-max formulations and also chance constrained formulations can be found. According to our experience the semi-infinite formulation discussed below has been most successful.

Consider the following robust version of the nonlinear programming problem written in the form of a semi-infinite optimization problem

$$\min \int_{\mathcal{O}} J(y(\cdot,\omega), p, \sigma(\cdot,\omega))\, d\mathcal{P}(\omega), \tag{6.19}$$
$$c(y(\cdot,\omega), p, \sigma(\cdot,\omega)) = 0 \quad \forall \omega \in \mathcal{O}, \tag{6.20}$$
$$h(y(\cdot,\omega), p, \sigma(\cdot,\omega)) \geq 0 \quad \forall \omega \in \mathcal{O}. \tag{6.21}$$

This definition of robustness can also be found in [205, 234]. Semi-infinite optimization problems have been treated directly so far only for rather small and weakly nonlinear problems; see, e.g., [128]. For the numerical treatment of complicated design tasks, one has to approximate the integral in the objective (6.19) resulting in the form of a multiple setpoint problem for the set points $\{\sigma_i\}_{i=1}^N$. We have

$$\min \sum_{i=1}^{N} w_i \, J(y_i, p, \sigma_i), \tag{6.22}$$

$$c(y_i, p, \sigma_i) = 0 \quad \forall i \in \{1, \ldots, N\}, \tag{6.23}$$

$$h(y_i, p, \sigma_i) \geq 0 \quad \forall i \in \{1, \ldots, N\}, \tag{6.24}$$

where w_i denote the quadrature weights. The approximation and discretization of the probability space using a goal-oriented KL expansion and an adaptive sparse-grid strategy in order to formulate the introduced multiple setpoint problem (6.22)–(6.24) are discussed in the following in more detail.

6.3.3 The Use of a Goal-Oriented KL Basis

In order to reduce the computational effort, the orthogonal basis functions $\{z_i\}$ of the KL expansion (cf. (6.9)) will be chosen goal-oriented; that is, the individual impact of the eigenvectors on the target functional will be taken into account. This method is well established in the model reduction methods of dynamic systems and the adaptive mesh refinement (cf. [28]). The idea is to develop an error indicator for the individual eigenvectors reflecting the influence on the drag. We discuss the use of sensitivity information to capture the local sensitivities of the drag with respect to the eigenvectors

$$\eta_i := \frac{\mathrm{d}J}{\mathrm{d}z_i} = -\lambda^\top \frac{\partial c}{\partial z_i} + \frac{\partial J}{\partial z_i} \qquad \forall i = 1, \ldots, d, \tag{6.25}$$

where λ solves the adjoint equation. The adjoint equation is independent of i; hence it has to be solved only once and the indicator η_i is numerically cheap to evaluate. Now, the reduced basis $\{\hat{z}_i\}$ can be automatically selected, and the eigenvector z_i with a large value η_i has to be kept in the reduced basis, whereas a small value indicates that the basis vector can be rejected from the basis.

6.3.4 Adaptive Sparse Grids for High-Dimensional Integration

We consider a dimension adaptive sparse-grid strategy in order to further reduce the number of grid points but conserve the approximation quality. First, a generalization of sparse grids will be introduced which allows us to weight the dimensions according to their importance on the target functional (cf. [141, 134, 97, 222]). The original sparse grid of order $\mathcal{J}$ combines all the incremental functions, defined as $\eta^i = w^{i+1} - w^i$, where w^j are the one-dimensional quadrature formulas of order j, which sum up to order $\mathcal{J}$. Further, [141] and [133] suggest allowing a more general index set which can then be adaptively chosen with respect to the importance of each dimension. An index set I is called *admissible* if for all $\mathbf{i} \in I$, we have

$$\mathbf{i} - e_j \in I \qquad \forall 1 \leq j \leq d, \; i_j > 1,$$

where $e_j \in \mathbb{R}^d$ is the jth unit vector. The generalized definition of sparse grids including the original sparse grid and the full tensor grid definition is then given as

$$S(\mathcal{J},d)(J) = \sum_{\mathbf{i}\in I} \left(\eta^{i_1} \otimes \cdots \otimes \eta^{i_d}\right)(J). \tag{6.26}$$

The algorithm computing the dimension adaptive sparse grid starts with the coarsest sparse grid, which means $I = \{(0,\ldots,0)\}$, and adds new indexes such that (1) the new index set remains admissible; and (2) the approximation error is reduced (cf. [141, 133]). The term $\eta_{\mathbf{i}} = \left(\eta^{i_1} \otimes \cdots \otimes \eta^{i_d}\right)(J)$ indicates the reduction in the approximated integral for each new added index, so that we directly use $\eta_{\mathbf{i}}$ as an error indicator for the adaptivity. The main advantage of the dimension adaptive refinement is the fact that the quadrature formulas can be chosen problem dependent. Considering geometric uncertainty described by a Gaussian random field, the Gauss–Hermite formulas are an appropriate choice for the quadrature. Since the nesting is a favorable feature constructing the sparse grid, only the Gauss–Hermite quadratures of orders $1,3,7,15$ are taken into account in the numerical results.

6.3.5 Numerically Computed Robust Aerodynamic Designs

We present results of numerical experiments using a numerical solution strategy which is based on the one-shot approach as outlined in Section 4.6 and in [183, 182, 312]. In the experiments that follow, an active set strategy has been employed, which uses the worst violation of the lift constraint as the active set in each iteration. The robust optimization under shape uncertainty of a transonic RAE2822 profile in Euler flow is considered. We use the flow solver TAU provided by DLR which allows the computation of gradients by the adjoint approach. The TAU code is a CFD software package for the prediction of viscous and inviscid flows about complex geometries from the low subsonic to the hypersonic flow regime employing hybrid unstructured grids. The profile is described by 129 surface grid points and the airfoil is parametrized by 21 Hicks–Henne functions. The geometric uncertainty is characterized by a Gaussian random field with $\mathbb{E}(\sigma)(x) = 0$ and $C_\sigma(x,y) = (0.005)^2 \cdot \exp(-\frac{\mathrm{arc}(x,y)^2}{(0.1)^2})$ for all $x,y \in \Gamma$. The distribution of the eigenvalues using the KL expansion of the given random field is shown in Figure 6.2.

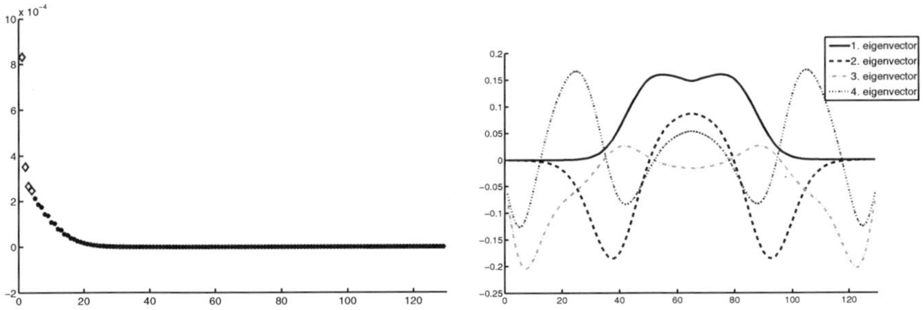

Figure 6.2. *Distribution of the eigenvalues and first four eigenvectors of the given random field σ. Reprinted with permission from A. Borzì, V. Schulz, C. Schillings, and G. von Winckel, On the treatment of distributed uncertainties in PDE-constrained optimization, GAMM-Mitteilungen,* 33(2) (2010), 230–246.

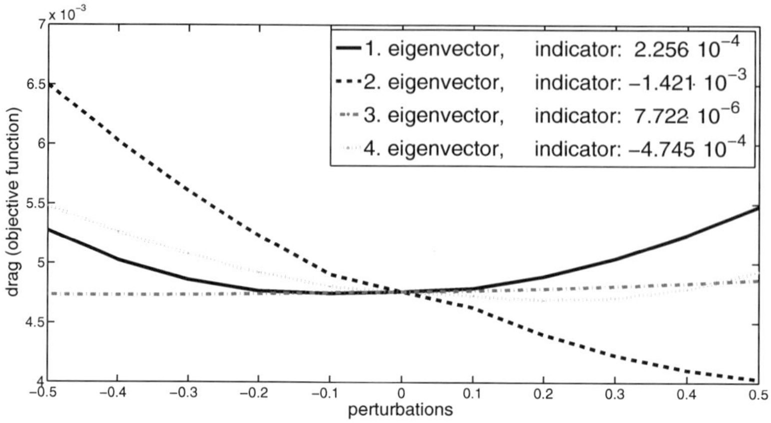

Figure 6.3. *Drag performance of the first four eigenvectors on the target functional. Reprinted with permission from A. Borzì, V. Schulz, C. Schillings, and G. von Winckel, On the treatment of distributed uncertainties in PDE-constrained optimization, GAMM-Mitteilungen,* 33(2) (2010), 230–246.

As stated before, the eigenvalues exponentially converge towards zero. For the numerical results below, we have considered only the first four eigenvalues and eigenvectors to represent the random field σ of the perturbations. The corresponding eigenvectors are shown in Figure 6.2. To further reduce the computational effort, we investigate the influence of the individual eigenvectors in order to reject those eigenvectors from the reduced basis which have no impact on the target functional. As Figure 6.3 shows, the third eigenvector has no impact on the objective function; hence it can be rejected from the KL basis and the dimension of the integral is reduced. This behavior is also reflected by the introduced indicator. The use of Gauss–Hermite formulas with an adaptive sparse-grid strategy results in a grid consisting of 21 points for a given error tolerance of 10^{-5}. Compared to a full tensor grid approach, a reduction by factor 10 can be observed; see Figure 6.4.

Figure 6.5 compares the results of the robust optimization and of the single setpoint optimization, i.e., without considering any uncertainty in the optimization.

The drag and lift performance is plotted against the 21 perturbed geometries, and the dashed line in Figure 6.5 indicates the mean value of the drag. The robust optimization improves the mean value of the target functional and also leads at the same time to a better lift performance over the whole range of perturbations, whereas the single setpoint optimization is infeasible in more than the half of the considered grid points. Summing it all up, it can be said that the robust optimization leads to a better lift-to-drag ratio than the single setpoint optimization and the resulting profile is more robust against small perturbations of the shape itself.

6.4 A Proper Orthogonal Decomposition Framework to Determine Robust Controls

In this section, we discuss a strategy for the fast computation of robust controls for PDE models with random coefficients [68]. This strategy is formulated in the framework of Section 6.2.1 and may be applied equally well to the case of robust shape design.

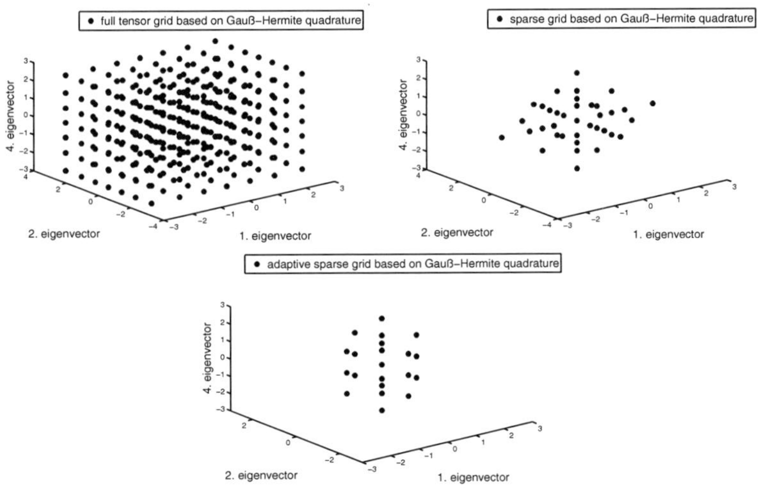

Figure 6.4. *Full tensor grid with* 343 *grid points, sparse grid with* 37 *grid points, and dimension adaptive sparse grid with* 21 *grid points. Reprinted with permission from A. Borzì, V. Schulz, C. Schillings, and G. von Winckel, On the treatment of distributed uncertainties in PDE-constrained optimization, GAMM-Mitteilungen,* 33(2) (2010), 230–246.

A robust control is defined as the control function that minimizes the expectation value of the objective over all coefficient configurations. However, a straightforward application of the full adjoint method on this problem results in a very large optimality system. In contrast, a fast method can be implemented where the expectation value of the objective is minimized with respect to a reduced proper orthogonal decomposition (POD) basis of the space of controls.

To illustrate this POD-based method, consider the following PDE optimization problem

$$\min J(y,u), \tag{6.27}$$

$$c(y,u,\omega) = 0, \tag{6.28}$$

where the PDE constraint has coefficients that are subject to randomness.

Now, assume an $\mathcal{N}$-dimensional and order $\mathcal{J}$ sparse-grid discretization of the probability space of the events $\omega \in \mathcal{O}$. Denote with χ the total number of sparse-grid points. We have that ω_i, $i = 1,\dots,\chi$, represent the grid points of the sparse grids that correspond to different events in the probability space. For each of these points we solve (6.27)–(6.28), thus obtaining the optimal control u_i corresponding to ω_i.

In the following, we discuss a framework that, by exploring the collection of controls, u_i, $i = 1,\dots,\chi$, allows us to determine a unique robust control for the governing random PDE model. We define this control as the minimizer $\hat{u}$ of the mean of the objective on the stochastic parameter space; recall the discussion in Sections 6.2 and 6.3. That is, we consider the following problem

$$\min_u \int_{\mathcal{O}} \hat{J}(u;\omega)\, d\mathcal{P}(\omega), \tag{6.29}$$

where $\hat{J}(u;\omega) = J(y(u;\omega),u)$ denotes the reduced objective corresponding to (6.27)–(6.28).

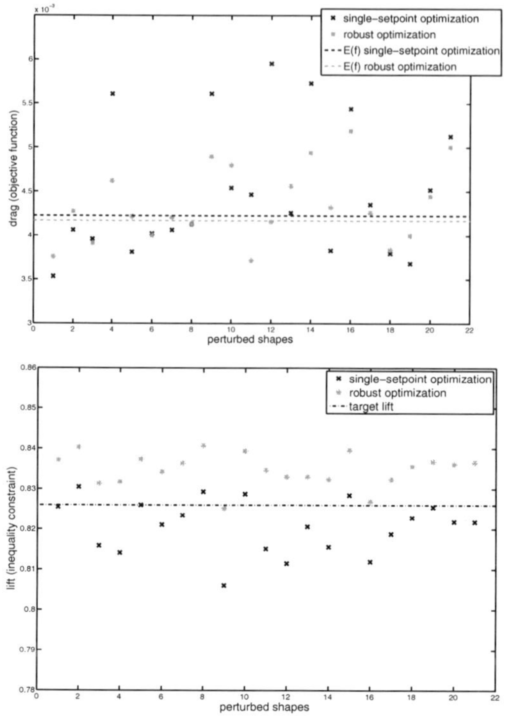

Figure 6.5. *Drag and lift performance of the* 21 *perturbed geometries. Reprinted with permission from A. Borzì, V. Schulz, C. Schillings, and G. von Winckel, On the treatment of distributed uncertainties in PDE-constrained optimization, GAMM-Mitteilungen,* 33(2) (2010), 230–246.

Now, after discretization of the stochastic space, an optimization problem with a very large number of PDE constraints must be solved. We have

$$\min_u \sum_{i=1}^{\chi} w_i \, \hat{J}(u;\omega_i), \tag{6.30}$$

where the w_i's denote the sparse-grid quadrature weights with $\sum_{i=1}^{\chi} w_i = 1$. This problem can be written in expanded form as follows

$$\begin{aligned} &\min_u \sum_{i=1}^{\chi} w_i \, J(y_i,u), \\ &c(y_1,u,\omega_1) = 0, \\ &c(y_2,u,\omega_2) = 0, \\ &\qquad \vdots \\ &c(y_\chi,u,\omega_\chi) = 0, \end{aligned} \tag{6.31}$$

where $y_i = y(u;\omega_i)$, $i = 1,\ldots,\chi$.

To characterize the solution to (6.31), we define the Lagrange function

$$L(y_1,y_2,\dots,y_\chi,p_1,p_2,\dots,p_\chi,u)=\sum_{i=1}^{\chi} w_i\, J(y_i,u)+\sum_{i=1}^{\chi} w_i\,(c(y_i,u,\omega_i),p_i).$$

We have that solving (6.31) results in the following optimality system consisting of a very large system of state and adjoint equations coupled through an optimality condition. We have

$$\begin{aligned}
c(y_1,u,\omega_1)=0,&\quad c_y^*(y_1,u,\omega_1)\,p_1=-J_y(y_1,u),\\
c(y_2,u,\omega_2)=0,&\quad c_y^*(y_2,u,\omega_2)\,p_2=-J_y(y_2,u),\\
\dots&\\
c(y_\chi,u,\omega_\chi)=0,&\quad c_y^*(y_\chi,u,\omega_\chi)\,p_\chi=-J_y(y_\chi,u),\\
&\sum_{i=1}^{\chi} w_i\,\big(c_u^*(y_i,u,\omega_i)\,p_i+J_u(y_i,u)\big)=0.
\end{aligned}\tag{6.32}$$

We call the solution of this problem the full solution.

The POD-based framework is different [68]. The sparse-grid collection of controls is computed and using POD, a reduced POD basis of this space is constructed. Then, the minimization problem (6.30) is solved in the span of the POD functions.

6.4.1 POD Analysis of the Control Space

Consider a positive random field σ representing a coefficient of our PDE governing model and assume that σ can be well approximated with a truncated KL expansion which allows a sparse-grid representation. Correspondingly, we obtain χ sparse-grid points, each of which represents a coefficient configuration function σ_i, $i=1,2,\dots,\chi$.

Let $u_i\in U$, $i=1,\dots,\chi$, represent an input collection of the controls corresponding to different coefficient configurations. The entire collection can be written as follows

$$\mathbb{U}=[u_1,\dots,u_\chi]\in U^\chi.$$

A reduced-order model of this collection is obtained with the following procedure.

Assume that each member of the collection can be written in terms of an nth order Galerkin finite element basis $\{\varphi_j\}_{j=1}^n$ and U is the corresponding finite element control space. Therefore, based on this representation, a finite-dimensional approximation of $\mathbb{U}$ results in an $n\times\chi$ matrix $\mathcal{U}$ where its m-column, $\mathcal{U}_{\cdot,m}$, contains the finite element coefficients of the mth member of the collection of controls. That is,

$$u_i=\sum_{j=1}^{n}\mathcal{U}_{ji}\,\varphi_j.$$

Now consider the problem of approximating all coordinate vectors $\mathcal{U}_{\cdot,m}$ simultaneously by a single normalized vector $\Psi_\cdot$ as well as possible. We assume that this vector contains the finite element coefficients of a function $\psi\in U$ that satisfies the following optimization problem

$$\max \sum_{i=1}^{\chi} |(u_i, \psi)_U|^2 \qquad \text{s.t. } \|\psi\|_U = 1. \tag{6.33}$$

The finite-dimensional approximation of this problem is obtained assuming that we can represent $\psi = \sum_{j=1}^{n} \Psi_j \varphi_j$ and denote with M the positive definite mass matrix with elements $M_{ij} = (\varphi_i, \varphi_j)_U$. With this setting the first-order optimality condition for the optimization problem (6.33) results in the following eigenvalue problem

$$\mathcal{U}\mathcal{U}^\top M \Psi = \rho \Psi.$$

This eigenvalue problem can be put in a more elegant form defining $\hat{\mathcal{U}} = M^{1/2}\mathcal{U}$ and $v = M^{1/2}\Psi$. This yields

$$\hat{\mathcal{U}}\hat{\mathcal{U}}^\top v = \mu^2 v. \tag{6.34}$$

Now, consider the normalized eigenvector v_1 of (6.34) corresponding to the largest eigenvalue. The first POD coordinate basis function is given by $\Psi_1 = M^{-1/2} v_1$. This is a column vector that defines the first POD basis function as follows

$$\psi_1 = \sum_{j=1}^{n} \Psi_{j1} \varphi_j.$$

The second POD basis function is found by minimizing (6.33) subject to the additional constraint $(\psi_1, \psi_2)_U = 0$. One obtains the coefficients $\Psi_2 = M^{-1/2} v_2$, where v_2 is the normalized eigenvector in (6.34) corresponding to the second largest eigenvalue. This process is repeated until all nonzero eigenvalues are obtained. If the matrix $\mathcal{U}$ has rank $\ell \leq \chi$, then we obtain a POD basis given by

$$\hat{\Psi} = [\psi_1, \ldots, \psi_\ell].$$

The advantage of the POD framework is that a much smaller number of POD basis functions can approximate the input collection in the case of rapidly decaying eigenvalues. In fact, given a fraction $\delta < 1$, there exists $m \ll \ell$ such that

$$\frac{\sum_{i=1}^{m} \mu_i^2}{\sum_{i=1}^{\ell} \mu_i^2} > \delta.$$

Then m gives the number of POD basis functions that guarantee an approximation with a mean square error less than $(1-\delta)\sum_{i=1}^{\ell} \mu_i^2$. That is, the first m POD basis functions retain 100δ percent of the structure of the input collection. Therefore, we use a truncated POD basis given by

$$\Psi = [\psi_1, \ldots, \psi_m].$$

Now, it becomes convenient to pursue the minimization of the expectation of the reduced objective in the truncated POD space. We have

$$\min_u \sum_{i=1}^{\chi} w_i \, \hat{J}(u; \omega_i),$$

where we use the representation

$$u = \sum_{i=1}^{m} q_i \psi_i = \Psi q,$$

where $q = (q_1, q_2, \ldots, q_m)^\top$.

6.4.2 A Robust Control for Elliptic Control Problems

In this section, we illustrate the POD framework [68] to determine robust controls in the case of a distributed elliptic control problem with random coefficients. Consider the following optimal control problem

$$\min J(y,u) := \frac{1}{2} \int_\Omega [(y-z)^2 + \gamma u^2] dx, \tag{6.35}$$

$$\{-\partial_x^2 + \sigma\} y = u \qquad \text{in } \Omega, \tag{6.36}$$

$$y = 0 \qquad \text{on } \partial\Omega, \tag{6.37}$$

where $\Omega = (0,1)$ and $z(x) = x^2 \sin(\pi x)$. We assume that for any configuration of the positive random field $\sigma \in L^\infty(\Omega)$, the state equation has a unique solution.

Now, for each $\sigma_i(x)$ coefficient function, we compute the corresponding optimal control u_i so that we form a collection $\mathcal{U} = \text{span}\{u_1, u_2, \ldots, u_\chi\}$. In the following experiments, the functions σ_i are piecewise constant in Ω. They are defined as follows. We choose the value of the sparse-grid dimension $\mathcal{N}$ and consider a subdivision of Ω in $\mathcal{N}$ subintervals. On the ℓth subinterval, the constant value of σ_i is the value of the ℓth coordinate of the ith sparse-grid point. That is, we have KL basis functions that are piecewise constant with random amplitudes on equally sized subregions.

Next, assume that a POD analysis of the space of controls has been performed and it results that this space is well approximated by retaining the first m largest eigenvalues and corresponding eigenvectors. Further, denote with S_i the self-adjoint elliptic operator as in (6.36)–(6.37) with σ_i as coefficient. With this setting, the expectation of the reduced cost functional, the corresponding gradient, and the reduced Hessian are given by

$$\hat{J}(u) = \frac{1}{2} \int_\Omega \left[\gamma u^2 + \sum_{i=1}^{\chi} w_i (S_i^{-1} u - z)^2 \right] dx,$$

$$\nabla \hat{J}(u) = \gamma u + \sum_{i=1}^{\chi} w_i S_i^{-1} (S_i^{-1} u - z),$$

$$\nabla^2 \hat{J}(u) = \gamma I + \sum_{i=1}^{\chi} w_i S_i^{-2}.$$

If the dimensionality of the KL expansion is large and one chooses a relatively high-order sparse-grid quadrature, there will be many solution operators S_i to invert, which can be quite expensive, as we will need a Krylov solver to compute the Newton step.

We have that the optimal control for a given σ_i satisfies the equation

$$[\gamma I + S_i^{-2}] u_i = -S_i^{-1} z.$$

The optimal control for the average reduced objective satisfies the following full problem

$$\left(\gamma I + \sum_{i=1}^{\chi} w_i\, S_i^{-2}\right) u = \sum_{i=1}^{\chi} w_i\, S_i^{-1}\, z. \tag{6.38}$$

Notice that in (6.38), in order to apply the Hessian, we must solve $2\,\chi$ boundary value problems.

We perform a POD decomposition on the set of control functions in order to obtain an orthogonal set of functions $[\psi_1, \psi_2, \ldots, \psi_m]$, where $m \ll \chi$. Therefore, we consider the control problem (6.38) using the POD expansion $u = \Psi\, q$. Further, we multiply both sides of (6.38) by $\Psi^\top$. We obtain

$$\Psi^\top \left(\gamma I + \sum_{i=1}^{\chi} w_i\, S_i^{-2}\right) \Psi\, q = \Psi^\top \sum_{i=1}^{\chi} w_i\, S_i^{-1}\, z.$$

Rearranging the Ψ and S_i^{-1} terms, we have

$$\left(\gamma I + \sum_{i=1}^{\chi} w_i\, (S_i^{-1}\Psi)^\top (S_i^{-1}\Psi)\right) q = \sum_{i=1}^{\chi} w_i\, (S_i^{-1}\Psi)^\top\, z.$$

Now, define $P_i = S_i^{-1}\,\Psi$ so that we obtain

$$\left(\gamma I + \sum_{i=1}^{\chi} w_i\, P_i^\top\, P_i\right) q = \sum_{i=1}^{\chi} w_i\, P_i^\top\, z. \tag{6.39}$$

We notice that in (6.39), we have to solve offline the χ boundary value problems $S_i\, P_i = \Psi$, $i = 1, \ldots, \chi$, with m right-hand sides each to obtain all the $P_i \in \mathbb{R}^{n \times m}$ matrices needed.

In fact, solving (6.38) requires us to solve 2χ boundary value problems, times the number of, e.g., conjugate gradient iterations to solve the Newton problem of size n. On the other hand, solving (6.39) requires the solution of $m\,\chi$ boundary value problems only once before the outer iteration starts.

In Figure 6.6, we compare the full solution obtained with (6.38) and with (6.39). We can see rapid changes of the control corresponding to the jumps in the reaction coefficient ($\mathcal{N} = 3$ on top and $\mathcal{N} = 5$ on bottom). In particular, we see that the POD robust control has larger gradients at the jump interfaces, which results in better minimizers. In Table 6.3, we see that the computational time of the POD method is an order of magnitude smaller than that of the full method. The CPU time to compute the input collection of controls and the POD basis results of approximately the same order of the POD solution process. All results with the POD method are obtained with $m = 10$ POD basis functions. The full problem (6.38) has been solved using a preconditioned conjugate gradient method.

6.5 Optimal Control of Probability Density Functions of Stochastic Processes

We remark that the discussion on the simulation and control of PDE models with data uncertainty focuses on deterministic models and the aim is to determine unique robust solutions to be used in application. On the other hand, a statistically meaningful solution would

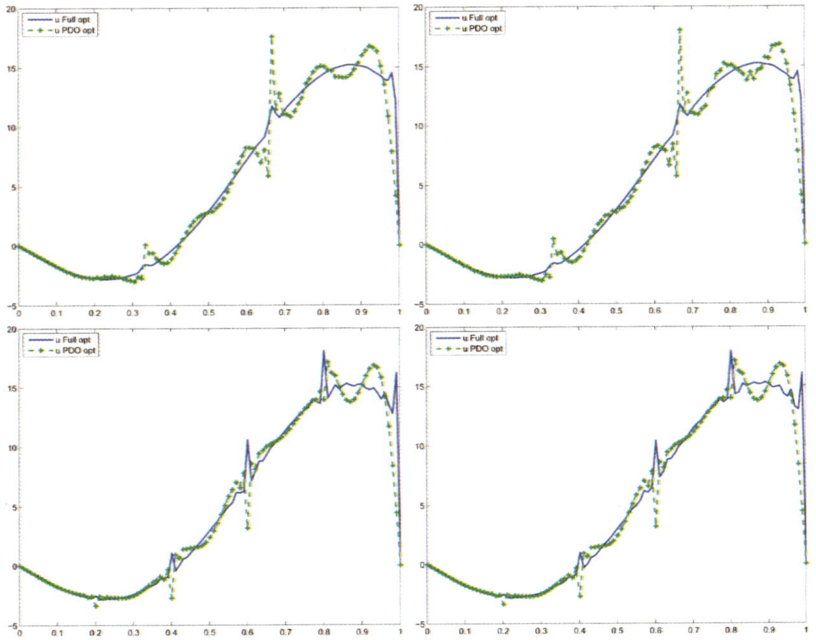

Figure 6.6. *Optimal controls computed with the full approach and the POD approach. Top:* $\mathcal{N} = 3$*; bottom:* $\mathcal{N} = 5$*. Left:* $\mathcal{J} = 3$*; right:* $\mathcal{J} = 5$*. From left to right and from top to bottom* $\chi = \{69, 441, 241, 2433\}$, $m = 10$*. Reprinted with permission from A. Borzì and G. von Winckel, A POD framework to determine robust controls in PDE optimization, Computing and Visualization in Science.*

Table 6.3. *Results with full and POD methods;* $n = 63$, $m = 10$*. CPU times include setup.*

$\mathcal{N}$	$\mathcal{J}$	χ	$\hat{J}_{pod}$	CPUsec	$\hat{J}_{full}$	CPUsec
3	3	69	$3.560\,10^{-2}$	$2.4\,10^{-2}$	$3.657\,10^{-2}$	34.2
3	5	441	$3.560\,10^{-2}$	$1.8\,10^{-1}$	$3.657\,10^{-2}$	221.6
5	3	241	$3.604\,10^{-2}$	$8.6\,10^{-2}$	$3.725\,10^{-2}$	225.6
5	5	2433	$3.605\,10^{-2}$	3.3	$3.725\,10^{-2}$	2291.4

require us to compute mean values and higher statistical moments depending on space and time. This is a very challenging problem setting especially in the infinite-dimensional functional framework where PDEs are defined. However, in the finite-dimensional case the problem of determining and controlling directly the probability density function (PDF) of a stochastic system is possible using the Fokker–Planck equation [288].

In this section, we discuss a Fokker–Planck framework for the formulation of an optimal control strategy for stochastic processes [7]. In this case, the governing model is a finite-dimensional Itô stochastic differential equation [144] and the PDE-based optimization theory enters this framework to define a control strategy of the PDF of the given stochastic process. In this framework, the control objectives are defined based on the PDF and the optimal control is obtained as the minimizer of the objective under the constraint given by the Fokker–Planck equation [288] that models the evolution of the PDF.

We focus on stochastic processes described by the following model

$$\begin{cases} dX_t = b(X_t,t;u)\,dt + \sigma(X_t,t)\,dW_t, \\ X_{t_0} = X_0, \end{cases} \tag{6.40}$$

where the state variable X_t is subject to deterministic infinitesimal increments of the first addend and to random increments proportional to a Wiener process dW_t having zero mean and unit variance. We have that $\sigma(X_t,t) > 0$ is the function of variance of the stochastic process and $b(X_t,t;u)$ is the drift term including the control mechanism.

We consider the action of a control $u = u(t)$ that allows us to drive the random process to approximate a desired evolution trajectory. In deterministic dynamics, the optimal control is achieved by finding the control law u that minimizes a given cost functional $J(X,u)$. In the present nondeterministic case, the state evolution X_t is random and represents an outcome of a probability space, so that a direct insertion of a stochastic process into a deterministic cost functional will result in a random variable. Therefore, when dealing with stochastic optimal control, usually the average of the cost function is considered. In particular, we have

$$J(X,u) = \mathbb{E}\left[\int_{t_0}^{t_1} L(t,X_t,u(t))\,dt + \Psi[X_T]\right]. \tag{6.41}$$

This is a Bolza-type cost functional in the finite horizon case, and it is supposed that the controller knows the state of the system at each instant of time (complete observations).

The common methodology to find an optimal controller of random processes consists in reformulating the problem from stochastic to deterministic. This is a reasonable approach when we consider the problem from a statistical point of view, with the perspective to find out the collective behavior of the process. In fact, the average of the functional of the process X_t is omnipresent in almost all stochastic optimal control problems considered in the scientific literature.

However, while the average of a stochastic process represents only a piece of information, the state of a stochastic process can be completely characterized in many cases by the shape of its statistical distribution, which is represented by the PDF. Therefore, a control methodology defined on the PDF would provide an accurate and flexible control strategy that could accommodate a wide class of objectives. For this reason, in [7, 129, 212, 215, 358] PDF control schemes were proposed, where the cost functional depends on the PDF of the stochastic state variable. In this way, a deterministic objective results and no average is needed.

Now, the important step is to recognize that the evolution of the PDF associated with the stochastic process (6.40) is characterized as the solution of the Fokker–Planck (also Fokker–Planck–Kolmogorov) equation; see, e.g., [279, 288]. This is a PDE of parabolic type with Cauchy data given by the initial PDF distribution. Therefore, the formulation of objectives in terms of the PDF and the use of the Fokker–Planck equation provide a consistent framework to formulate an optimal control strategy of stochastic processes.

In this framework, we consider a stochastic process in a time interval, with a given initial PDF and the objective of approximating a desired final PDF target with the actual

PDF of the state variable. We assume that the control is a constant function in the time interval to be determined by the optimal control scheme once and for all the evolution of the process in the time interval. The cost functional consists of a terminal-time tracking objective and the control cost. The resulting open-loop optimal control problem is formulated as the problem of finding a controller that minimizes this cost function within the time interval under the constraint provided by the Fokker–Planck equation. In [7], this control strategy is applied to a sequence of time subintervals to construct a fast closed-loop control scheme of the stochastic process based on the receding-horizon (RH) model predictive control (MPC) approach [244, 250]. Notice that the RH-MPC approach does not optimize a true performance index; see [155] for a method to quantify the performance degradation. Nevertheless, these schemes provide robust controllers that apply equally well to linear and nonlinear models and allow us to accommodate different control and state constraints [155, 207]. For this reason, RH-MPC schemes are among the most widely used control techniques in process control.

6.5.1 A Fokker–Planck Optimal Control Formulation

Denote with $f(x,t)$ the PDF function of finding the process at x at time t. Further, let $\hat{f}(x,t;y,s)$ denote the transition density probability distribution for the stochastic process to move from y at time s to x at time t, which means that $\hat{f}(x,s;y,s) = \delta(x-y)$. Both $f(x,t)$ and $\hat{f}(x,t;y,s)$ are nonnegative functions and the following holds

$$\int_\Omega f(x,t)dx = 1 \quad \forall t \geq s. \tag{6.42}$$

This is the conservation condition. If $\rho(y,s)$ is the given initial density probability of the process at time s, then we have that the density probability of the process at time $t > s$ is given by the following

$$f(x,t) = \int_\Omega \hat{f}(x,t;y,s)\rho(y,s)dy. \tag{6.43}$$

Notice that ρ should be nonnegative and normalized: $\int_\Omega \rho(y,s)dy = 1$.

First, we consider the stochastic process in the time interval (t_k, t_{k+1}), and domain Ω is chosen large enough such that $f(x,t) = 0$ on $\Sigma = \Omega \times (t_k, t_{k+1})$. We assume to know the initial value of the process at time t_k, in the sense that we give the probability density $\rho(x,s)$ at time $s = t_k$. Our problem is to determine a control u such that starting with an initial distribution ρ the process evolves such that a desired target probability density $f_d(x,t)$ at time $t = t_{k+1}$ is matched as close as possible. For this purpose, we consider the following control problem in $Q = \Omega \times (t_k, t_{k+1})$. We have

$$\min J(f,u) := \frac{1}{2}\|f(\cdot,t_{k+1}) - f_d(\cdot,t_{k+1})\|^2_{L^2(\Omega)} + \frac{\nu}{2}|u|^2, \tag{6.44}$$

$$\partial_t \hat{f}(x,t;y,t_k) - \frac{1}{2}\partial_x^2(\sigma(x,t)^2 \hat{f}(x,t;y,t_k)) + \partial_x(b(x,t;u)\hat{f}(x,t;y,t_k)) = 0, \tag{6.45}$$

$$\hat{f}(x,t_k;y,t_k) = \delta(x-y), \tag{6.46}$$

where (6.45) is the Fokker–Planck equation, i.e., the *forward Kolmogorov* equation, for the transition density probability distribution $\hat{f}(x,t;y,t_k)$ of the stochastic process X_t, and

$y \in \Omega$. Notice that the set of equations (6.45)–(6.46) can be interpreted as the fundamental solution of the stochastic process such that for any given ρ the corresponding f is obtained. This problem is as difficult to compute as the Green function of a PDE. However, in our case we assume that the initial distribution ρ is given, and hence we can reformulate (6.44)–(6.46) as follows

$$\min J(f,u) := \frac{1}{2}\|f(\cdot,t_{k+1}) - f_d(\cdot,t_{k+1})\|^2_{L^2(\Omega)} + \frac{\nu}{2}|u|^2, \tag{6.47}$$

$$\partial_t f(x,t) - \frac{1}{2}\partial_x^2(\sigma(x,t)^2 f(x,t)) + \partial_x(b(x,t;u)f(x,t)) = 0, \tag{6.48}$$

$$f(x,t_k) = \rho(x), \tag{6.49}$$

where we dropped s in the initial distribution. We consider solutions of the Fokker–Planck equation that are sufficiently regular [144], i.e., $f(x,t)$ has continuous first derivative in time, and continuous second derivative in space, jointly to the conservation condition.

6.5.2 An RH-MPC Scheme

The optimal control problem (6.47)–(6.49) is formulated for a stochastic process in the time interval (t_k, t_{k+1}) with a given terminal observation. Now, we consider the problem of controlling the PDF of a stochastic process to track a given sequence of desired PDFs in time. Let $(0,T)$ be the time interval where this process is considered. We assume time windows of size $\Delta t = T/N$, with N a positive integer. Let $t_k = k\Delta t$, $k = 0,1,\ldots,N$. At time t_0, we have a given initial PDF denoted with ρ and with $f_d(\cdot,t_k)$, $k = 1,\ldots,N$, we denote the sequence of desired PDFs. Our scheme starts at time t_0 and solves the minimization problem $\min_u J(f(u),u)$ defined in the interval (t_0,t_1). Then, with the PDF f resulting at $t = t_1$ solving the optimal control problem, we define the initial PDF for the subsequent optimization problem defined in the interval (t_1,t_2). This procedure is repeated by receding the time horizon until the last time window is reached. This is an instance of the class of RH-MPC schemes [250, 244] that is widely used in engineering applications to design closed-loop algorithms. One important aspect of this approach is that it can be applied to infinite-dimensional evolution systems [207], which is the case of the Fokker–Planck model.

The RH-MPC procedure is summarized in the following algorithm.

ALGORITHM 6.1. RH-MPC control.
Set $k = 0$; assign the initial PDF, $f(x,t_k) = \rho(x)$ and the targets $f_d(\cdot,t_k)$, $k = 0,\ldots,N-1$;

1. In (t_k, t_{k+1}), solve $\min_u J(f(u),u)$.
2. With the optimal solution u compute $f(\cdot,t_{k+1})$.
3. Assign this PDF as the initial condition for the Fokker–Planck problem in the next time window.
4. If $t_{k+1} < T$, set $k := k+1$, go to 1. and repeat.
5. End.

In the following, we report results of numerical experiments where we compute the control u that solves the control problem corresponding to a representative stochastic process. The classical problem of a particle immersed in a viscous fluid and subject to random Brownian fluctuations due to interaction with other particles is modeled by the Ornstein–Uhlenbeck process. We focus on the control of the PDF of this process. In this model, we have $b(X_t,t;u) = -\gamma X_t + u$, $\sigma(X_t,t) = \sigma$, where X_t represents the velocity of the particle and u is the momentum induced by an external force field defining the control mechanism.

In this case, the solution to the Fokker–Planck equation is well known to be a Gaussian distribution with mean

$$\mu(t;y,t_k;u) = u/\gamma + (y - u/\gamma)e^{-\gamma(t-t_k)}$$

and variance

$$\bar{\sigma}^2(t,t_k) = \frac{\sigma^2}{2\gamma}(1 - e^{-2\gamma(t-t_k)}).$$

Therefore the solution is as follows

$$\hat{f}(x,t;y,t_k;u) = \frac{1}{\sqrt{2\pi\bar{\sigma}^2(t,t_k)}} \exp\left(-\frac{(x-\mu(t;y,t_k;u))^2}{2\bar{\sigma}^2(t,t_k)}\right). \tag{6.50}$$

This solution defines a mapping $\hat{f} = \hat{f}(u)$. Now, assuming an initial distribution ρ at $t = t_k$, and having $\hat{f}(x,t_{k+1};y,t_k)$, the final distribution $f(x,t_{k+1})$ is given by integration as defined in (6.43). This procedure provides a mapping $f = f(u)$, and thus the following reduced cost functional is obtained

$$J(f(u),u) = \frac{1}{2}\|(f(u))(\cdot,t_{k+1}) - f_d(\cdot,t_{k+1})\|^2_{L^2(\Omega)} + \frac{\nu}{2}|u|^2.$$

As a result, one obtains the following one-parameter optimization problem

$$\min_u J(f(u),u).$$

This problem can be solved efficiently by a bisection minimization procedure [265].

In Figure 6.7, we show the optimal solution for the Ornstein–Uhlenbeck process. This solution is obtained solving (6.48) with an appropriate conservative numerical scheme [7], and these results are almost identical to those obtained from the direct quadrature of (6.43) with (6.50). In this experiment, the parameters are $\gamma = 1$, $\sigma = 0.8$, $\nu = 0.1$. The initial distribution (black dotted) is a Gaussian with zero mean and variance 0.1; the target is also Gaussian with mean value following the law $2\sin(\pi t_k/5)$ and variance 0.2. The optimal distribution (solid line) is calculated at time windows of $\Delta t = 0.5$ until time $T = 5$. Other successful applications of this methodology are given in [7], where also a geometric-Brownian process with additive drift control and a controlled Shiryaev process are considered.

We remark that the complexity of this Fokker–Planck approach increases exponentially as the dimensionality of the random process X_t increases. Also the idea of solving the reduced minimization problem $\min_u J(f(u),u)$ by bisection may be too inefficient when increasing the dimensionality of the process and of the control space. In all these cases, it

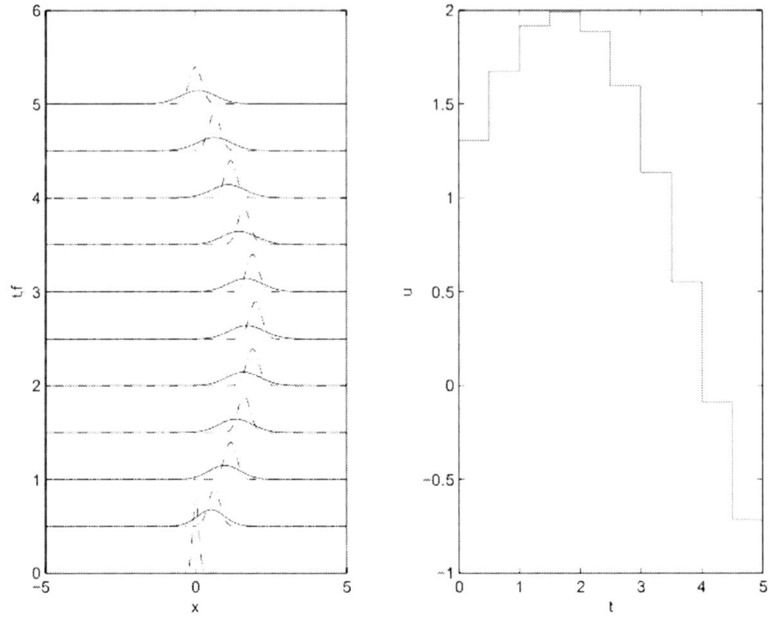

Figure 6.7. *The optimal control of the Ornstein–Uhlenbeck process. Left-hand side: computed PDF (solid line) and desired PDF (dotted-dashed line) at different time windows. Right-hand side: the optimal control function. Reprinted with permission from M. Annunziato and A. Borzì, Optimal control of probability density functions of stochastic processes, Math. Model. Anal.,* 15(4) (2010), 393–407.

becomes necessary to obtain gradient information by solving the following Fokker–Planck optimality system

$$\left\{\begin{array}{rcl} \partial_t f(x,t) - \frac{1}{2}\partial_x^2(\sigma(x,t)^2 f(x,t)) + \partial_x(b(x,t;u)f(x,t)) & = & 0, \\ f(x,t_k) & = & f_k(x), \\ -\partial_t p(x,t) - \frac{1}{2}\sigma(x,t)^2\,\partial_x^2\, p(x,t) + b(x,t;u)\,\partial_x p(x,t) & = & 0, \\ p(x,t_{k+1}) & = & f_d(x,t_{k+1}) - f(x,t_{k+1}), \\ \nu u + \left(\partial_x(\frac{\partial b}{\partial u} f), p\right) & = & 0. \end{array}\right.$$

6.6 Bayesian Uncertainty Quantification

Central to this section is the concept of the state of information over a parameter set or a parameter function. We assume to describe such a state of information as a probability density over the parameter space. Therefore also the results corresponding to such parameters and the information on the physical correlations between observable and model parameters are all described using probability densities. Here, model parameters play the role of the optimization function in the sense that the mathematical link between observable and model parameters is established through the formulation of an optimization problem. Moreover, we consider models defined by PDEs. Notice that the Bayesian approach has mainly been investigated with inverse problems [336] and it is only recently that this framework started to be extended to inverse problems governed by PDE models; see [127, 247, 336].

6.6.1 Statistical Inverse Problems

In the statistical approach to inverse problems all variables included in the model are modeled as random variables. The randomness describes our degree of information and uncertainty concerning model realizations, and these values are coded in the probability distributions. In this framework, the solution of an inverse problem is represented by posterior probability distributions.

Assume that a quantity $y \in \mathbb{R}^q$ represents (part of) the state of the system under consideration and it is also the quantity that we measure (data) in order to determine another quantity $x \in \mathbb{R}^n$ which is not directly accessible to measurements. This assumption requires the existence of a governing model that describes how these two quantities are related. The model itself may be inaccurate and it may contain parameters that are not well known. Furthermore, the measured quantity y may be affected by noise. Therefore, within this framework it is reasonable to consider the following functional relationship

$$y = F(x, \epsilon), \tag{6.51}$$

where $F : \mathbb{R}^n \times \mathbb{R}^k \to \mathbb{R}^q$ is the model function and $\epsilon \in \mathbb{R}^k$ is the vector containing all model uncertainty as well as the measurement noise. One possible approach to treat the presence of noise by appropriately chosen calibration measurements is to determine those components of ϵ that do not change from measurement to measurement. The noise, however, may be different from one instant to the other, and regularization methods are applied to cope with this part.

The statistical formulation of the inverse problem represented by (6.51) leads to the following

$$Y = F(X, E).$$

This is a parameter-to-observable map linking together the three random variables X, Y, and E, and consequently their probability distributions.

We call the directly observable random variable Y the measurement, and any of its realization $y_{obs} \in Y$ is the actual measurement producing the data. The nonobservable random variable X is called the unknown model. Here, $x \in X$ is a realization of X. In Bayesian theory, it is assumed that any information on the model which is available a priori can be coded into a probability density $\pi_{prior}(x)$ called the prior density. It expresses what we know about the model prior to the measurement. Further, the probability density $\pi_{noise}(\epsilon)$ describes the modeling error and the observation noise. We also introduce the likelihood function $\pi(y|x)$, which describes the relationship between the observable y and the model parameter x. It is the statistical representation of the (forward) model.

To be more specific, we illustrate the following few examples of PDF functions related to the present discussion. We do not discuss the appropriate normalization by constants and give the PDFs in terms of proportionality.

In the case where we have no a priori information on x, the $\pi_{prior}(x)$ equals a normalized constant function. If $x \in X$ is a realization of a Gaussian random field X with given mean x_{prior}, we have

$$\pi_{prior}(x) \propto \exp\left(-\frac{1}{2}(x - x_{prior})^T \, C_X^{-1} \, (x - x_{prior}) \right),$$

where C_X is the covariance matrix of X.

Concerning the modeling error and the measuring data error, for simplicity assume an additive noise such that the following model can be considered: $y = f(x) + \epsilon$. Then the theoretical model uncertainty described by a Gaussian probability density with covariance matrix C_M results in the following

$$\pi(y|x) \propto \exp\left(-\frac{1}{2}(y - f(x))^T \, C_M^{-1} \, (y - f(x)) \right).$$

On the other hand, if there is no theoretical modeling uncertainty, we have

$$\pi(y|x) \propto \delta(y - f(x)),$$

where δ is the Dirac delta. In this latter case, we can only have measurement errors and we obtain

$$\pi_{noise}(\epsilon) = \pi_{noise}(y_{obs} - f(x)).$$

Now, we state Bayes' theorem that is central to the present discussion.

Theorem 6.1 (Bayes' theorem). *Given two events A and B, we have that*

$$P(B|A)P(A) = P(A|B)P(B).$$

Assume we know the (unconditional) probabilities $P(A)$ and $P(B)$ and the conditional probability $P(A|B)$ for the effect A given the cause B. If the effect A is observed, the Bayes formula gives the probability $P(B|A)$ for B being the cause. In particular, if A and B are independent, we have

$$P(A \cup B) = P(A)P(B);$$

then $P(B|A) = P(B)$ and $P(A|B) = P(A)$ and the statement becomes obvious. This latter remark means that the conditional probabilities equal the unconditional ones, and hence the term "independent" for A and B.

In our context, in terms of probability densities, Bayes' theorem states that the prior probability density, the likelihood function, and the data can be combined to form the posterior probability density on the model parameters X as follows

$$\pi_{post}(x) := \pi(x|y_{obs}) = \pi_{prior}(x)\pi(y_{obs}|x)/\pi(y_{obs}) \propto \pi_{prior}(x)\pi(y_{obs}|x). \qquad (6.52)$$

By definition, $\pi_{post}(x)$ is the solution of the inverse problem in the statistical sense.

6.6.2 A Fast Scheme for Large-Scale Linear Inverse Problems

We consider the case that the prior probability density of X and the probability density of error E are both Gaussian. With this setting, we discuss a methodology for the fast solution of large-scale linear inverse problems presented in [127]. This methodology applies also in the case of high-dimensional parameter spaces, which arise when an infinite-dimensional parameter field is discretized.

Consider the following probability densities

$$\pi_{prior}(x) \propto \exp\left(-\frac{1}{2}(x - x_{prior})^T C_X^{-1} (x - x_{prior})\right)$$

and

$$\pi_{noise}(\epsilon) \propto \exp\left(-\frac{1}{2}(\epsilon - \epsilon_o)^T C_E^{-1} (\epsilon - \epsilon_o)\right),$$

where ϵ_o represents the mean error.

We insert these distributions in (6.52) assuming no modeling uncertainty and obtain

$$\pi_{post}(x) \propto \exp\left(-\frac{1}{2}\|x - x_{prior}\|^2_{C_X^{-1}} - \frac{1}{2}\|y_{obs} - f(x) - \epsilon_o\|^2_{C_E^{-1}}\right). \tag{6.53}$$

Notice that since $f(x)$ may be nonlinear in x, the posterior probability density may not be Gaussian. If we assume linearity of this map, $f(x) = Ax$, then $\pi_{post}(x)$ is Gaussian, with mean x_{post} corresponding to the maximum point of $\pi_{post}(x)$, or equivalently

$$x_{post} = \underset{x}{\operatorname{argmin}} \left(\frac{1}{2}\|x - x_{prior}\|^2_{C_X^{-1}} + \frac{1}{2}\|y_{obs} - Ax - \epsilon_o\|^2_{C_E^{-1}}\right). \tag{6.54}$$

We remark that this is similar to a regularized least-squares problem. It is clearly possible to see that the inverse covariance matrix of the posterior probability density of model parameters is given by the Hessian of the least-squares objectives. Therefore, we have

$$C_{post}^{-1} = A^T C_E^{-1} A + C_X^{-1}, \tag{6.55}$$

where $H_{misfit} = A^T C_E^{-1} A$ is the Hessian of the data misfit and $H_{prior} = C_X^{-1}$ is the Hessian of the prior. We denote $H_{post} = H_{misfit} + H_{prior}$.

This means that we have a complete knowledge of the solution to the inverse problem if we are able to estimate the mean x_{post} and the covariance matrix C_{post}, since then

$$\pi_{post}(x) \propto \exp\left(-\frac{1}{2}\|x - x_{post}\|^2_{C_{post}^{-1}}\right).$$

As discussed in [127], the task of determining the mean of the posterior density of the parameters can be found readily by solving (6.54) using state-of-the-art techniques from large-scale PDE-constrained optimization. However, computing the posterior covariance matrix of the parameters using (6.55) is a challenge. To solve this problem, in [127] a low-rank spectral approximation of the prior-preconditioned Hessian of the data misfit is computed at a cost that is a small multiple (independent of the parameter dimension) of the cost of solving the underlying forward PDEs. Then, with this low-rank approximation, the final expression for approximation of the posterior covariance is easily obtained. This approach is illustrated in the following. For all details see [127].

The first step in the development of this approach is to recognize that typically, H_{misfit} is spectrally similar to a discrete compact operator with the eigenvalues decaying rapidly to zero. This fact is used to construct a low-rank approximation of H_{misfit} by retaining only a certain number of the largest eigenvalues and corresponding eigenvectors as follows.

Denote with λ_i and v_i the eigenvalues and eigenvectors of the prior-preconditioned Hessian of the data misfit given by

$$\tilde{H}_{misfit} = C_X^{1/2} \left(A^T C_E^{-1} A \right) C_X^{1/2}.$$

Further, let Λ be the diagonal matrix of the eigenvalues λ_i and let V be the matrix of the corresponding normalized eigenvectors, such that $\tilde{H}_{misfit} = V \, \Lambda \, V^T$.

Now, when the λ_i decay rapidly to zero, one can retain the first r largest eigenvalues and obtain the following low-rank approximation

$$\tilde{H}_{misfit} \approx V_r \, \Lambda_r \, V_r^T,$$

where Λ_r and V_r denote the truncated eigenvalue and eigenvector matrices. Specifically, the r columns of V_r correspond to the r eigenvectors of the first r largest eigenvalues with which Λ_r is constructed.

Next, an important step in [127] is to use the Sherman–Morrison–Woodbury formula to write the following expansion

$$\left(I + \tilde{H}_{misfit} \right)^{-1} = I - V_r \, D_r \, V_r^T + \mathcal{O}\left(\sum_{i=r+1}^{n} \frac{\lambda_i}{\lambda_i + 1} \right),$$

where $D_r = \operatorname{diag}\left(\frac{\lambda_i}{\lambda_i+1}\right) \in \mathbb{R}^{r \times r}$. Notice that this expression also provides the estimate of the error due to truncation. For rapidly decaying eigenvalues the rest sum quickly approaches zero by increasing the order of truncation.

With this preparation, we are now ready to construct a low-rank approximation to C_{post}. We have

$$\begin{aligned} C_{post} &= \left(A^T C_E^{-1} A + C_X^{-1} \right)^{-1} = C_X^{1/2} \left(A^T C_E^{-1} A + I \right)^{-1} C_X^{1/2} \\ &\approx C_X^{1/2} \left(I - V_r \, D_r \, V_r^T \right) C_X^{1/2} \\ &= C_X - C_X^{1/2} \, V_r \, D_r \, V_r^T \, C_X^{1/2}. \end{aligned} \tag{6.56}$$

The importance of this formula lies in the fact that usually the choice of a small r is reasonable and independent of problem size. This is particularly true when the prior is of smoothing type.

As shown in [127] one cannot explicitly construct $\tilde{H}_{misfit}$ and apply a truncated singular value decomposition algorithm to obtain Λ_r and V_r. The viable approach is to use a matrix-free Lanczos method to find the dominant eigenvalues and corresponding eigenvectors. The application of the Lanczos scheme requires only a matrix-vector product with $\tilde{H}_{misfit}$ at each iteration. By definition of $\tilde{H}_{misfit}$, this process can be broken down in a few steps: multiplications with C_X, solving for C_E, and action of A and A^T on vectors. The last involve forward and adjoint PDE solutions, which dominate the cost of the computation when one considers a PDE forward model.

6.6.3 An Inverse Parabolic Problem

In this section, we illustrate the Bayesian framework in the case of an inverse parabolic problem. Our purpose is also to provide a clear link to the PDE-based deterministic optimization framework considered in previous sections.

Consider the following parabolic inverse problem:

$$\begin{aligned} \min_{y_0 \in L^2(\Omega)} \quad J(y, y_0) &:= \frac{\beta_{noise}}{2} \|y(\cdot, T) - y_{obs}\|^2_{L^2(\Omega)} + \frac{\beta_{prior}}{2} \|y_0\|^2_{L^2(\Omega)}, \\ -\partial_t y + \Delta y &= 0 \qquad \text{in } Q = \Omega \times (0, T), \\ y &= y_0 \qquad \text{in } \Omega \times \{t = 0\}. \end{aligned} \tag{6.57}$$

That is, find the initial configuration y_0 that represents the unknown distributed parameter x, based on the observations at the final time T that provide the data y_{obs}.

Notice that the discretized functional $J(y, y_0)$ provides the minus exponent of the noise and prior PDEs, respectively. We have

$$C_X^{-1} = \beta_{prior}\, h^2\, I \qquad \text{and} \qquad C_E^{-1} = \beta_{noise}\, h^2\, I,$$

where h is the space mesh size. Clearly, $(y_0)_{prior} = 0$.

We denote with A the discrete parabolic operator such that $y_T = A\, y_0$, where y_T is the solution of the discretized parabolic problem in (6.57) with y_0 as initial condition. Therefore, discretization of the minimization problem (6.57) results in the following posterior mean:

$$(y_0)_{post} = \operatorname*{argmin}_{y_0} \left(\frac{h^2}{2}\, \beta_{prior}\, y_0^T\, y_0 + \frac{h^2}{2}\, \beta_{noise}\, (y_{obs} - A\, y_0)^T\, (y_{obs} - A\, y_0) \right).$$

The choice of prior and noise distributions given above result in the following:

$$\tilde{H}_{misfit} = \frac{\beta_{noise}}{\beta_{prior}}\, A^T\, A,$$

where A is the forward operator and A^T is the adjoint operator. On a square domain and homogeneous Dirichlet boundary conditions, one can easily verify that the eigenvalues λ_i of $\tilde{H}_{misfit}$ are linearly proportional to the ratio $\beta_{noise}/\beta_{prior}$ and decay exponentially; see [127].

Chapter 7

Applications

7.1 Introduction

The range of applications of PDE optimization is vast and quickly widening to all fields of scientific computing. An outlook of some application problems, for which the design of algorithms with optimal computational complexity is ongoing, is the purpose of this chapter. We discuss PDE optimization problems with representative models of mathematical physics, such as the Navier–Stokes equations, the Schrödinger equation, and the Maxwell equations.

Based on the Navier–Stokes model, we discuss shape design problems where the geometry of the model becomes an optimization variable, and therefore the need arises of defining a "calculus" with respect to this variable. In particular, we focus on the modern powerful approach of nonparametric shape sensitivity analysis that allows us to consider challenging shape optimization problems.

The models described by Schrödinger-type equations are receiving much attention because they are central in nanosciences. At the nanoscale and beyond, we are in the quantum world and the Schrödinger equation becomes the governing model with a large range of applications including quantum optics, quantum chemistry, and design of nanostructures. We discuss quantum control problems where PDE optimization is used to design control potentials to optimally drive quantum systems to perform given tasks.

The Maxwell equations provide one of the most representative models related to inverse problems formulated within the PDE optimization framework. In particular, we consider noninvasive imaging which is very important in medical diagnosis and material probing. In this case, inverse scattering problems are formulated where given incident electromagnetic waves penetrate a medium and based on the scattered waves one determines the presence and the properties of objects hidden in the medium.

7.2 Aerodynamic Shape Design Supported by the Shape Calculus

7.2.1 Overview on Shape Sensitivity Analysis

Shape calculus defines the mathematical framework, when the shape of an object is the unknown quantity and should be determined in accordance with a certain (differentiable) goal. For a gradient-based optimization approach the sensitivity information of an objective function is needed with respect to the shape of the boundary on which it is defined, which is often termed "shape sensitivity analysis." While these methods have been used quite successfully to derive theoretical results for optimal shapes, such as the optimal supersonic Haak body [160] or the rugby-ball ogive shapes in a Stokes flow [276], most numerical schemes are actually based on smooth ansatz functions such as the very popular Hicks–Henne functions [189] or other curve parameterizations such as b-splines.

However, we consider the nonparametric shape sensitivity approach to numerically conduct large-scale shape optimization. In the following, let the domain under consideration be given by $\Omega \subset \mathbb{R}^3$ compact with sufficiently smooth boundary $\Gamma := \partial\Omega$. We seek to compute the derivative or sensitivity information of general objective functions such as

$$J_1(\Omega) := \int_{\Omega} f \, dA, \tag{7.1}$$

$$J_2(\Omega) := \int_{\Gamma} g \, dS \tag{7.2}$$

with respect to the domain Ω, where f and g are sufficiently smooth functions. To this end, a deformed domain

$$\Omega_t = T_t(\Omega) = \{T_t(x) : x \in \Omega\}$$

is considered. The family of bijective mappings T_t is usually chosen as the perturbation of identity

$$T_t(x) = x + tV(x)$$

or the speed method, where T_t is the solution of the flow equation

$$\frac{\partial x}{\partial t} = V(t,x), \quad x(0) = x_0 \in \Gamma.$$

Here, V is a sufficiently smooth vector field that takes the role of the differentiation direction. For the volume case (7.1), it is easy to see that

$$\begin{aligned}\int_{\Omega_t} f \, dA_t &= \int_{\Omega} f(T_t(x))\sqrt{\det DT_t^T \, DT_t(x)} \, dA(x) \\ &= \int_{\Omega} f(T_t(x)) |\det DT_t(x)| \, dA(x),\end{aligned}$$

which can easily be differentiated with respect to t, taking into account that physically meaningful vectorfields V preserve the orientation:

$$\begin{aligned} dJ_1(\Omega)[V] &= \int_\Omega \frac{d}{dt}\bigg|_{t=0} f(T_t(x))\det(DT_t(x))\, dA(x) \\ &= \int_\Omega \langle \nabla f(x), V(x)\rangle + f(x)\operatorname{div} V(x)\, dA(x) \\ &= \int_\Omega \operatorname{div}(f(x)V(x))\, dA(x). \end{aligned} \tag{7.3}$$

A similar computation for the surface objective function results in

$$\begin{aligned} dJ_2(\Omega)[V] &= \frac{d}{dt}\bigg|_{t=0} \int_{\Gamma_t} g\, dS_t \\ &= \int_\Gamma \langle \nabla g, V\rangle + g\cdot(\operatorname{div} V - \langle DVn, n\rangle)\, dS \\ &= \int_\Gamma \langle \nabla g, V\rangle + g \operatorname{div}_\Gamma V\, dS, \end{aligned} \tag{7.4}$$

where $\operatorname{div}_\Gamma$ is the tangential or surface divergence operator. Thus, the sensitivity equations (7.3) and (7.4) can be interpreted as the directional derivatives in direction V.

7.2.2 The Hadamard Formula

The Hadamard theorem or Hadamard–Zolésio structure theorem [321, 110] states that under some regularity assumptions, the sensitivity equations (7.3) and (7.4) can be expressed as a scalar product of the normal component of the perturbation field V with some shape gradient on the surface of the geometric object to be optimized, i.e.,

$$dJ(\Omega)[V] = dJ(\Gamma)[\langle V, n\rangle n].$$

For the volume case (7.3) applying the divergence theorem results in

$$dJ_1(\Omega)[V] = \int_\Gamma \langle V, n\rangle\, f\, dS.$$

The surface case, (7.4), requires elements from tangential calculus, especially the tangential Stokes formula. However, one can show that

$$dJ_2(\Omega)[V] = \int_\Gamma \langle V, n\rangle \left[\frac{\partial g}{\partial n} + \kappa g\right] dS,$$

where $\kappa := \operatorname{div} n$ is the additive mean curvature. Thus, formulas are created which can be very efficiently evaluated numerically allowing a large-scale shape optimization. The need to compute mesh sensitivity Jacobians is completely circumvented. Instead, some geometric quantities such as the curvature κ must be computed, which can be conveniently done using discrete differential geometry [292].

For the application in computational fluid dynamics and aerodynamic design, the above formula becomes more complex, because the usual objective functions such as fluid forces in addition to a dependence on the PDE fluid state also depend on the geometry and these quantities must be differentiated accordingly. For a dependence on the normal, one has

$$J(g,\Gamma) := \int_\Gamma g(\varphi, D\varphi, n)\, dS,$$

where $g : \mathbb{R}^d \times \mathbb{R}^{d\times d} \times \mathbb{R}^d \to \mathbb{R}, (\varphi, \zeta, \psi) \mapsto g(\varphi, \zeta, \psi)$ is a sufficiently smooth functional. The shape derivative of the above expression is then given by

$$\begin{aligned} dJ(g,\Gamma)[V] = \int_\Gamma \langle V, n\rangle \Big[& D_\varphi g D\varphi\, n + D_\zeta g D^2\varphi\, n \\ & + \kappa\left(g - D_\psi g\, n\right) + \mathrm{div}_\Gamma \left(D_\psi g\right)^T \Big]\, dS. \end{aligned}$$

7.2.3 Shape Optimization and the Incompressible Navier–Stokes Equations

The incompressible Navier–Stokes equations are considered in a general setting:

$$\begin{aligned} \min_{(u,p,\Omega)} J(u,p,\Omega) := & \int_\Omega f(u, Du, p)\, dA \\ & + \int_{\Gamma_0} g(u, D_n u, p, n)\, dS \end{aligned}$$

subject to

$$\begin{aligned} -\mu\Delta u + \rho u \nabla u + \nabla p = \rho G \quad & \text{in} \quad \Omega, \\ \mathrm{div}\, u = 0, & \\ u = u_+ \quad & \text{on} \quad \Gamma_+, \\ u = 0 \quad & \text{on} \quad \Gamma_0, \\ pn - \mu \frac{\partial u}{\partial n} = 0 \quad & \text{on} \quad \Gamma_-, \end{aligned}$$

where Γ_+ is the inflow boundary, Γ_- is the outflow boundary, and Γ_0 is the fluid obstacle of which the shape is to be optimized. Here, $f : \mathbb{R}^3 \times \mathbb{R}^{3\times 3} \times \mathbb{R} \to \mathbb{R}$ and $g : \mathbb{R}^3 \times \mathbb{R}^3 \times \mathbb{R} \times \mathbb{R}^3 \to \mathbb{R}$ are assumed to be continuously differentiable in each argument. In order to keep the notation readable, components of the Jacobian are denoted as follows

$$\begin{aligned} Du &=: \left[a_{ij}\right]_{ij} \in \mathbb{R}^{3\times 3}, \\ D_n u = Du \cdot n &= \frac{\partial u}{\partial n} =: [b_i]_i \in \mathbb{R}^3. \end{aligned}$$

The surface part g is chosen such that there exists a functional $\lambda : \Omega \to \mathbb{R}^d$ satisfying the following conditions on Γ_0

$$\begin{aligned} \lambda_i &= \frac{1}{\mu}\frac{\partial g}{\partial b_i}\; i = 1,2,3, \\ \langle \lambda, n\rangle &= -\frac{\partial g}{\partial p}. \end{aligned}$$

A formal differentiation of the above leads to the gradient expression [304]

$$
\begin{aligned}
&dJ(u,p,\Omega)[V] \\
&= \int_{\Gamma_0} \langle V,n\rangle f(u,Du,p)\,dS \\
&+ \int_{\Gamma_0} \langle V,n\rangle \left[D_{(u,b,p)}g(u,D_n u,p,n)\cdot n + \kappa g(u,D_n u,p,n)\right] dS \\
&+ \int_{\Gamma_0} \langle V,n\rangle \left[-\sum_{i=1}^{d}\left(\frac{\partial g}{\partial u_i} + \mu\frac{\partial \lambda_i}{\partial n} + \sum_{j=1}^{d}\frac{\partial f}{\partial a_{ij}} n_j\right)\frac{\partial u_i}{\partial n}\right] dS \\
&+ \int_{\Gamma_0} \langle V,n\rangle \left[(\mathrm{div}_\Gamma \nabla_n g) - \kappa\langle \nabla_n g, n\rangle\right] dS,
\end{aligned}
$$

where $\nabla_n g$ denotes the vector consisting of components $\frac{\partial g}{\partial n_i}$ and λ and λ_p solve the adjoint incompressible Navier–Stokes equations

$$
\begin{aligned}
-\mu\Delta\lambda_i - \rho\sum_{j=1}^{d}\left(\frac{\partial \lambda_j}{\partial x_i}u_j + \frac{\partial \lambda_i}{\partial x_j}u_j\right) - \frac{\partial \lambda_p}{\partial x_i}& \\
= \frac{\partial f}{\partial u_i} - \sum_{j=1}^{d}\frac{\partial}{\partial x_j}\frac{\partial f}{\partial a_{ij}} &\quad \text{in } \Omega, \\
\mathrm{div}\,\lambda = \frac{\partial f}{\partial p}&
\end{aligned}
$$

with boundary conditions

$$
\begin{aligned}
\lambda &= 0 && \text{on } \Gamma_+, \\
\lambda_i &= \frac{1}{\mu}\frac{\partial g}{\partial b_i} && \text{on } \Gamma_0, \\
\langle \lambda, n\rangle &= -\frac{\partial g}{\partial p} && \text{on } \Gamma_0, \\
\mu\frac{\partial \lambda_i}{\partial n} + \rho\left(\sum_{j=1}^{d}\lambda_j u_j n_i + \lambda_i u_j n_j\right) + \lambda_p n_i &= 0 && \text{on } \Gamma_-.
\end{aligned}
$$

7.2.4 Shape Hessian Approximation and Operator Symbols

Very efficient optimization methods are usually variants of the SQP or Newton's method, which results in the need of higher-order derivative information. However, shape Hessians are very complex objects and usually need to be approximated in some way. The above Navier–Stokes problem can be made into an excellent problem for studying shape Hessians. By choosing the constant density $\rho = 0$, the nonlinearity vanishes. Coupled with the minimization of the kinetic energy dissipation, one arrives at the following problem, where the fluid is modeled by the Stokes equations

$$\min_{(u,p,\Omega)} J(u,p,\Omega) := \int_\Omega \mu \sum_{i,j=1}^{3} \left(\frac{\partial u_i}{\partial x_j}\right)^2 dA$$

subject to

$$\begin{aligned} -\mu\Delta u + \nabla p &= 0 && \text{in} \quad \Omega, \\ \operatorname{div} u &= 0, && \\ u &= u_+ && \text{on} \quad \Gamma_+, \\ u &= 0 && \text{on} \quad \Gamma_0, \\ pn - \mu\frac{\partial u}{\partial n} &= 0 && \text{on} \quad \Gamma_-, \\ \mathrm{Vol} &= V_0. && \end{aligned}$$

The shape derivative of this problem is given by

$$dJ(u,p,\Omega)[V] = -\mu \int_{\Gamma_0} \langle V, n\rangle \sum_{k=1}^{3} \left(\frac{\partial u_k}{\partial n}\right)^2 dS, \tag{7.5}$$

which means the problem is self-adjoint and the Hessian can be studied without the need to consider variations in the adjoint multiplicators λ.

Shape gradients can be split in a geometric part, which stems from variations of the objective with respect to the geometry, and a dynamic part, which stems from a variation of the PDE state (u,p) due to a change of the domain on which the PDE is defined. The idea presented here from [305, 302] is to study the symbol of the control to state mapping of the dynamic part only. Considering a sinusoidal perturbation $\tilde{q}(x) = \hat{q}e^{i\omega x}$ of some control q, the pseudodifferential operator nature of the Hessian H can be seen by comparing the input $\tilde{q}$ with the output $H\tilde{q}$. For example, if

$$H\tilde{q} = i\omega\hat{q}e^{i\omega x} = i\omega\tilde{q},$$

then $\Sigma(\omega) := i\omega$ is the symbol of the Hessian, and this corresponds to a classical differential operator of order +1. If, for example, one has

$$H\tilde{q} = -\omega^2\hat{q}e^{i\omega x} = -\omega^2\tilde{q},$$

then $\Sigma(\omega) := -\omega^2$ is the symbol of the Hessian, and this corresponds to a classical differential operator of order +2. However, if

$$H\tilde{q} = |\omega|\hat{q}e^{i\omega x} = |\omega|\tilde{q},$$

then H is a pseudodifferential operator of order +1.

Assuming flow over a flat plate, the domain Ω is considered to be given by

$$\Omega = \{(x_1,x_2,x_3) \in \mathbb{R}^3 : x_3 \geq 0\}.$$

Thus, at the boundary $\Gamma = \{(x_1,x_2,x_3) \in \mathbb{R}^3 : x_3 = 0\}$, the outer normal is given by

$$n = (0,0,1)^T,$$

and a complex-valued oscillation of the two-dimensional flat plate is described by

$$x_3 = \alpha(x_1, x_2) := e^{i(\omega_1 x_1 + \omega_2 x_2)},$$

where i is the imaginary unit. Using this setting, the disturbed gradient from (7.5) is given by

$$\begin{aligned}\tilde{G} &:= -2\mu \sum_{k=1}^{3} \frac{\partial u_k}{\partial n} \frac{\partial u_k'[\alpha]}{\partial n} \\ &= -2\mu \sum_{k=1}^{3} \frac{\partial u_k}{\partial x_3} \frac{\partial u_k'[\alpha]}{\partial x_3}\end{aligned}$$

due to the flat initial domain under consideration. To identify the symbol of the Hessian, the mapping

$$S\alpha := -2\mu \sum_{k=1}^{3} \frac{\partial u_k}{\partial x_3} \frac{\partial u_k'[\alpha]}{\partial x_3}$$

must now be characterized by making α appear explicitly in the right-hand side.

Furthermore, the perturbed states $u'[\alpha]$ and $p'[\alpha]$ are also considered to be oscillatory:

$$\begin{aligned}u_k'[\alpha] &= \hat{u}_k e^{i(\omega_1 x_1 + \omega_2 x_2)} e^{\omega_3 x_3}, \\ p'[\alpha] &= \hat{p} e^{i(\omega_1 x_1 + \omega_2 x_2)} e^{\omega_3 x_3}.\end{aligned}$$

The no-slip boundary condition on $x_3 = 0$ leads to

$$u_k'[\alpha] = \hat{u}_k \alpha e^0,$$

which results in

$$\hat{u}_k = -\frac{\partial u_k}{\partial x_3} \neq 0. \tag{7.6}$$

However, the linearized Stokes PDE must also be solved inside the domain. Applying the Laplace, the gradient, and the divergence operators on the disturbances $u'[\alpha]$ and $p'[\alpha]$ is equivalent to

$$A \begin{pmatrix} \hat{u}_1 \\ \hat{u}_2 \\ \hat{u}_3 \\ \hat{p} \end{pmatrix} \alpha(x_1, x_2) e^{\omega_3 x_3} = 0,$$

where the matrix A can be thought of as the Stokes equation in the Fourier space. This is noncontradictory to the consequences of the boundary condition (7.6) only if the linearized system matrix A does not have full rank, which means the determinant of A must vanish:

$$\det(A) = \left[\mu\left(\omega_1^2+\omega_2^2-\omega_3^2\right)\right]^2\left(\omega_1^2+\omega_2^2-\omega_3^2\right) \stackrel{!}{=} 0$$
$$\Rightarrow \omega_3 = \pm\sqrt{\omega_1^2+\omega_2^2}.$$

Hence, it is possible to remove ω_3 from the equations and the local shape derivative of the velocity is given by

$$u_k'[\alpha] = -\frac{\partial u_k}{\partial x_3}\alpha e^{\pm\sqrt{\omega_1^2+\omega_2^2}x_3}$$
$$\Rightarrow \left.\frac{\partial u_k'[\alpha]}{\partial x_3}\right|_{x_3=0} = \left[-\frac{\partial^2 u_k}{\partial x_3^2} \mp \frac{\partial u_k}{\partial x_3}\sqrt{\omega_1^2+\omega_2^2}\right]\alpha,$$

and the operator S is given by

$$S = -2\mu\sum_{k=1}^{3}\frac{\partial u_k}{\partial x_3}\left[-\frac{\partial^2 u_k}{\partial x_3^2} \pm \frac{\partial u_k}{\partial x_3}\sqrt{\omega_1^2+\omega_2^2}\right]. \tag{7.7}$$

In two dimensions, one can assume $\omega_2 \equiv 0$ and the symbol becomes

$$\pm|\omega_1|,$$

which is the symbol of a pseudodifferential operator of order +1 closely related to the Dirichlet-to-Neumann map. For a true pseudodifferential operator with symbol $|\omega|$, a discretization using finite differences or surface finite elements is not straightforward. A symbol which can much more easily be approximated is ω^2, and this corresponds to the tangential Laplace operator, also known as Laplace–Beltrami operator Δ_Γ. Thus, the discrete Hessian is approximated by

$$H_h \approx k\Delta_\Gamma^h + I, \tag{7.8}$$

where $I \in \mathbb{R}^{m\times m}$ is the identity matrix for m surface mesh nodes and k is an appropriately chosen smoothing parameter.

7.2.5 Aerodynamic Design Using Shape Calculus

We conclude this section with the application of the above to a large-scale aerodynamic design problem [306, 307]. The single set-point shape optimization of the very efficient large aircraft (VELA) blended wing-body aircraft is discussed. Rather than using the full Navier–Stokes equations, we employ the compressible Euler equations as a simplified model for the inviscid fluid and the shape is to be optimized such that the pressure or wave drag in a transonic cruise of Mach 0.85 is minimized while the lift is kept constant. Thus, the optimization problem reads as

$$
\begin{aligned}
\min_{(\rho,u,E,\Omega)} \; & C_D(\rho,u,E,\Omega) \\
\text{subject to} \quad & \int_\Omega \operatorname{div}(\rho u)\, dA = 0, \\
& \int_\Omega \sum_{j=1}^{3} \left[\frac{\partial(\rho u_i u_j)}{\partial x_j} \right] + \frac{\partial p}{\partial x_i}\, dA = 0, \\
& \int_\Omega \operatorname{div}(\rho H u)\, dA = 0, \\
& \langle u, n\rangle = 0 \quad \text{on} \quad \Gamma_0, \\
& C_L = C_{L_0}, \\
& \mathrm{Vol} = V_0.
\end{aligned}
$$

Here, the first three constraints are the Euler equations, and the pressure p is linked to the conserved variables

$$U := (\rho, \rho u, \rho E)^T$$

by the perfect gas law

$$p = (\gamma - 1)\rho \left(E - \frac{1}{2}\|u\|^2 \right)$$

and H is the fluid enthalpy. The condition $\langle u, n\rangle = 0$ is the Euler slip boundary condition on the aircraft surface Γ. The farfield boundary conditions are usually more complex, involving characteristics, and are thought of as treated by the flow solver discretely.

Since the gradient-based optimization method needs sensitivity information for both the objective function and the lift constraint, the generalized inviscid fluid forces

$$J(U,\Omega) = \int_{\Gamma_0} \langle p \cdot a, n\rangle \; dS \tag{7.9}$$

are considered, where a is a constant vector describing the rotation of the coordinate system, such that drag, lift, and the appropriate angle of attack are considered simultaneously. The shape derivative of (7.9) is given by [303]

$$
\begin{aligned}
& dJ(U,\Omega)[V] \\
&= \int_{\Gamma_0} \langle V, n\rangle \left[\frac{\partial(p \cdot a)}{\partial n} n - \lambda U_H \left\langle \frac{\partial u}{\partial n}, n \right\rangle \right. \\
&\qquad\qquad \left. + \operatorname{div}_\Gamma (p \cdot a - \lambda U_H u) \right] dS \qquad (7.10) \\
&= \int_{\Gamma_0} \langle V, n\rangle \left[\frac{\partial(p \cdot a)}{\partial n} n - \lambda U_H \left\langle \frac{\partial u}{\partial n}, n \right\rangle + \kappa \langle p \cdot a, n\rangle \right] \\
&\qquad\qquad + \langle p \cdot a - \lambda U_H u, dn[V]\rangle \; dS. \qquad (7.11)
\end{aligned}
$$

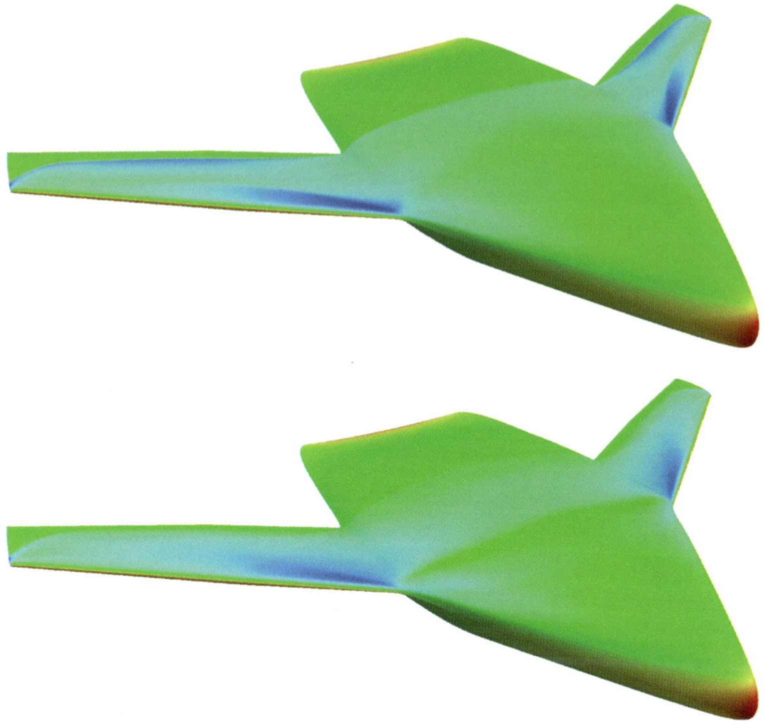

Figure 7.1. *Initial and optimized VELA aircraft. Color denotes pressure. Reprinted with permission from C. Schillings, S. Schmidt, and V. Schulz, Efficient shape optimization for certain and uncertain aerodynamic design, Comput. & Fluids,* 46(1) (2011), 78–87.

Additionally, U_H is the vector of conserved variables with the last component replaced by ρH. The adjoint variables λ are given as the solution of [146, 136]

$$-\frac{\partial}{\partial x_1}(A_1^T\lambda) - \frac{\partial}{\partial x_2}(A_2^T\lambda) - \frac{\partial}{\partial x_3}(A_3^T\lambda) = 0 \qquad \text{in} \quad \Omega,$$

$$\langle(\lambda_2,\lambda_3,\lambda_4)^T, n\rangle = -\langle a, n\rangle \quad \text{on} \quad \Gamma_0,$$

where the A_i are the Euler flux Jacobians.

The drag minimization problem with constant lift is solved by the application of the one-shot method presented in Section 4.6, where the reduced Hessian approximation B is chosen according to (7.8). The VELA mesh consists of 115,673 surface nodes and 5,859,435 tetrahedra in the volume. Thus, there are approximately $29.3 \cdot 10^6$ unknowns for the fluid, which are computed using the DLR flow solver TAU [131, 140, 317]. The planform of the aircraft is kept constant, such that 113,956 surface nodes of the CFD mesh are the design unknowns for the shape. At cruise condition of Mach 0.85 and 1.8° angle of attack, the initial drag value is $C_D = 4.77 \cdot 10^{-3}$ and the initial lift coefficient is $C_L = 1.787 \cdot 10^{-1}$. The initial and optimized shapes are shown in Figure 7.1. The upper-side transonic shock wave is completely removed, such that the optimized aircraft features a drag value

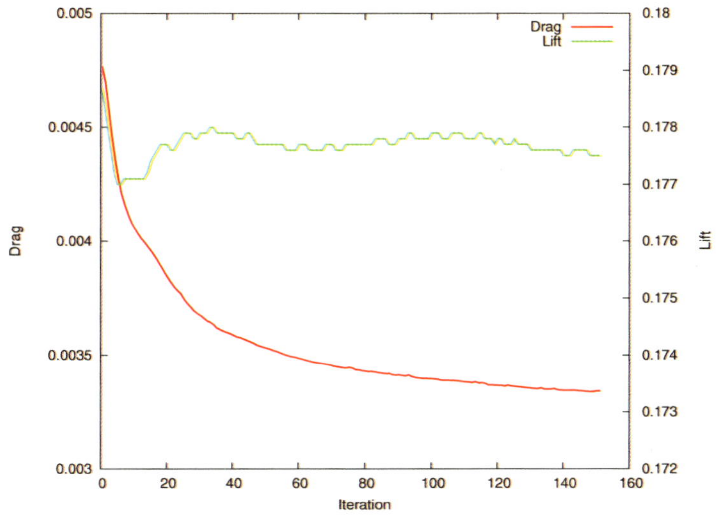

Figure 7.2. *Optimization history of the single set-point VELA aircraft. Note that the lift values are stored with four digits only. Reprinted with permission from C. Schillings, S. Schmidt, and V. Schulz, Efficient shape optimization for certain and uncertain aerodynamic design, Comput. & Fluids,* 46(1) (2011), 78–87.

of $C_D = 3.342 \cdot 10^{-3}$, an improvement by 30.06%, and the lift with $C_L = 1.775 \cdot 10^{-1}$ is kept within 0.67% of the original value. The optimization history is shown in Figure 7.2.

7.3 Quantum Control Problems

7.3.1 Introduction to Quantum Control Problems

Control of quantum phenomena is becoming central in a variety of systems with present and perspective applications ranging from quantum optics and quantum chemistry to semiconductor nanostructures, including the emerging fields of quantum computation and quantum communication; see, e.g., [16, 95, 100, 108, 198, 283, 282, 360] and the report of the AMO 2010 study undertaken by the U.S. National Research Council [5]. In most cases, quantum control is aiming at quantum devices where there is the need to manipulate quantum states with the highest possible precision. The successful application of optimal control theory in this field together with the enormous effort towards nanosciences explains the large growing interest of the mathematical and physical communities towards quantum control problems.

In the realm of quantum control problems, we can roughly identify classes of problems ranging from the realization of laser pulses to break a bond in a molecule or to drive a certain chemical reaction [360] to the control of photochemical processes [329] and the control of quantum dots [198]. More generally, control may be required to drive state transitions, maximize observable expectation, and obtain best performance of quantum operators.

In most of the envisioned applications, it is important to define fast control mechanisms that cannot be constructed based on perturbation theory strategies or on a priori parameterized control fields. This fact has motivated the increasing interest in the extension of the optimal control theory framework to the quantum world, resulting in many recent successful results [54, 63, 66, 70, 241, 243, 242, 356]. A pioneering work in this field was done by Peirce, Dahleh, and Rabitz [273], who investigated the optimal dipole control of a diatomic molecule represented by a Morse potential. The focus in this and in other early papers (see the references in [95, 100]) was to validate the ability of the optimal control framework to provide suitable quantum control mechanisms.

However, computational difficulties arising from solving optimality systems with Schrödinger-type equations, the bilinear control structure of the quantum control mechanism, and the resulting nonconvexity of the optimization problem led to research focusing more on finite-dimensional Schrödinger equations; see, e.g., [63, 70, 243]. In this case, the computational schemes of choice have been the monotonic iterative scheme [243, 283] and accelerated versions of the gradient scheme [63, 66]. These schemes perform well for finite-level quantum systems and provide acceptable results [242, 354] when applied to infinite-dimensional systems. More recently, second-order Newton schemes have been proposed [355, 356] by carefully taking into account the variational formulation of quantum optimization problems in complex Hilbert spaces and constructing accurate and stable discretization schemes with time-varying potentials. The development of the MGOPT method to solve infinite-dimensional quantum control problems is discussed in [54].

7.3.2 Finite-Level Quantum Systems

Quantum control problems that are genuinely finite dimensional are problems of controlling spin systems, and in molecular dynamics there are infinite-dimensional systems in interaction with external fields that can be well described by finite-dimensional models.

We show that a finite-level quantum system is modeled by a Schrödinger equation for an n-component wave function $\psi : [0,T] \to \mathbb{C}^n$ as follows

$$i\dot{\psi}(t) = H(u(t))\psi(t), \qquad \psi(0) = \psi_0, \tag{7.12}$$

for $t \in (0,T]$, and $T > 0$ is a given terminal time. The Hamiltonian matrix $H \in \mathbb{C}^{n\times n}$ depends on the external control field $u : [0,T] \to \mathbb{C}$ and $\psi_0 \in \mathbb{C}^n$ is a given initial condition with $|\psi_0|^2 = 1$. The Hamiltonian $H = H_0 + H_c(u)$ has two constitutive components: the constant free Hamiltonian $H_0 \in \mathbb{C}^{n\times n}$ describing the unperturbed (uncontrolled) system and the control Hamiltonian $H_c \in \mathbb{C}^{n\times n}$ modeling the coupling of the quantum state to an external control field, typically a laser (electromagnetic) field. In particular, a dipole control mechanism can be written as $H_c(u) = H_d\, u$, where H_d is a Hermitian dipole matrix. Less investigated is the case of dipole and polarization control terms, $H_c(u) = H_d\, u + H_p\, u^2$; see, e.g., [106].

The choice of T is a modeling issue motivated by physical considerations. Based on the quantum indeterminacy principle $\Delta\lambda\,\Delta t \geq \hbar$ (where $\hbar$ is the Planck constant that we set equal to one) we can state the following: For small T a highly energetic ($\Delta\lambda \gg 1$) optimal control results, thus involving many energy levels. On the other hand, for large T additional decoherence channels become important, which should be avoided in order to have an efficient control process; see, e.g., [198].

Strictly speaking, the wavefunction description given in (7.12) is appropriate for an isolated quantum system and in that case the governing Hamiltonian H is Hermitian. For a more realistic nonisolated system with environment couplings and subject to control, a more general density-matrix description would be required. Alternatively, a non-Hermitian Hamiltonian as in [66] can be constructed to account for environment losses.

Consider the Hermitian case that is very important in applications and that is also the most studied. Notice that being H Hermitian the evolution of ψ is unitary. This means that (7.12) defines an evolution in the unit sphere of $\mathbb{C}^n$. Further, we assume that the spectrum of H_0 is nondegenerate, with real eigenvalues $\lambda_1, \ldots, \lambda_n$, called energy levels, and normalized eigenvectors $\phi_1, \ldots, \phi_n$, called eigenstates, representing an orthonormal basis.

Once the eigenvalues and eigenvectors are known, the time evolution of a free state is computed as follows. If $\psi_0 = \sum_{j=1}^n c_0^j \phi_j$, then

$$\psi(t) = \sum_{j=1}^{n} c^j(t)\phi_j \qquad \text{with} \qquad c^j(t) = c_0^j \exp(-\lambda_j t),$$

where $|\psi(t)|^2 = \sum_{j=1}^n |c^j(t)|^2 = 1$. In this framework, $|c^j(t)|^2$ is the probability that making a measurement of energy of the system at time t, we obtain λ_j as its value.

Now, denote with $\langle \cdot, \cdot \rangle$ the complex scalar product in $\mathbb{C}^n$ and assume the system at time $t = 0$ is in an ϕ_i-eigenstate in the sense that $|\langle \psi(0), \phi_i \rangle|^2 = 1$. A typical state transition problem is to design control functions $u_1, \ldots, u_m$ such that at time T the system is in another prescribed ϕ_j-eigenstate such that $|\langle \psi(T), \phi_j \rangle|^2 = 1$. Here, we have $H_c(u) = \sum_{k=1}^m H_k u_k$. Notice that because of the complex absolute function $|\cdot|$, the value of $|\langle \psi, \phi_j \rangle|^2 = 1$ is invariant under a phase shift $\psi \to \psi\, e^{i\theta}$.

For finite-level quantum systems with n energy levels, the state space is the unit sphere $S^{2n-1} \subset \mathbb{C}^n$. These problems are right-invariant control problems on the Lie group of special unitary matrices $SU(n)$, and the quantum controllability problem of proving that for every couple of points in $SU(n)$ one can find controls steering the system from one point to the other is well understood. Indeed the system is controllable if and only if [108, 284] we have

$$\mathrm{Lie}\{i H_0, i H_1, \ldots, i H_m\} = su(n),$$

where $su(n)$ denotes the subalgebra of zero trace matrices in $u(n)$, the Lie algebra of skew-Hermitian $n \times n$ matrices considered as a Lie algebra over the real field.

However, verifying the above condition is not always straightforward, and alternative necessary and sufficient conditions for controllability have been formulated; see, e.g., [283, 301]. On the other hand, much less is known on the controllability of infinite-dimensional (infinite-level) quantum systems; see [20, 27, 103, 342] and the references given therein.

In the following, we focus on the problem of determining an optimal control field $u \in L^2(0, T; \mathbb{C})$, such that (7.12) is fulfilled and a number of optimality criteria are met. The most frequent objectives encountered in quantum control problems require that the control sequence drive the system at a time T close to a desired target configuration $\psi_d \in \mathbb{C}^n$. In addition, limited laser resources and smoothness of the shape of the laser field are accounted for through a minimization of appropriate norms of the control function.

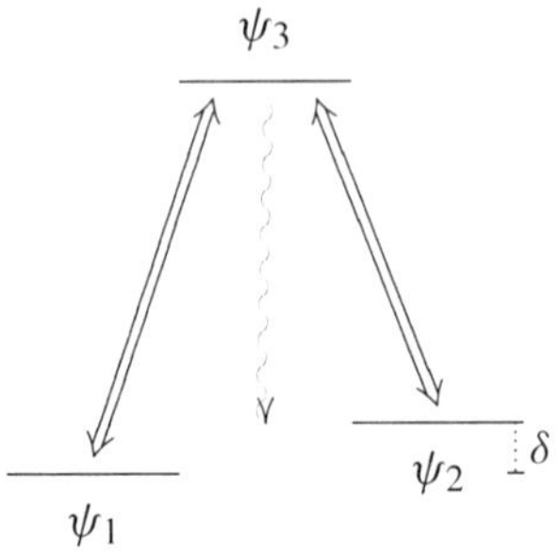

Figure 7.3. *Prototypical Λ-type three-level scheme: ψ_1 and ψ_2 are long-lived states, whereas ψ_3 is a short-lived state which is optically coupled to both ψ_1 and ψ_2; wiggled line indicates relaxation and dephasing of state ψ_3. Reprinted with permission from A. Borzì, J. Salomon, and S. Volkwein, Formulation and numerical solution of finite-level quantum optimal control problems, J. Comput. Appl. Math.,* 216(1) (2008), 170–197.

Further, one can require suppressing population of intermediate states which suffer strong environment losses. All these requirements are realized in a cost functional as

$$\begin{aligned} J(\psi,u) &= \frac{1}{2}\,|\psi(T)-\psi_d|^2_{\mathbb{C}^n} + \frac{\gamma}{2}\,\|u\|^2_{L^2(0,T;\mathbb{C})} + \frac{\mu}{2}\,\|\dot{u}\|^2_{L^2(0,T;\mathbb{C})} \\ &\quad + \frac{1}{2}\sum_{j\in I}\alpha_j\,\|\psi_j\|^2_{L^2(0,T,\mathbb{C})}, \end{aligned} \tag{7.13}$$

where the constants $\gamma, \mu \geq 0$, $\gamma+\mu > 0$, are regularization parameters, which allow us to vary the relative importance of the objectives represented by the various terms. The goal of the first term of the cost functional is to track the state ψ close to a given terminal state at $t = T$. The second and third terms are for the regularization of the problem so that existence of at least one optimal control is ensured. In the last term of (7.13), which penalizes the occupation of certain states ψ_j, the set $I \subset \{1,\dots,n\}$ denotes a subset of possible state indices and $\alpha_j \geq 0$, $j \in I$, are weighting factors.

In [63, 66] a quantum control problem with (7.12)–(7.13) is considered. The optimality system is discretized by implicit second-order CN schemes and solved using cascadic NCG schemes, as discussed in Chapter 4, and monotonic schemes. Numerical results show that this optimization problem has very flat minima and the resulting optimal control is sensitive to the tolerance on the norm of the reduced gradient. In the following, we give details of this application.

Consider a three-level quantum system whose configuration is represented by $\psi = (\psi_1,\psi_2,\psi_3) \in H^1(0,T;\mathbb{C}^3)$ which consists of two long-lived states ψ_1 and ψ_2, which are energetically separated by some amount δ, and a state ψ_3, which has a finite lifetime because of environment coupling (wiggled line); see Figure 7.3. Such Λ-type configurations have a long-standing history in quantum optics, and more recently, similar configurations have received increasing attention also in semiconductor quantum dots; see, e.g., [198] and the references given therein.

Table 7.1. *Optimization results depending on optimization parameters,* $tol_{abs} = 10^{-7}$.

γ	μ	α	$\lvert\psi(T) - \psi_d\rvert_{\mathbb{C}^3}$	J
10^{-7}	10^{-7}	0.05	$8.6 \cdot 10^{-4}$	$2.37 \cdot 10^{-3}$
10^{-7}	10^{-9}	0.05	$3.7 \cdot 10^{-4}$	$5.46 \cdot 10^{-4}$
10^{-7}	0	0.05	$6.9 \cdot 10^{-5}$	$1.41 \cdot 10^{-4}$
10^{-7}	0	0	$1.2 \cdot 10^{-3}$	$2.33 \cdot 10^{-6}$
10^{-4}	10^{-4}	0.05	$3.3 \cdot 10^{-2}$	$6.52 \cdot 10^{-2}$
10^{-4}	10^{-6}	0.05	$4.4 \cdot 10^{-3}$	$9.03 \cdot 10^{-3}$
10^{-4}	0	0.05	$2.7 \cdot 10^{-3}$	$5.68 \cdot 10^{-3}$
10^{-4}	0	0	$8.3 \cdot 10^{-3}$	$3.34 \cdot 10^{-4}$

Time evolution of this finite-level quantum system is governed by the following non-Hermitian Hamiltonian

$$H_0 = \frac{1}{2}\begin{pmatrix} -\delta & 0 & 0 \\ 0 & \delta & 0 \\ 0 & 0 & -i\gamma_0 \end{pmatrix}, \tag{7.14}$$

where the term $-i\gamma_0$ accounts for environmental losses (e.g., spontaneous photon emissions). The external control field is assumed complex, $u(t) \in \mathbb{C}$, which corresponds to two real control fields. The coupling of this field to the system is modeled as follows

$$H_c(u) = -\frac{1}{2}\begin{pmatrix} 0 & 0 & \mu_1 u \\ 0 & 0 & \mu_2 u \\ \mu_1 u^* & \mu_2 u^* & 0 \end{pmatrix}, \tag{7.15}$$

where μ_1 and μ_2 describe the coupling strengths of states ψ_1 and ψ_2 to the interconnecting state ψ_3 (e.g., optical dipole matrix elements). Typical initial and final states are given by

$$\psi_0 = \begin{pmatrix} 1 \\ 0 \\ 0 \end{pmatrix} \quad \text{and} \quad \psi_d = \begin{pmatrix} 0 \\ 1 \\ 0 \end{pmatrix},$$

respectively.

Now, we discuss the effect of different choices of values of the optimization parameters using the results reported in Table 7.1 and Figure 7.4. These results are obtained with a cascadic NCG which appears to be the most robust and efficient scheme for this class of problems.

In Table 7.1, we see that smaller values of $|\psi(T) - \psi_d|_{\mathbb{C}^3}$ are attained for smaller γ. We remark that $\gamma = 10^{-7}$ is quite small, and that makes the problem quite stiff and ill-conditioned. The NCG algorithm appears to be robust with respect to changes of γ. We also can see the effect of the regularization parameter μ. As μ increases, $|\psi(T) - \psi_d|_{\mathbb{C}^3}$ increases, demonstrating that the additional smoothness of the control function (slightly) reduces the capability of tracking. Apparently, larger μ makes the problem behaving better, resulting in less computational effort. Concerning the parameter $\alpha_3 = \alpha$ we obtain better

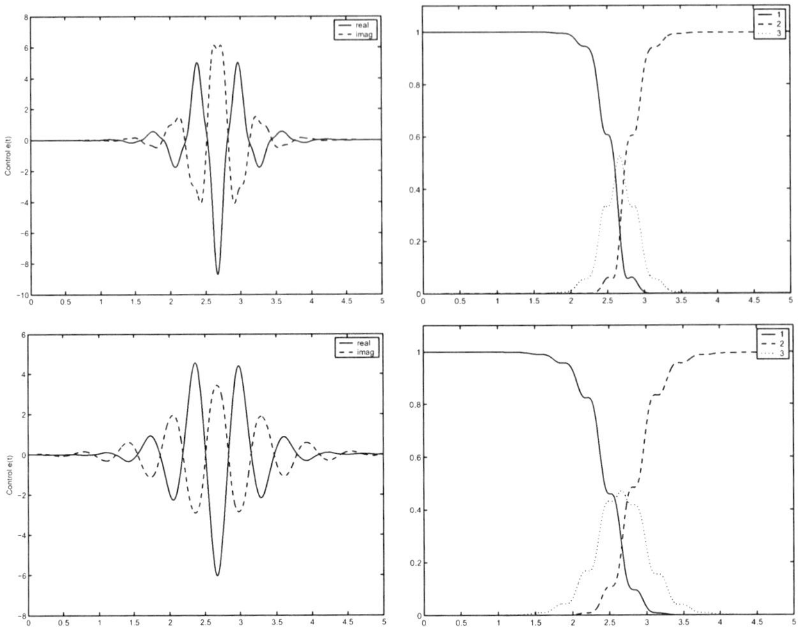

Figure 7.4. *Optimal solutions for* $\gamma = 10^{-4}$ *and* $\mu = 0$ *(top) and* $\mu = 10^{-6}$ *(bottom):* $\alpha_3 = 0.01$. *Left: the control field; right: the wave function. Reprinted with permission from A. Borzì, J. Salomon, and S. Volkwein, Formulation and numerical solution of finite-level quantum optimal control problems, J. Comput. Appl. Math.*, 216(1) (2008), 170–197.

tracking for nonzero α. This is expected since we have $|\psi_d|_{\mathbb{C}^3} = 1$, whereas $|\psi(T)|_{\mathbb{C}^3} < 1$ whenever $\gamma_0 > 0$, because of dissipation. By taking $\alpha > 0$ dissipation is reduced and therefore better $|\psi(T) - \psi_d|_{\mathbb{C}^3}$ is possibly achieved.

7.3.3 Infinite-Dimensional Quantum Systems

Optimization with infinite-dimensional quantum systems is one of the most challenging research fields in computational sciences. In fact, the complexity of multiparticle quantum systems and their foreseen applications outmatch our present computational capabilities [294] and our understanding of physical processes at nanoscales [5].

A fundamental model to start with the investigation of optimization problems of infinite dimensional quantum systems is the control problem of one-particle state transition in a confining potential. The term infinite dimensional refers to the fact that the constraint is a PDE and the free Hamiltonian is unbounded. We have $H_0 = -\Delta + V_0$, where the minus Laplacian represents the kinetic energy operator of the particle and V_0 the confining potential.

In quantum mechanics, the quantum state of a particle is described by a wavefunction $\psi : \Omega \times (0,T) \to \mathbb{C}$ that is governed by the following time-dependent Schrödinger equation

(TDSE) [101]

$$i\,\partial_t \psi(x,t) = \{-\Delta + V(x,t)\}\,\psi(x,t), \tag{7.16}$$

where we choose the scaling of the Planck constant $\hbar = 1$ and the mass $m = 1/2$. We consider the potential $V(x,t)$ consisting of a stationary part $V_0(x)$ and a time varying control part.

Notice that quantum systems are defined in unbounded spaces while the presence of a confinement potential results in wavefunctions ψ whose support is localized in a bounded region. Therefore, with Ω we represent a spatial domain that is large enough to represent ψ during evolution. In this case, periodic or homogeneous boundary conditions may be appropriate; see [8] for more sophisticated boundary conditions to solve numerically the TDSE on unbounded domains.

The next representative infinite-dimensional quantum system discussed in this section is the Bose–Einstein condensate (BEC) model that describes the state of matter formed by a cloud of bosons (e.g., helium-4, rubidium) cooled to temperatures very near to absolute zero. Under such conditions, the atoms with magnetic spin collapse into the lowest quantum state sharing the same wavefunction. This wavefunction describes the mean-field dynamics of a coherent BEC. It is modeled by the following Gross–Pitaevskii equation [71, 109]

$$i\,\partial_t\,\psi(x,t) = \left\{-\Delta + V(x,u(t)) + g\,|\psi(x,t)|^2\right\}\psi(x,t), \tag{7.17}$$

where $x \in \Omega$ and $t \in (0,T)$, with g a coupling constant related to the scattering length of the atoms, density, and transversal confinement. In (7.17), the confining potential $V(x,u(t))$ is produced by magnetic microtraps whose variation is described by a control function $u : (0,T) \to \mathbb{R}$.

Among models with higher complexity, we find multiparticle Schrödinger-type equations that allow us to describe quantum systems like molecules. In particular, we mention the time-dependent Kohn–Sham equation [126, 223] and the n-particle Schrödinger equation; see, e.g., [368] and the references given therein. For these two classes of systems, optimization and control are topics of future research efforts.

In the following, we discuss in detail the solution of optimal control problems governed by (7.16) and (7.17). By doing this, we address some numerical optimization issues which arise in quantum control problems. In particular, we encounter the formulation of optimization operators in complex Hilbert spaces. We start discussing a control problem with (7.16) and a dipole control mechanism.

A Dipole Quantum Control Problem

In the quantum mechanical framework, a dynamically stable system like an atom or a molecule exists in the presence of a stationary confining potential with a "well" envelope [101]. The states of this system are defined based on the following eigenproblem

$$\left\{-\Delta + V_0(x) - \lambda_j\right\}\phi_j(x) = 0, \quad j = 1,2,\ldots. \tag{7.18}$$

The eigenfunctions ϕ_j represent the stationary states and the eigenvalues λ_j represent the corresponding energy. Like in the finite-dimensional case, the time evolution of these states is formally given by $\psi_j(x,t) = \phi_j(x)\exp(-i\lambda_j t)$. A representative stationary potential with various applications in semiconductor nanostructures, which is meaningful in one spatial dimension, with $\Omega = (-\ell/2, \ell/2)$, is the following infinite barrier well potential

$$V_0(x) = 0 \quad \text{for } x \in \Omega \quad \text{and} \quad V_0(\pm\ell/2) = +\infty.$$

The infinite barrier condition is equivalent to homogeneous Dirichlet boundary conditions for the wavefunction, and thus we have $\lambda_j = \frac{j^2\pi^2}{\ell^2}$ and $\phi_j(x) = \cos(j\pi x/\ell)$.

To discuss some mathematical properties of the present framework, we give some definitions. We write $Re(z)$ and $\Im m(z)$ for the real and imaginary part of a complex $z \in \mathbb{C}$. Moreover, z^* stands for the complex conjugate of z and $|z|_{\mathbb{C}} = \sqrt{z^* z}$ for its absolute value. We define $\mathcal{H} = L^2(\Omega;\mathbb{C})$, the Hilbert space endowed with the inner product

$$(\varphi, \psi)_{\mathcal{H}} = \int_\Omega \varphi(x)^* \psi(x)\, dx \qquad \text{for } \varphi, \psi \in \mathcal{H}$$

and the induced norm $\|\varphi\|_{\mathcal{H}}$ for $\varphi \in \mathcal{H}$. The Hilbert space $\mathcal{V} = H_0^1(\Omega;\mathbb{C})$ is given by

$$\mathcal{V} = \left\{ \varphi \in \mathcal{H} \,\middle|\, \|\varphi\|_{\mathcal{V}} = \left(\int_\Omega |\nabla\varphi(x)|_{\mathbb{C}}^2\, dx \right)^{1/2} < \infty, \qquad \varphi = 0 \text{ on } \partial\Omega \right\},$$

supplied with the inner product $(\varphi, \psi)_{\mathcal{V}} = (\nabla\varphi, \nabla\psi)_{\mathcal{H}}$ for $\varphi, \psi \in \mathcal{V}$ and the induced norm $\|\varphi\|_{\mathcal{V}}$. We also need the Hilbert space

$$\mathcal{W} = L^2(0,T; H_0^1(\Omega;\mathbb{C}) \cap H^2(\Omega;\mathbb{C})) \cap H^1(0,T; L^2(\Omega;\mathbb{C})).$$

For more details on the above Lebesgue and Sobolev spaces see, e.g., [124].

Now, consider the TDSE with an initial state of the quantum system given by $\psi_0(x) \in \mathcal{V}$ at $t = 0$. One recognizes that the Schrödinger evolution operator is time-reversible (nondissipative) and therefore ψ cannot have better regularity than ψ_0 [71]. We also see that with a time-varying potential there is no energy conservation. In fact, we want to change the energy of the system. However, we have mass conservation as stated by the following proposition; see, e.g., [101].

Proposition 7.1. *Let $V(x,t) = V_0(x) + u(t)x$ and $\|\psi_0(\cdot)\|_{\mathcal{H}} = 1$; then we have $\|\psi(\cdot,t)\|_{\mathcal{H}} = 1$ for all $t \in [0,T]$.*

An important result of perturbation theory is that in a long time horizon a time-harmonic control $u(t)$ is able to induce transition between two eigenstates $i \to j$ if its frequency equals the difference of energy of the two states; see, e.g., Chapter 15 of [101]. Therefore, it is relatively easy to control state transitions for long time intervals. However, the problem becomes very difficult if short time intervals are considered. Here short means that $T \approx 2\pi/\omega$, where $\omega = \lambda_j - \lambda_i$, and in this case the perturbation theory is inapplicable and the control $u(t)$ deviates greatly from an harmonic function. In this case, an optimal control approach is unavoidable.

To formulate the optimal control problem, we have to decide in which functional space the control is sought. From the previous discussion, it appears that the control space $\mathcal{U} = H_0^1(0,T;\mathbb{R})$ is the most appropriate for dipole controls, as it means that the laser pulse cannot change instantaneously and it accommodates sinusoidal functions for long time controls. In the optimal control framework, this choice means that the objective of the optimization has a regularization term of the form $\|u\|^2_{\mathcal{U}}$, where this norm is induced by the following inner product

$$(u,v)_{\mathcal{U}} = \int_0^T (u(t)v(t) + \alpha\, \dot{u}(t)\dot{v}(t))\, dt \qquad \text{for } u,v \in \mathcal{U}, \tag{7.19}$$

with $0 < \alpha \ll 1$. Notice that with this norm the control is continuous since $H_0^1(0,T)$ is compactly embedded in $C^0([0,T])$ in one dimension. Use of smaller values of α allows for controls with larger rates of change. We have that the control is zero at the beginning and at the end of the time interval which is the maximum time window for the laser pulse.

The present control problem requires finding a control $u \in \mathcal{U}$ such that a quantum system initially in the state ψ_0 evolves with (7.16) to a state $\psi(\cdot,T)$ that is as close as possible to a desired target configuration ψ_d. This aim is formulated by requiring us to minimize the objective given by the cost functional

$$\min J(\psi,u) := \frac{1}{2}\left(1 - \|P\psi(\cdot,T)\|^2_{\mathcal{H}}\right) + \frac{\gamma}{2}\,\|u\|^2_{\mathcal{U}}, \tag{7.20}$$

under the constraint given by the TDSE, including the initial condition, denoted as

$$c(\psi,u) := \{i\,\partial_t - H(u)\}\,\psi = 0, \tag{7.21}$$

where $H(u) = -\Delta + V_0(x) + u(t)x$ and we introduce the projector $P\psi = (\psi_d,\psi)_{\mathcal{H}}\,\psi_d$. The goal of the first term of the cost functional is to track the given terminal state ψ_d up to a global phase $e^{i\varphi T}$ which cannot be specified.

A control suitable for fast quantum state transition can be obtained in the optimal control formulation given by (7.20) and (7.21). To characterize the solution to this problem, we introduce the following Lagrangian

$$L(\psi,u,p) = J(\psi,u) + Re\int_0^T\int_\Omega p^*(x,t)c(\psi,u)(x,t)\,dx\,dt, \tag{7.22}$$

where p is the Lagrange multiplier. In [356] it is proved that any minima of (7.20) and (7.21) correspond to an extremal point of the Lagrangian; see also [235, 239]. Therefore, taking the Fréchet derivatives of $L(\psi,u,p)$ with respect to the optimization variables gives the following first-order optimality system that characterizes the optimal solution. We have

$$\begin{aligned} \{i\,\partial_t + \Delta - V_0(x) - u(t)x\}\,\psi(x,t) &= 0,\\ \{i\,\partial_t + \Delta - V_0(x) - u(t)x\}\,p(x,t) &= 0,\\ -\gamma\, u(t) + \gamma\alpha\,\ddot{u}(t) + Re\int_\Omega p^*(x,t)\,x\,\psi(x,t)\,dx &= 0. \end{aligned} \tag{7.23}$$

This system consists of the state equation, the adjoint equation, and the optimality condition, respectively, with homogeneous Dirichlet boundary conditions, and initial and terminal conditions given by

$$\begin{aligned} \psi(x,0) &= \psi_0(x),\\ p(x,T) &= i\,(\psi_d(\cdot),\psi(\cdot,T))_{\mathcal{H}}\,\psi_d(x),\\ u(0) &= 0, \quad u(T) = 0. \end{aligned} \tag{7.24}$$

In [356] it is proved that there exists at least one solution to (7.23)–(7.24). In this reference it is also proved that if γ is sufficiently large and the projection $\|P\psi(T)\|^2_{\mathcal{H}}$ is sufficiently close to 1, then the second-order sufficient optimality condition holds and the optimization problem is locally strictly convex.

Notice that the control is a function of time only, and the state and adjoint variables can be seen as implicit functions of the control. Therefore the dimensionality of the optimization problem can be reduced significantly, introducing a reduced cost functional $\hat{J}(u) = J(\psi(u), u)$. The corresponding gradient is given by

$$\left(\nabla \hat{J}(u)\right)(t) = \gamma\, u(t) - \gamma\alpha\, \ddot{u}(t) - Re \int_\Omega p^*(x,t)\, x\, \psi(x,t)\, dx. \tag{7.25}$$

Therefore we have that $\nabla \hat{J}(u) \in H^{-1}(0,T;\mathbb{R})$, which is problematic with a gradient-based approach because the gradient is not in the same space of the solution and thus it does not provide an update to the control along the descent direction. It has been shown [354] that this problem can be solved by using the Riesz representation of the gradient in the $H^1_0(0,T)$ space as a means of Sobolev smoothing.

We remark that a main difficulty in the analysis of quantum control problems is that they may admit multiple solutions (as most bilinear control problems). This is the case of the present control problem. In fact, we have the following proposition [356].

Proposition 7.2. *Let the initial and target states be eigenfunctions and the stationary potential be symmetric. Then the reduced cost functional does not have a unique minimizer. In particular, if $u^*(t)$ is a minimizer, then so is $-u^*(t)$ and consequently $\hat{J}(u)$ is nonconvex independently of the values of the regularization parameters.*

The Discrete Dipole Quantum Control Problem and Numerical Results

We recall the introductory discussion in Chapter 3 and pursue the approach of discretize-before-optimize in order to avoid any discrepancy between the directional derivative and its approximation, and to guarantee a symmetric Hessian; see [16, 99]. In our experience, gradient inconsistency is usually not negligible in the case of hyperbolic- and Schrödinger-type equations with bilinear controls.

Regarding discretization of the Schrödinger equation, the time-discretization schemes discussed in Chapter 3 are not appropriate since they do preserve unitarity. Furthermore, in the case of time-varying Hamiltonian, also the CN scheme is not norm-preserving. For this reason, in [356] a modified Crank–Nicolson (MCN) method is proposed which is proved to be second-order accurate and unconditionally stable also in the case of time-varying potentials. Let N_t be the number of time steps $t_k = k\,\delta t$, $k = 0, 1, \ldots, N_t$, where the size $\delta t = \frac{T}{N_t}$, and let N_x be the number of grid points of the Ω discretization including the boundary points. The TDSE discretized by the MCN scheme results in the following

$$\psi_k - \psi_{k-1} = -\frac{i\,\delta t}{4}[H(t_k) + H(t_{k-1})][\psi_k + \psi_{k-1}]. \tag{7.26}$$

Let $\mathbf{y}_k \in \mathbb{C}^{N_x - 2}$ be a vector which contains the values of ψ on the interior grid points at the kth time step. Let $N_d = (N_x - 2)(N_t - 1)$ be the total number degrees of freedom of the state variable, which excludes the initial and boundary points.

Based on [355], we discuss a formulation in complex spaces of the discrete equality constraint (7.26) and of the optimization operators involved. It results in a more compact notation than the real-imaginary matrix notation used in [356]. We introduce the following matrices

$$\mathbf{A}_1 = \mathbf{I} + \frac{i\delta t}{2}\mathbf{H}_0, \quad \mathbf{A}_2 = \mathbf{I} - \frac{i\delta t}{2}\mathbf{H}_0, \quad \mathbf{B} = \frac{i\delta t}{4}\mathbf{X}, \tag{7.27}$$

where $\mathbf{H}_0$ is the discretization of the stationary Hamiltonian $H_0 = -\Delta + V_0(x)$, using linear finite elements on a uniform grid, and $\mathbf{X}$ is the discretized position operator. The kth discrete equality constraint may now be written as follows

$$\begin{aligned} \mathbf{c}_k(\mathbf{y},\mathbf{u}) = {} & [\mathbf{A}_1 - (u_k + u_{k-1})\mathbf{B}]\mathbf{y}_k \\ & - [\mathbf{A}_2 + (u_k + u_{k-1})\mathbf{B}]\mathbf{y}_{k-1}. \end{aligned} \tag{7.28}$$

To take variations using the Wirtinger calculus [346], we write the cost in terms of the discretized wavefunction and its complex conjugate

$$J(\mathbf{y},\bar{\mathbf{y}},\mathbf{u}) = \frac{1}{2}(1 - \bar{\mathbf{y}}^\top \mathbf{P}\mathbf{y}) + \frac{\gamma}{2}\mathbf{u}^\top \mathbf{W}\mathbf{u}, \tag{7.29}$$

where $^\top$ means transpose, and $\mathbf{P} \in \mathbb{R}^{N_x \times N_d}$ is the discrete projection operator, defined by

$$\mathbf{P}\mathbf{y} = \begin{pmatrix} 0 \\ \vdots \\ 0 \\ (\mathbf{y_d}^\top \mathbf{y}_{N_t})\mathbf{y_d} \end{pmatrix}, \tag{7.30}$$

where $\mathbf{W}$ is the finite element approximation to the Helmholtz operator $I - \alpha\,\partial_t^2$.

We consider the discrete Lagrangian, with Lagrange multipliers $\mathbf{p}_1,\ldots,\mathbf{p}_{N_t}$ and $\bar{\mathbf{p}}_1,\ldots,\bar{\mathbf{p}}_{N_t}$:

$$\begin{aligned} L(\mathbf{y},\bar{\mathbf{y}},\mathbf{u},\mathbf{p},\bar{\mathbf{p}}) = {} & J(\mathbf{y},\bar{\mathbf{y}},\mathbf{u}) \\ & + \sum_{k=1}^{N_t} \mathbf{p}_k^\top \mathbf{c}_k(\mathbf{y},\mathbf{u}) + \bar{\mathbf{p}}_k^\top \bar{\mathbf{c}}_k(\mathbf{y},\mathbf{u}). \end{aligned} \tag{7.31}$$

Here, the variables are understood to be column vectors. The components of the derivatives of the Lagrange function are row vectors given as follows

$$L_{\mathbf{y}} = -\frac{1}{2}\bar{\mathbf{y}}^\top \mathbf{P} + \mathbf{p}^\top \mathbf{c}_{\mathbf{y}}, \tag{7.32}$$

$$L_{\bar{\mathbf{y}}} = -\frac{1}{2}\mathbf{y}^\top \mathbf{P} + \bar{\mathbf{p}}^\top \bar{\mathbf{c}}_{\bar{\mathbf{y}}}, \tag{7.33}$$

$$L_{\mathbf{u}} = \gamma\mathbf{u}^\top \mathbf{W} + \mathbf{p}^\top \mathbf{c}_{\mathbf{u}} + \bar{\mathbf{p}}^\top \bar{\mathbf{c}}_{\mathbf{u}}, \tag{7.34}$$

$$L_{\mathbf{p}} = \mathbf{c}^\top, \tag{7.35}$$

$$L_{\bar{\mathbf{p}}} = \bar{\mathbf{c}}^\top. \tag{7.36}$$

With this notation, the discrete equality constraint $\mathbf{c}^\top \in \mathbb{R}^{1\times N_d}$ is a row vector given by $\mathbf{c}^\top = \begin{pmatrix} \mathbf{c}_1^\top & \mathbf{c}_2^\top & \cdots & \mathbf{c}_{N_t}^\top \end{pmatrix}$.

We collect the components (7.32)–(7.36) to define the full gradient

$$\nabla L = \begin{pmatrix} L_{\mathbf{y}}, & L_{\bar{\mathbf{y}}}, & L_{\mathbf{u}}, & L_{\mathbf{p}}, & L_{\bar{\mathbf{p}}} \end{pmatrix}.$$

The Hessian is defined as the Jacobian of the Hermitian conjugate of this gradient, which we denote with $\nabla^2 L = \nabla(\nabla L)^*$. This approach results in the following Hermitian Hessian

$$\nabla^2 L = \begin{pmatrix} L_{\mathbf{yy}} & 0 & L_{\mathbf{yu}} & 0 & {\mathbf{c_y}}^* \\ 0 & L_{\bar{\mathbf{y}}\bar{\mathbf{y}}} & L_{\bar{\mathbf{y}}\mathbf{u}} & {\mathbf{c_y}}^\top & 0 \\ L_{\mathbf{yu}}^* & L_{\mathbf{yu}}^\top & L_{\mathbf{uu}} & {\mathbf{c_u}}^\top & {\mathbf{c_u}}^* \\ 0 & \bar{\mathbf{c}}_{\mathbf{y}} & \bar{\mathbf{c}}_{\mathbf{u}} & 0 & 0 \\ \mathbf{c_y} & 0 & \mathbf{c_u} & 0 & 0 \end{pmatrix}. \tag{7.37}$$

To make the notion clear, $\mathbf{c_y} \in \mathbb{C}^{N_d \times N_d}$ is the part of the Jacobian of the equality constraint obtained by differentiating with respect to the state $\mathbf{y}$, ${\mathbf{c_y}}^\top$ is its transpose without complex conjugation, ${\mathbf{c_y}}^*$ is the Hermitian conjugate, and $\bar{\mathbf{c}}_{\mathbf{y}}$ is its complex conjugate without transposition. These operators are given in expanded matrix form in the following

$$\mathbf{c_u} = -\begin{pmatrix} \mathbf{B}(\mathbf{y}_1+\mathbf{y}_0) & & & & \\ \mathbf{B}(\mathbf{y}_2+\mathbf{y}_1) & \mathbf{B}(\mathbf{y}_2+\mathbf{y}_1) & & & \\ & \mathbf{B}(\mathbf{y}_3+\mathbf{y}_2) & \mathbf{B}(\mathbf{y}_3+\mathbf{y}_2) & & \\ & & \ddots & \ddots & \\ & & & & \mathbf{B}(\mathbf{y}_{N_t}+\mathbf{y}_{N_t-1}) \end{pmatrix},$$

$$\mathbf{c_y} = \begin{pmatrix} \mathbf{A}_1-(u_1+u_0)\mathbf{B} & & & \\ -\mathbf{A}_2-(u_2+u_1)\mathbf{B} & \mathbf{A}_1-(u_1+u_0)\mathbf{B} & & \\ & & \ddots & \\ & & -\mathbf{A}_2-(u_{N_t}+u_{N_t-1})\mathbf{B} & \mathbf{A}_1-(u_{N_t}+u_{N_t-1})\mathbf{B} \end{pmatrix},$$

$$L_{\mathbf{yu}} = -\begin{pmatrix} \mathbf{B}^*(\bar{\mathbf{p}}_1+\bar{\mathbf{p}}_2) & \mathbf{B}^*\bar{\mathbf{p}}_2 & & & \\ \mathbf{B}^*\bar{\mathbf{p}}_2 & \mathbf{B}^*(\bar{\mathbf{p}}_2+\bar{\mathbf{p}}_3) & \mathbf{B}^*\bar{\mathbf{p}}_2 & & \\ & & \ddots & \ddots & \\ & & & \mathbf{B}^*\bar{\mathbf{p}}_{N_t-1} & \mathbf{B}^*(\bar{\mathbf{p}}_{N_t-1}+\bar{\mathbf{p}}_{N_t}) \\ & & & & \mathbf{B}^*\bar{\mathbf{p}}_{N_t} \end{pmatrix},$$

$$L_{\mathbf{yy}} = L_{\bar{\mathbf{y}}\bar{\mathbf{y}}} = -\begin{pmatrix} 0 & \cdots & \cdots & 0 \\ \vdots & \ddots & & \vdots \\ \vdots & & 0 & 0 \\ 0 & \cdots & 0 & -\frac{1}{2}\mathbf{P} \end{pmatrix},$$

$$L_{\mathbf{uu}} = \gamma \mathbf{W}.$$

It is important to recognize that the complex differentiation and transposition operators do not commute, so the order of operations is critical and this notation is employed to make clear that in each case differentiation occurs first and then conjugation second.

Using the fact that $\mathbf{c} = 0$ and $L_{\mathbf{y}} = 0$ by construction, the full KKT Hessian and gradient equations allow us to compute the differential change in the state and adjoint variables due to a differential change in the control

$$\delta \mathbf{y} = -\mathbf{c}_{\mathbf{y}}^{-1} \mathbf{c_u} \delta \mathbf{u}, \tag{7.38}$$

$$\delta \mathbf{p} = -{\mathbf{c_y}}^{-\top}(L_{\bar{\mathbf{y}}\bar{\mathbf{y}}}\delta \bar{\mathbf{y}} + L_{\bar{\mathbf{y}}\mathbf{u}}\delta \mathbf{u}). \tag{7.39}$$

Using the Schur reduction on the KKT system allows us to write the Newton method in terms of the reduced cost functional as follows

$$\nabla^2 \hat{J}(\mathbf{u})\delta\mathbf{u} = -\nabla \hat{J}^*(\mathbf{u}),$$

where $\nabla \hat{J}^*(\mathbf{u}) = L_{\mathbf{u}}^* = L_{\mathbf{u}}^\top$, and the application of the reduced Hessian on $\delta\mathbf{u}$ is given by

$$\nabla^2 \hat{J}(\mathbf{u})\delta\mathbf{u} = \mathbf{L_{uu}}\delta\mathbf{u} + 2Re[\mathbf{c_u}^\top \delta\mathbf{p} + \mathbf{L_{yu}}^\top \delta\mathbf{y}]. \tag{7.40}$$

Based on this framework, a globalized Krylov–Newton scheme and the corresponding code are presented in [355]. This methodology is discussed in detail in Section 4.2. In the following, we report results obtained with the Krylov–Newton scheme.

In Figure 7.5, optimal controls for different state transitions are depicted and in Table 7.2, we report the decrease of the norm of the gradient for the method of steepest descent and for the Krylov–Newton method for successive iterations. From this table, typical Newton's second-order convergence can be seen. The Krylov–Newton method attains a gradient which is zero to machine precision in just 8 steps. For the test of Table 7.2, 400 time steps and 200 spatial grid points were used with $T = 1$, $\alpha = 10^{-2}$, $\gamma = 10^{-3}$, $\psi_0(x) = \cos(\pi x/2)$, $\psi_d(x) = \frac{1}{\sqrt{10}}[\cos(\pi x/2) + 3\cos(\pi x)]$.

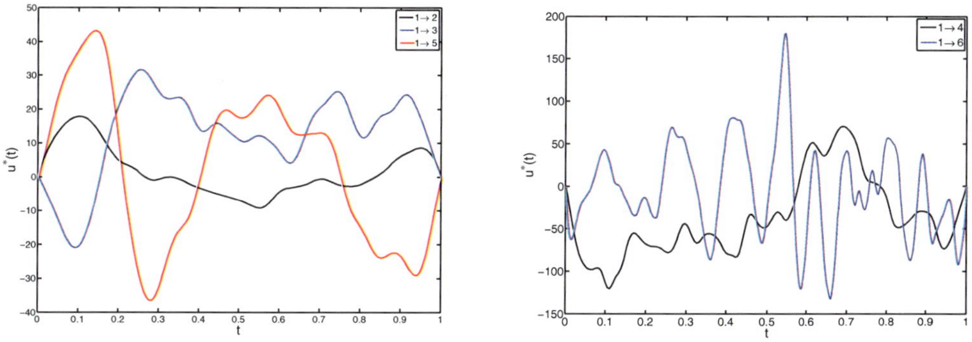

Figure 7.5. *Left: Optimal controls for transitions from the first state to the second, the third, and the fifth states. Right: Optimal controls for transitions from the first state to the fourth and the sixth states. This figure first appeared in G. von Winckel, A. Borzì, and S. Volkwein, A globalized Newton method for the accurate solution of a dipole quantum control problem, SIAM J. Sci. Comput.,* 31(6) (2009), 4176–4203.

Optimal Control of BECs

Quantum optimal control of transport of BECs in magnetic microtraps was proposed [199] to solve the problem for a trapped-atom interferometer setup where a dilute BEC should be split from a single to a double-well ground state. We recall that a coherent BEC is described by the following

$$i\,\partial_t \psi(x,t) = \left\{ -\Delta + V(x,u(t)) + g\,|\psi(x,t)|^2 \right\} \psi(x,t), \tag{7.41}$$

where $x \in \Omega$ and $t \in [0,T]$. We measure mass in units of the atom mass and length in units of micrometers. We assume that the quantum state wavefunction $\psi(x,t)$ is normalized to

Table 7.2. *Convergence of the steepest descent scheme and of the Krylov–Newton scheme with respect to* $\|\nabla \hat{J}(u)\|$.

Iteration	$\|\nabla \hat{J}_{SD}(u)\|$	$\|\nabla \hat{J}_{KN}(u)\|$
1	1.8615×10^{-4}	1.8615×10^{-4}
2	6.5263×10^{-5}	6.5263×10^{-5}
3	6.0031×10^{-5}	2.4732×10^{-5}
4	2.3535×10^{-5}	1.5557×10^{-5}
5	2.8106×10^{-5}	1.2316×10^{-6}
6	1.5703×10^{-5}	1.0977×10^{-8}
7	1.7062×10^{-5}	3.5480×10^{-13}
8	1.0322×10^{-5}	2.0009×10^{-17}
9	1.3312×10^{-5}	0
50	1.9114×10^{-7}	0

one, $\int_\Omega |\psi(x,t)|^2 dx = 1$, $t \geq 0$, and therefore g in (7.41) incorporates the number of atoms N_A. The fact that V is a confinement potential results in a wavefunction ψ whose support is localized in a bounded region. Therefore, with Ω we represent a spatial domain that is large enough to represent the support of ψ during evolution and we use periodic boundary conditions.

The magnetic control potential $V_u(x,t) = V(x,u(t))$ is modulated by a control function $u(t)$, and the initial and final potential configurations are given; therefore we require that $u(t)$ take initial and final values of zero and one, respectively. These two extremal values correspond to the case where the potential V_u is convex and to the case where it has a double-well structure. Furthermore we require that V_u be spatially symmetric with respect to the origin of coordinates.

Suppose that initially the system is in the ground state ψ_0 for the potential $V(x,0)$. Upon varying $u(t)$ in the time interval $t \in [0,T]$ from zero to one, the system will pass through a sequence of states and will end up in the final state $\psi(T)$. Our purpose is to determine an optimal control time evolution that allows us to channel the system from the initial state ψ_0 at time zero to a desired state ψ_d at final time T and ψ_d represents the ground state for the potential $V(x,1)$ at time T. The ground state for a given potential $V(x,u)$ is defined as the stationary state $\phi(x)$ with $\int_\Omega |\phi(x)|^2 dx = 1$ that minimizes the following energy [22]

$$E_u(\phi) = \frac{1}{2}\int_\Omega \left(|\nabla\phi(x)|^2 + V_u(x)\,|\phi(x)|^2 + \frac{g}{2}\,|\phi(x)|^4 \right) dx. \tag{7.42}$$

See Figure 7.6 for stationary solutions and corresponding initial target states.

We choose the control potential proposed in [231] to create condensates of trapped atoms coupled with a radio frequency fields. We have that

$$V(x,u) = -\frac{u^2 d^2}{8c}x^2 + \frac{1}{c}x^4, \tag{7.43}$$

where $c = 40$ and d is a parameter corresponding to twice the distance of the two minima in the double-well potential.

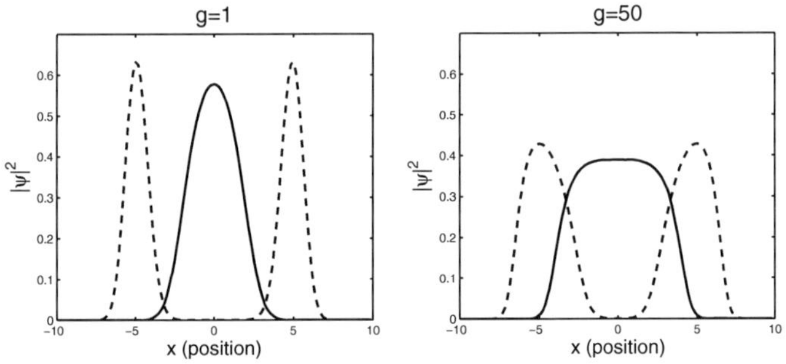

Figure 7.6. *Stationary solutions of the Gross–Pitaevskii equation for nonlinearity strengths $g = 1$ and $g = 50$. The initial state ψ_0 is shown with the solid line and the target state ψ_d is represented by the dashed line. Reprinted with permission from G. von Winckel and A. Borzì, Computational techniques for a quantum control problem with H^1-cost, Inverse Problems,* 24 (2008), 034007.

For our purpose, in order to define a well-defined control problem, we consider the cost functional [199]

$$J(\psi,u) = \frac{1}{2}\left(1 - \left|(\psi_d, \psi(T))\right|^2\right) + \frac{\gamma}{2}\int_0^T (\dot{u}(t))^2\,dt \tag{7.44}$$

with $(u,v) = \int_\Omega u(x)^* v(x)\,dx$ the usual inner product in complex spaces and $\|u\| = (u,u)^{1/2}$. The second term in the cost functional represents an H^1 cost and aims at penalizing fast varying confinement potentials that are more difficult to realize in real experiments.

The control problem under consideration is therefore to minimize $J(\psi,u)$ subject to the condition that $\psi(x,t)$ fulfills the BEC equation (7.41) with given initial conditions. As in the dipole control problem, to solve this problem we introduce a Lagrange function where $p(x,t)$ is the Lagrange multiplier, and we obtain the following optimality system characterizing the solution to the BEC optimal control problem. We have [54]

$$i\frac{\partial \psi}{\partial t} = \left(-\Delta + V_u + g|\psi|^2\right)\psi, \tag{7.45}$$

$$i\frac{\partial p}{\partial t} = \left(-\Delta + V_u + 2g|\psi|^2\right)p + g\,\psi^2 p^*, \tag{7.46}$$

$$\gamma\frac{d^2u}{dt^2} = -Re\left(\psi, \frac{\partial V_u}{\partial u}p\right), \tag{7.47}$$

which has to be solved together with the initial and terminal conditions

$$\psi(0) = \psi_0, \tag{7.48}$$

$$ip(T) = -(\psi_d, \psi(T))\,\psi_d, \tag{7.49}$$

$$u(0) = 0, \quad u(T) = 1. \tag{7.50}$$

Because of H^1 regularization we have a natural setting to impose the required Dirichlet boundary conditions on the control function, $u(0) = 0$ and $u(T) = 1$. See [54] for a discussion concerning second-order optimality conditions.

We have that (7.45) is uniquely solvable for every $u \in H^1(0,T;\mathbb{R})$ such that V_u is a symmetric double-well potential; see [296]. Thus, it is meaningful to introduce the reduced cost functional $\hat{J} : H^1(0,T;\mathbb{R}) \to \mathbb{R}$ given by $\hat{J}(u) = J(\psi(u),u)$, where $\psi(u)$ denotes the unique solution to (7.45) for given u. One can show that the (L^2) gradient of $\hat{J}$ with respect to u is given by

$$\nabla \hat{J}(u) = -\gamma \frac{d^2 u}{dt^2} - Re\left(\psi, \frac{\partial V_u}{\partial u} p\right), \tag{7.51}$$

where ψ and p solve the state and the adjoint equations with given u.

Now, the Taylor series of $\hat{J}(u)$ in a Hilbert space is

$$\hat{J}(u+\epsilon\varphi) = \hat{J}(u) + \epsilon\left(\nabla \hat{J}(u),\varphi\right)_X + \frac{\epsilon^2}{2}\left([\nabla^2 \hat{J}(u)]\varphi,\varphi\right)_X + O(\epsilon^3).$$

The actual gradient will depend on the choice of which inner product space is used. If we choose the space $X = L^2(0,T;\mathbb{R})$, we have the gradient given by (7.51).

In the case of the $X = H^1(0,T;\mathbb{R})$ formulation, we can determine the formula for the gradient because the Taylor series must be identical term-by-term regardless of the choice of X. Since we know the gradient in L^2 given by (7.51), we can determine the gradient in H^1 by requiring

$$\left(\nabla \hat{J}_{H^1}(u),\varphi\right)_{H^1} = \left(\nabla \hat{J}_{L^2}(u),\varphi\right)_{L^2}.$$

Using the definition of the H^1 inner product and integrating by parts shows that the H^1 gradient must satisfy the one-dimensional Poisson equation with homogeneous Dirichlet conditions. That is,

$$\frac{d^2}{dt^2}[\nabla \hat{J}(u)] = \gamma \frac{d^2 u}{dt^2} + Re\left(\psi, \frac{\partial V_u}{\partial u} p\right), \tag{7.52}$$

with $[\nabla \hat{J}(u)](0) = 0$ and $[\nabla \hat{J}(u)](T) = 0$. Two important differences between the L^2 and H^1 formulations are immediately apparent. First, the L^2 gradient does not vanish at $t = T$, and second, the H^1 gradient possesses the same degree of smoothness as u, while the L^2 gradient does not.

The discussion on which optimization space is used is very important also for the numerical implementation. As pointed out in [354], not only is the definition of the gradient also space dependent but also the construction of the optimization schemes. In [354], optimization methods in L^2 and H^1 spaces are discussed with a focus on the nonlinear conjugate gradient scheme of Hager and Zhang [181] and on the BFGS algorithm. These schemes are illustrated in Chapter 4. We remark that our BEC control problem has been successfully solved in [54] by applying the MGOPT optimization strategy.

For the discretization of the BEC control problem, we use an unconditionally stable explicit second-order norm-preserving time-splitting spectral scheme (TSSP); see [23, 54, 354] for more details. We take $\Omega = (-\ell/2, \ell/2) \subset \mathbb{R}$, where ℓ is large enough so that the support of the state and adjoint variables is well within the domain. The spatial mesh size is $h = \ell/N$ with $N = 128$. The time interval $(0,T)$ is divided in M subintervals; thus the time step size is given by $\delta t = T/M$. The function ψ is periodic in Ω in the sense that $\psi(-\ell/2^+) = \psi(\ell/2^-)$.

Some of the results of numerical experiments are summarized in Figure 7.7, which shows a series of optimal control curves corresponding to different time horizons, obtained

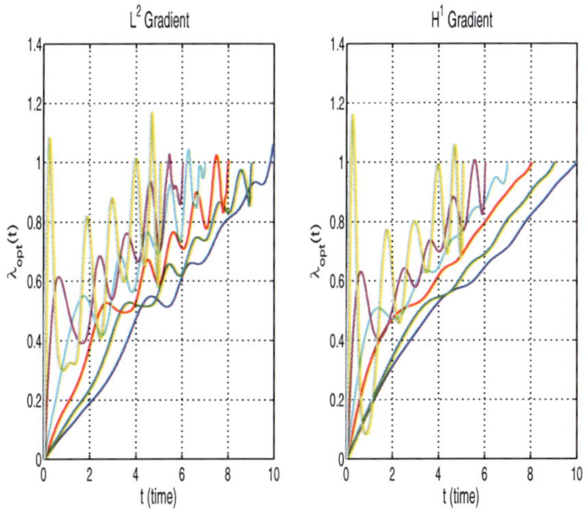

Figure 7.7. *The optimal control function in the L^2 and H^1 settings for decreasing time horizons. As T becomes smaller, u becomes a more oscillating function; here $\gamma = 10^{-4}$. Reprinted with permission from G. von Winckel and A. Borzì, Computational techniques for a quantum control problem with H^1-cost, Inverse Problems,* 24 (2008), 034007.

by minimizing in the L^2 and H^1 spaces. We see that, as T becomes shorter, a more oscillatory optimal control results. However, the solutions obtained with H^1 space minimization are less oscillating. Indeed we see that with the same setting, the controls obtained in the two functional space settings are different.

The regularization parameter γ plays an important role in the L^2-based optimization, as it forces the solution to have a finite H^1 norm, while this will always be the case by construction with the H^1-based formulation. In the L^2 space, the limit as $\gamma \to 0$ small gives a control function $u(t)$ which has a steep gradient at $t = T$. On the other hand, since the H^1-based scheme produces $u \in H^1$ at every optimization step, the precise value of the regularization term is not essential to obtain a smooth control. Rather in this case the γ term serves only to penalize highly oscillatory solutions. In the limit as $\gamma \to 0$, the H^1-based solution approaches a smooth curve which gives a small cost functional.

Additional results are reported in Tables 7.3 through 7.6, which allow a more detailed discussion. These tables show the computational performance of the HZ-NCG and BFGS schemes using the L^2 and H^1 spaces minimization, respectively. With the nonlinearity strength $g = 10$ and final time $T = 6$, this is a difficult optimization problem, the optimal control being far away from the original linear function. The CPU usage in both spaces is comparable. We see that, at convergence and similar computational effort, the H^1-based approach produces a smaller objective and substantially smaller norm of the gradient on the finest mesh than the L^2-based approach.

The advantage of the H^1-based minimization on the L^2-based scheme is more evident in Tables 7.5 and 7.6, which show the performance of the BFGS scheme in the L^2 and H^1 spaces, respectively. The L^2-based BFGS method requires progressively more iterations to reach the minimum, whereas the H^1 approach requires fewer iterations on the

Table 7.3. *Results with L^2-based HZ-NCG minimization with $g = 10$ and $T = 6$.*

M	$\hat{J}$	$\|\|\nabla\hat{J}\|\|_{L^2}$	Iterations	CPU time (sec)
400	2.9676×10^{-3}	4.4182×10^{-2}	81	1.6271×10^2
800	1.6492×10^{-3}	2.113×10^{-2}	110	4.1789×10^2
1600	1.8103×10^{-3}	1.1400×10^{-2}	155	1.1932×10^3
3200	3.4755×10^{-3}	1.13815×10^{-1}	76	1.1985×10^3

Table 7.4. *Results with H^1-based HZ-NCG minimization with $g = 10$ and $T = 6$.*

M	$\hat{J}$	$\|\|\nabla\hat{J}\|\|_{L^2}$	Iterations	CPU time (sec)
400	3.0295×10^{-2}	2.7397×10^{-1}	24	3.3911×10^1
800	2.2888×10^{-3}	2.9977×10^{-2}	217	7.9421×10^2
1600	1.1995×10^{-3}	1.8310×10^{-2}	153	1.1986×10^3
3200	2.4388×10^{-3}	4.5315×10^{-3}	74	1.1883×10^3

Table 7.5. *Results with L^2-based BFGS minimization with $g = 10$ and $T = 6$.*

M	$\hat{J}$	$\|\|\nabla\hat{J}\|\|_{L^2}$	Iterations	CPU time (sec)
400	1.8922×10^{-3}	5.8348×10^{-2}	25	6.1990×10^1
800	1.2472×10^{-3}	9.7978×10^{-2}	34	1.7211×10^2
1600	3.5128×10^{-4}	1.3702×10^{-2}	105	9.6031×10^2
3200	1.1394×10^{-4}	5.8215×10^{-3}	500	8.1292×10^3

Table 7.6. *Results with H^1-based BFGS minimization with $g = 10$ and $T = 6$.*

M	$\hat{J}$	$\|\|\nabla\hat{J}\|\|_{L^2}$	Iterations	CPU time (sec)
400	1.6605×10^{-2}	1.4288×10^{-1}	15	3.8407×10^1
800	5.5963×10^{-4}	4.5284×10^{-2}	62	2.8107×10^2
1600	2.9634×10^{-4}	1.0733×10^{-2}	30	3.6334×10^2
3200	1.0562×10^{-4}	3.6378×10^{-3}	37	9.6153×10^2

finest mesh, similar to NCG. Compared to the previous two tables, the BFGS scheme provides at least one order of magnitude better results in terms of the value of the objective and concerning CPU times. See [354] for additional results.

Figure 7.8 shows results for a wavefunction splitting for an optimized and a linear [199] control function [54]. One can see that the wavefunction becomes split along the time evolution and is transported in the respective minima. Excited vibrational states that originate during the initial splitting process can be seen plotting the absolute value of the wavefunction profiles at final time. These are evident in the case of a linearly varying u where the split is not complete and part of the population remains localized between the two wells.

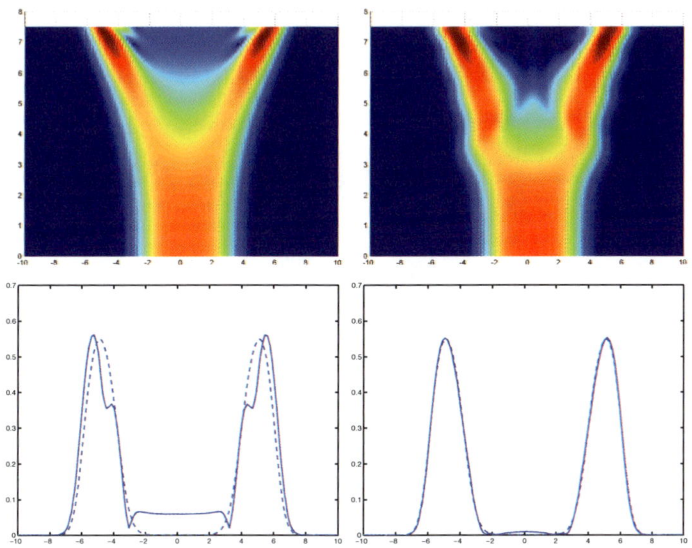

Figure 7.8. *The function* $|\psi(x,t)|$ *on the space-time domain (top) for the linear (left) and optimized (right) controls. The corresponding profiles at final time (bottom, continuous line) compared to the desired state (dashed line). Mesh* 128×1250*;* $\gamma = 10^{-4}$. *This figure first appeared in A. Borzì and U. Hohenester, Multigrid optimization schemes for solving Bose–Einstein condensate control problems, SIAM J. Sci. Comput.*, 30(1) (2008), 441–462.

7.4 Electromagnetic Inverse Scattering

7.4.1 Introduction

In this section, we discuss electromagnetic inverse scattering problems in the PDE-based optimization framework. These problems arise in many different practical application fields and are among the most representative classes of inverse problems. We provide details of the formulation of the inverse scattering problem as a PDE optimization problem and discuss some aspects of its numerical implementation.

Consider an incident electromagnetic wave which propagates in a known background medium where there are objects with electromagnetic properties different from the background. Because of this difference, these objects scatter the electromagnetic waves, and with this scenario, we can distinguish two kinds of scattering problems. The first category is that of direct scattering problems where the incident waves, the background medium, and the characteristics of the objects are given and computation of the scattered field generated by the interaction of the incident waves with the objects is required. The inverse scattering problems represent the second category of problems where the incident waves, the scattered waves, and the medium are given and we have the problem of deducing the features of the unknown objects. These problems are very important in applications concerning noninvasive imaging. In fact, because electromagnetic waves can penetrate various media where objects may be placed, the solution of inverse scattering problems allows us to determine the presence and the properties of these objects hidden in the medium.

We focus on electromagnetic inverse scattering problems in the space-time domain [31], trying to determine the electromagnetic properties of objects in the medium, and the geometric structure and position of these scatterers. We consider the time-domain formulation that naturally accommodates data corresponding to a large spectrum of possible frequencies of the incident field [159, 161, 163, 164, 246, 269, 287, 335]. Notice that the Maxwell equations could be considered in the frequency domain with monochromatic fields and one would need to process the data in the frequency domain with particular techniques; see, e.g., [298]. In the discussion that follows, a two-dimensional spatial domain with a model given by the Maxwell curl equations is considered. In particular, we consider a two-dimensional scattering scenario and the governing Maxwell curl equations including initial conditions, and radiation conditions. The objective includes observation of the electric and magnetic fields' data misfit and regularization terms of zero- and first-order Tikhonov type [105] for the electromagnetic properties that are sought. To characterize the solution of the optimization problem, the first-order optimality system is discussed, including the formulation of the radiation conditions for the adjoint equations. The implementation of radiation conditions is important to avoid unphysical wave reflections at the boundary of the computation domain. However, see [161, 164] for simplified boundary conditions.

To solve the optimality system, the finite difference time domain (FDTD) method is illustrated in the one-dimensional case, and we report some details concerning numerical dispersion, stability, and accuracy of FDTD solutions. Notice that the scientific literature of discretization schemes for Maxwell problems is vast; see, e.g., [257] and the references given therein. Further, we discuss the perfectly matched layer scheme to implement free propagation in the free space of the scattered waves. This section is completed with numerical results with an inversion problem.

7.4.2 The Scattering Problem in the Time Domain

We formulate a time-domain electromagnetic inverse scattering problem in the framework of optimization with PDE constraints. We start introducing a general two-dimensional scattering scenario and the governing Maxwell curl equations with initial and radiation boundary conditions. Our objective is to minimize a functional of electromagnetic field data misfit, including a regularization functional for the electromagnetic properties that are sought. To characterize the solution of the resulting optimization problem, we derive the corresponding first-order optimality system.

Consider the two-dimensional scenario illustrated in Figure 7.9.

This scenario consists of a domain V which can be unbounded. If it is bounded, we denote with ∂V its boundary. The domain V represents also a free space background medium, characterized by the electric permittivity ε_0 and the magnetic permeability μ_0. In a subdomain $D \subset V$, called scatterer domain, we imagine that we have a spatial distribution of the electromagnetic properties $\varepsilon(\vec{r})$, $\mu(\vec{r})$, and $\sigma(\vec{r})$, which are different from those of the background medium. Outside the domain D, we have electromagnetic sources, i.e., an antenna that generates an electromagnetic field that propagates in the domain V and also in the domain D where the scattering occurs.

In the direct scattering problem, we know the sources, the background medium, and also the distribution of electromagnetic properties in the scatterer domain D. The purpose is to calculate the fields $(\vec{E}(\vec{r},t), \vec{H}(\vec{r},t))$ in V by solving the following Maxwell curl

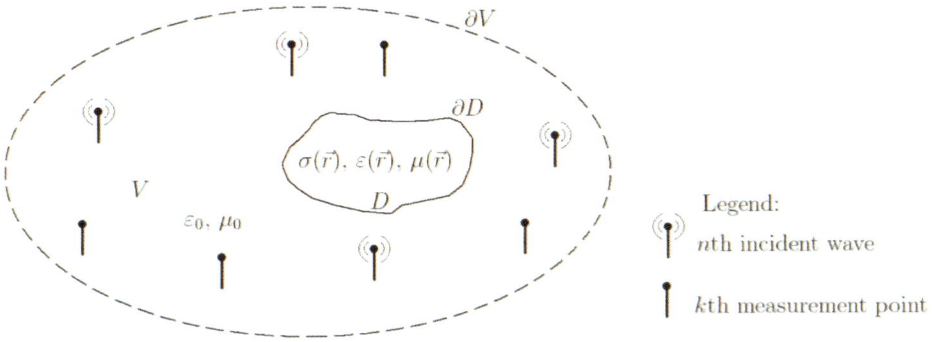

Figure 7.9. *A scenario of direct and inverse scattering problems.*

equations

$$\begin{cases} \nabla \times \vec{E} + \mu \dfrac{\partial \vec{H}}{\partial t} = 0, \\ \nabla \times \vec{H} - \varepsilon \dfrac{\partial \vec{E}}{\partial t} - \sigma \vec{E} - J = 0, \end{cases} \tag{7.53}$$

where we have used the compact notation $(\vec{E}, \vec{H})$ instead of $(\vec{E}(\vec{r},t), \vec{H}(\vec{r},t))$. This set of equation needs in addition to specify the initial conditions and the boundary conditions that the evolving field must satisfy. The initial conditions set the values of the electromagnetic field at the starting time, $t = 0$. The boundary conditions prescribe the behavior of the fields on the boundary of the domain V for all times. Given the system (7.53) and the initial and boundary conditions, we can solve the wave equations for the fields.

In an inverse scattering problem, we have antennas and receivers placed in certain positions in V, external to the scatterer domain D, which generate and receive and record the fields during the time interval $[0, T]$. With this data, we apply our inversion procedure to determine the properties $\mathbf{p} = (\varepsilon(\vec{r}), \mu(\vec{r}), \sigma(\vec{r}))$ in D, knowing the sources and the background medium. Since the presence and positioning of the scatterers in D are unknown, a procedure that is defined on all the subdomain D must be applied. In this procedure, the domain D is illuminated by N different incident waves and, for every incidence wave, we measure the fields in K different measurement points during the time interval $[0, T]$.

To solve the inverse scattering problem, one defines a functional that contains the discrepancies between the measured fields and the calculated fields, and the aim is at minimizing the misfit by varying the unknown electromagnetic properties which define the optimization variables. Because the ill-posedness of this problem, a regularization term is required and the following functional results

$$J\left(\mathbf{p}, \vec{E}, \vec{H}\right) = \frac{1}{2} \sum_{n=1}^{N} \sum_{k=1}^{K} \int_0^T \left(\| \vec{E}_{nk} - \vec{E}_{nk}^m \|^2 + \eta_0^2 \| \vec{H}_{nk} - \vec{H}_{nk}^m \|^2 \right) dt + R(\mathbf{p}), \tag{7.54}$$

where $\vec{E}_{nk}$ and $\vec{H}_{nk}$ are the calculated fields, whereas $\vec{E}_{nk}^m$ and $\vec{H}_{nk}^m$ are the measured fields. These four fields are evaluated in the kth measurement points for the nth incident wave.

Here, $R(\mathbf{p})$ represents the regularization term given by the following

$$\begin{aligned} R(\mathbf{p}) =& \frac{\beta}{2} \int_D \left(|\varepsilon_0^{-1} \nabla \varepsilon|^2 + |\nabla \sigma|^2 + |\mu_0^{-1} \nabla \mu|^2 \right) dv \\ &+ \frac{\gamma}{2} \int_D \left(|\varepsilon - \overline{\varepsilon}|^2 + |\sigma - \overline{\sigma}|^2 + |\mu - \overline{\mu}|^2 \right) dv. \end{aligned} \tag{7.55}$$

In this regularization functional we distinguish two different terms. The first term represents a first-order Tikhonov regularization scheme, as in [161, 287]. A regularization term similar to (7.55) is considered in [164]. Minimizing the functional J with this type of regularization means that we search the model's parameters $\mathbf{p}$ in a smooth space with spatially slow varying functions. A result of this choice is that we obtain the edges of the reconstructed objects smeared out by diffusion. The other regularization term represents a zero-order Tikhonov regularization, and it corresponds to the requirement that the ε sought be as close as possible to $\overline{\varepsilon}$, and similarly σ be as close as possible to $\overline{\sigma}$, and μ be as close as possible to $\overline{\mu}$. This regularization term can be used even if we do not have a priori information with known $\overline{\varepsilon}, \overline{\mu}, \overline{\sigma}$. In this case we choose these values to be zero and the resulting regularization term penalizes large parameters values.

The objective defined by (7.54) is to be minimized under the constraints given by the Maxwell curl equations for each incidence field. We have

$$\begin{cases} \nabla \times \vec{E}_n + \mu \dfrac{\partial \vec{H}_n}{\partial t} = 0, \\ \nabla \times \vec{H}_n - \varepsilon \dfrac{\partial \vec{E}_n}{\partial t} - \sigma \vec{E}_n - J_n = 0, \end{cases} \tag{7.56}$$

$$\vec{E}_n \big|_{t=0} = 0, \qquad \vec{H}_n \big|_{t=0} = 0, \tag{7.57}$$

$$\lim_{r \to \infty} r \left[\nabla \times \begin{pmatrix} \vec{E}_n \\ \vec{H}_n \end{pmatrix} + \frac{1}{c_0} \hat{n}_{V_\infty} \times \frac{\partial}{\partial t} \begin{pmatrix} \vec{E}_n \\ \vec{H}_n \end{pmatrix} \right] = 0, \tag{7.58}$$

$$\Sigma(\vec{E}_n, \vec{H}_n, \vec{J}_n) = f(\vec{r}, t). \tag{7.59}$$

The system (7.56) is constituted by the Maxwell curl equations and the initial conditions (7.57), and (7.59) represents the (hard and soft) sources. The limit (7.58) represents the Sommerfeld radiation condition at infinity, where $k_0 = \omega\sqrt{\mu_0 \varepsilon_0} = 2\pi f \sqrt{\mu_0 \varepsilon_0}$ and $\hat{n}_{V_\infty}$ is the outward-pointing vector normal to ∂V, the surface boundary of V_∞. This constraint means that, for every estimated $\mathbf{p}$ and for the nth incident wave, the calculated fields have to satisfy the Maxwell curl equations.

To solve this infinite-dimensional constrained optimization problem in the framework of PDE optimization, we introduce the following Lagrange functional

$$\begin{aligned} L\left(\mathbf{p}, \vec{E}, \vec{H}, \vec{e}, \vec{h}\right) =& J\left(\mathbf{p}, \vec{E}, \vec{H}\right) \\ &+ \sum_{n=1}^{N} \int_0^T \int_V \left(\vec{e}_n \cdot \left(\nabla \times \vec{H}_n - \varepsilon \frac{\partial \vec{E}_n}{\partial t} - \sigma \vec{E}_n - \vec{J}_n \right) \right) dt dv \\ &+ \sum_{n=1}^{N} \int_0^T \int_V \left(\vec{h}_n \cdot \left(\nabla \times \vec{E}_n + \mu \frac{\partial \vec{H}_n}{\partial t} \right) \right) dt dv, \end{aligned} \tag{7.60}$$

where $\vec{e}_n$ and $\vec{h}_n$ represent the Lagrange multipliers.

The characterization of the solution of this constrained minimization problem results in the optimality conditions of the first order that are formally given by the conditions for extrema of the Lagrange functional. For this purpose, we compute the derivatives of the Lagrange functional (7.60) with respect to $\vec{E}$, $\vec{H}$, $\vec{e}$, ε, μ, σ, with the inner product given by

$$\left(\vec{\varphi},\vec{\vartheta}\right)=\sum_{n=1}^{N}\int_0^T\int_V\left(\vec{\varphi}\cdot\vec{\vartheta}\right)\,dvdt.$$

We also use the L^2-norm in space which is denoted as follows

$$||\vec{\varphi}||^2=\int_V(\vec{\varphi}\cdot\vec{\varphi})\,dv. \tag{7.61}$$

We have that the functional derivative with respect to the Lagrange multipliers gives the Maxwell curl equations. Further, the derivatives with respect to the fields give the adjoint equations and the derivatives with respect to the electromagnetic properties provide the optimality conditions; see [246] for details. The following optimality system is obtained:

- The direct problem

$$\begin{cases}\nabla\times\vec{E}_n+\mu\dfrac{\partial\vec{H}_n}{\partial t}=0,\\[2mm] \nabla\times\vec{H}_n-\varepsilon\dfrac{\partial\vec{E}_n}{\partial t}-\sigma\vec{E}_n-J_n=0,\end{cases}$$

$$\vec{E}_n\big|_{t=0}=0,\qquad \vec{H}_n\big|_{t=0}=0,$$

$$\lim_{r\to\infty}r\left[\hat{n}_{V_\infty}\times\left(\nabla\times\begin{pmatrix}\vec{E}_n\\ \vec{H}_n\end{pmatrix}\right)-jk_0\begin{pmatrix}\vec{E}_n\\ \vec{H}_n\end{pmatrix}\right]=0,$$

$$\Sigma(\vec{E}_n,\vec{H}_n,\vec{J}_n)=f(\vec{r},t);$$

- The adjoint problem

$$\begin{cases}\nabla\times\vec{e}_n-\mu\dfrac{\partial\vec{h}_n}{\partial t}+\eta_0^2\displaystyle\sum_{k=1}^{K}\left(\vec{H}_n-\vec{H}_n^m\right)\delta(\vec{r}-\vec{r}_k)=0,\\[2mm] \nabla\times\vec{h}_n+\varepsilon\dfrac{\partial\vec{e}_n}{\partial t}-\sigma\vec{e}_n+\displaystyle\sum_{k=1}^{K}\left(\vec{E}_n-\vec{E}_n^m\right)\delta(\vec{r}-\vec{r}_k)=0,\end{cases}$$

$$\vec{e}_n\big|_{t=T}=0,\qquad \vec{h}_n\big|_{t=T}=0,$$

$$\lim_{r\to\infty}r\left[\hat{n}_{V_\infty}\times\left(\nabla\times\begin{pmatrix}\vec{e}_n\\ \vec{h}_n\end{pmatrix}\right)+jk_0\begin{pmatrix}\vec{e}_n\\ \vec{h}_n\end{pmatrix}\right]=0;$$

- The optimality conditions with respect to μ, ε, and σ, respectively

$$\begin{cases} \displaystyle\sum_{n=1}^{N} \int_0^T \vec{h}_n \cdot \frac{\partial \vec{H}_n}{\partial t} dt - \beta \mu_0^{-2} \nabla^2 \mu + \gamma(\mu - \overline{\mu}) = 0 \\ \displaystyle -\left(\sum_{n=1}^{N} \int_0^T \vec{e}_n \cdot \frac{\partial \vec{E}_n}{\partial t} dt + \beta \varepsilon_0^{-2} \nabla^2 \varepsilon - \gamma(\varepsilon - \overline{\varepsilon}) \right) = 0 \\ \displaystyle -\left(\sum_{n=1}^{N} \int_0^T \vec{e}_n \cdot \vec{E}_n dt + \beta \nabla^2 \sigma - \gamma(\sigma - \overline{\sigma}) \right) = 0. \end{cases}$$

Notice that the Lagrange multipliers represent ingoing waves whereas the fields $\vec{E}$ and $\vec{H}$ are outgoing waves. The Lagrange multipliers have the same nature as electromagnetic field variables. The sources that drive these fields are the discrepancies between the measured and calculated fields at the measurement points.

The residuals of the optimality conditions provide the minus gradient of the objective functional with respect to the optimization variables, and therefore they represent the descent direction that is required in any gradient-based optimization procedure as those discussed in Chapter 4. We present results obtained with a steepest descent scheme; see [161, 164] concerning the application of quasi-Newton schemes to electromagnetic inverse problems.

7.4.3 The Maxwell Curl Equations

In order to illustrate some approximation aspects of the Maxwell curl equations and of the corresponding adjoint equations, we discuss the FDTD discretization of these equations with different spatial dimensions as follows.

Consider the following four Maxwell equations

$$\begin{cases} \nabla \cdot \vec{B} = 0 & \text{Gauss's law for the magnetic field,} \\ \nabla \cdot \vec{D} = 0 & \text{Gauss's law for the electric field,} \\ \nabla \times \vec{E} = -\dfrac{\partial \vec{B}}{\partial t} - \vec{M} & \text{Faraday's law,} \\ \nabla \times \vec{H} = \dfrac{\partial \vec{D}}{\partial t} + \vec{J} & \text{Ampere's law.} \end{cases} \tag{7.62}$$

In a linear, isotropic, and nondispersive medium, we can write

$$\vec{D} = \varepsilon \vec{E} = \varepsilon_0 \varepsilon_r \vec{E}, \qquad \vec{B} = \mu \vec{H} = \mu_0 \mu_r \vec{H}$$

and

$$\vec{J} = \vec{J}_s + \sigma \vec{E}, \qquad \vec{M} = \vec{M}_s + \sigma^* \vec{H}.$$

Thus, the equations in (7.62) become

$$\begin{cases} \dfrac{\partial \vec{H}}{\partial t} = -\dfrac{1}{\mu} \nabla \times \vec{E} - \dfrac{1}{\mu} \left(\vec{M}_s + \sigma^* \vec{H} \right), \\ \dfrac{\partial \vec{E}}{\partial t} = \dfrac{1}{\varepsilon} \nabla \times \vec{H} - \dfrac{1}{\varepsilon} \left(\vec{J}_s + \sigma \vec{E} \right). \end{cases} \tag{7.63}$$

In Cartesian coordinates, the equations (7.63) have the following components

$$\begin{cases} \dfrac{\partial H_x}{\partial t} = \dfrac{1}{\mu}\left(\dfrac{\partial E_y}{\partial z} - \dfrac{\partial E_z}{\partial y} - \left(M_{sx} + \sigma^* H_x\right)\right), \\ \dfrac{\partial H_y}{\partial t} = \dfrac{1}{\mu}\left(\dfrac{\partial E_z}{\partial x} - \dfrac{\partial E_x}{\partial z} - \left(M_{sy} + \sigma^* H_y\right)\right), \\ \dfrac{\partial H_z}{\partial t} = \dfrac{1}{\mu}\left(\dfrac{\partial E_x}{\partial y} - \dfrac{\partial E_y}{\partial x} - \left(M_{sz} + \sigma^* H_z\right)\right), \end{cases} \tag{7.64}$$

$$\begin{cases} \dfrac{\partial E_x}{\partial t} = \dfrac{1}{\varepsilon}\left(\dfrac{\partial H_z}{\partial y} - \dfrac{\partial H_y}{\partial z} - \left(J_{sx} + \sigma E_x\right)\right), \\ \dfrac{\partial E_y}{\partial t} = \dfrac{1}{\varepsilon}\left(\dfrac{\partial H_x}{\partial z} - \dfrac{\partial H_z}{\partial x} - \left(J_{sy} + \sigma E_y\right)\right), \\ \dfrac{\partial E_z}{\partial t} = \dfrac{1}{\varepsilon}\left(\dfrac{\partial H_y}{\partial x} - \dfrac{\partial H_x}{\partial y} - \left(J_{sz} + \sigma E_z\right)\right). \end{cases} \tag{7.65}$$

These six coupled PDEs are the fundamental equations in the three-dimensional case.

The two-dimensional case is obtained setting $\partial/\partial z = 0$ in (7.64) and in (7.65). Then, we can group the six equations in the following two independent set of equations

$$\begin{cases} \dfrac{\partial H_x}{\partial t} = \dfrac{1}{\mu}\left(-\dfrac{\partial E_z}{\partial y} - \left(M_{sx} + \sigma^* H_x\right)\right), \\ \dfrac{\partial H_y}{\partial t} = \dfrac{1}{\mu}\left(\dfrac{\partial E_z}{\partial x} - \left(M_{sy} + \sigma^* H_y\right)\right), \\ \dfrac{\partial E_z}{\partial t} = \dfrac{1}{\varepsilon}\left(\dfrac{\partial H_y}{\partial x} - \dfrac{\partial H_x}{\partial y} - \left(J_{sz} + \sigma E_z\right)\right) \end{cases} \tag{7.66}$$

and

$$\begin{cases} \dfrac{\partial E_x}{\partial t} = \dfrac{1}{\varepsilon}\left(\dfrac{\partial H_z}{\partial y} - \left(J_{sx} + \sigma E_x\right)\right), \\ \dfrac{\partial E_y}{\partial t} = \dfrac{1}{\varepsilon}\left(-\dfrac{\partial H_z}{\partial x} - \left(J_{sy} + \sigma E_y\right)\right), \\ \dfrac{\partial H_z}{\partial t} = \dfrac{1}{\mu}\left(\dfrac{\partial E_x}{\partial y} - \dfrac{\partial E_y}{\partial x} - \left(M_{sz} + \sigma^* H_z\right)\right). \end{cases} \tag{7.67}$$

The set in (7.66) is called the transverse magnetic mode (TM_z), and the set in (7.67) is called the transverse electric mode (TE_z).

For completeness, we write the equations for the one-dimensional case. We impose $\partial/\partial y = 0$ in both modes and obtain

$$\begin{cases} \dfrac{\partial H_x}{\partial t} = -\dfrac{1}{\mu}\left(M_{sx} + \sigma^* H_x\right), \\ \dfrac{\partial H_y}{\partial t} = \dfrac{1}{\mu}\left(\dfrac{\partial E_z}{\partial x} - \left(M_{sy} + \sigma^* H_y\right)\right), \\ \dfrac{\partial E_z}{\partial t} = \dfrac{1}{\varepsilon}\left(\dfrac{\partial H_y}{\partial x} - \left(J_{sz} + \sigma E_z\right)\right) \end{cases}$$

and

$$\begin{cases} \dfrac{\partial E_x}{\partial t} = -\dfrac{1}{\varepsilon}(J_{sx} + \sigma E_x), \\ \dfrac{\partial E_y}{\partial t} = \dfrac{1}{\varepsilon}\left(-\dfrac{\partial H_z}{\partial x} - (J_{sy} + \sigma E_y)\right), \\ \dfrac{\partial H_z}{\partial t} = \dfrac{1}{\mu}\left(-\dfrac{\partial E_y}{\partial x} - (M_{sz} + \sigma^* H_z)\right). \end{cases}$$

Now, we pose M_{sx} and J_{sx} zero for every t and also H_x E_x zero in $t = 0$. Consequently, we have $E_x = 0$ and $H_x = 0$ for every t. Then, we obtain the following two sets of equations

$$\begin{cases} \dfrac{\partial H_y}{\partial t} = \dfrac{1}{\mu}\left(\dfrac{\partial E_z}{\partial x} - (M_{sy} + \sigma^* H_y)\right), \\ \dfrac{\partial E_z}{\partial t} = \dfrac{1}{\varepsilon}\left(\dfrac{\partial H_y}{\partial x} - (J_{sz} + \sigma E_z)\right) \end{cases} \tag{7.68}$$

and

$$\begin{cases} \dfrac{\partial E_y}{\partial t} = \dfrac{1}{\varepsilon}\left(-\dfrac{\partial H_z}{\partial x} - (J_{sy} + \sigma E_y)\right), \\ \dfrac{\partial H_z}{\partial t} = \dfrac{1}{\mu}\left(-\dfrac{\partial E_y}{\partial x} - (M_{sz} + \sigma^* H_z)\right). \end{cases} \tag{7.69}$$

The system (7.68) is called TEM mode x-directed and z-polarized. The system (7.69) is called TEM mode x-directed and y-polarized.

7.4.4 The FDTD Discretization

To evaluate the functional gradient and therefore define an optimization step, we need to solve the direct Maxwell problem and the adjoint Maxwell equations. This solution procedure can be formulated in the framework of FDTD methods; see [120, 333] for a detailed introduction to this methodology. We remark that the Maxwell equations are first-order PDEs. In this case, it is known that stable finite difference schemes are obtained in a staggered grid. It was Yee [366] who first proposed such a scheme defined on a staggered grid that is now called the Yee grid. The Yee algorithm is based upon the discretization of the Maxwell curl equations in time and space domain with central difference formula on a staggered grid. The scheme does not enforce the Gauss laws, because they are a consequence of the two Maxwell curl equations. The electric and magnetic fields are sampled on different discrete spatial points and at different discrete temporal points.

To illustrate the Yee scheme in the three-dimensional case, consider a cubic spatial domain in $\mathbb{R}^3$, divided in N_x, N_y, and N_z cells in the x, y, and z directions, respectively. In this lattice we can identify a spatial point as $(i, j, k) = (i\Delta x, j\Delta y, k\Delta z)$, where i, j, k are integers and Δx, Δy, and Δz are the spatial increments on the x, y, and z axes, respectively. The generic single cell in the position (i, j, k) is called the Yee cell and is illustrated in Figure 7.10.

Also the time is discretized, with an index n and time increment Δt, uniform over the observation interval. A function u in a three-dimensional grid on the point (i, j, k) and at the nth discretized time instant is individuated by

$$u(i\Delta x, j\Delta y, k\Delta z, n\Delta t) = u^n_{ijk}.$$

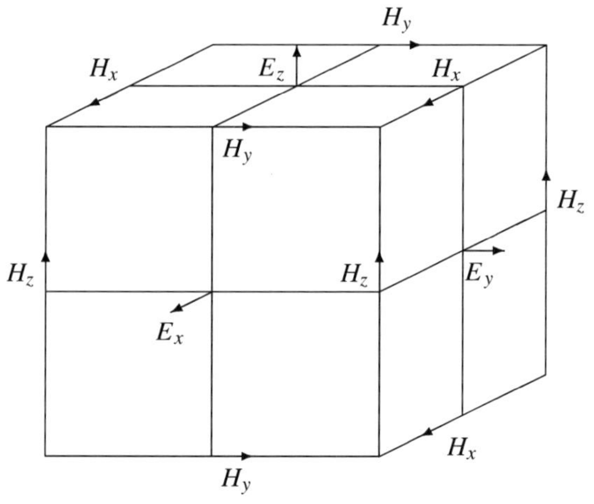

Figure 7.10. *The Yee cell.*

The sampling instants of the electric and magnetic fields are not the same. The magnetic field is sampled at the instants $0, \Delta t, 2\Delta t, \ldots n\Delta t, \ldots$, and the electric field is sampled at $1/2\Delta t, (1+1/2)\Delta t, \ldots (n+1/2)\Delta t, \ldots$. Second-order accurate centered differences are used for the first-order derivatives [333].

Consider the one-dimensional case, where we have increments only in the x and the t axis. We start with the discretization of (7.68), more precisely with the first equation. We imagine evaluating the equation in the point $(i, n+1/2)$. We have

$$\frac{H_{y|i}^{n+1} - H_{y|i}^{n}}{\Delta t} = \frac{1}{\mu_i}\left(\frac{E_{z|i+1/2}^{n+1/2} - E_{z|i-1/2}^{n+1/2}}{\Delta x} - M_{s\,y|i}^{n+1/2} - \sigma_i^* H_{y|i}^{n+1/2}\right).$$

With the hypothesis that the field H is evaluated every time step Δt, we do not have its value in $(n+1/2)\Delta t$. Because of this, we use a semi-implicit approximation as follows

$$H_{y|i}^{n+1/2} = \frac{H_{y|i}^{n} + H_{y|i}^{n+1}}{2}.$$

Substituting this and manipulating the equation, we obtain

$$H_{y|i}^{n+1} = \frac{1 - \frac{\Delta t \sigma_i^*}{2\mu_i}}{1 + \frac{\Delta t \sigma_i^*}{2\mu_i}} H_{y|i}^{n} + \frac{\frac{\Delta t}{\mu_i \Delta x}}{1 + \frac{\Delta t \sigma_i^*}{2\mu_i}}\left(E_{z|i+1/2}^{n+1/2} - E_{z|i-1/2}^{n+1/2} - \Delta x M_{s\,y|i}^{n+1/2}\right).$$

In the same manner, we discretize the second equation of (7.68) on $(i-1/2, n)$ as follows

$$\frac{E_{z|i-1/2}^{n+1/2} - E_{z|i-1/2}^{n-1/2}}{\Delta t} = \frac{1}{\varepsilon_{i-1/2}}\left(\frac{H_{y|i}^{n} - H_{y|i-1}^{n}}{\Delta x} - J_{s\,z|i-1/2}^{n} - \sigma_{i-1/2} E_{z|i-1/2}^{n}\right).$$

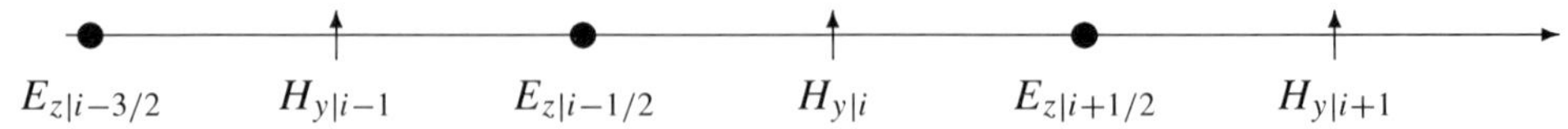

Figure 7.11. *Spatial arrangement for E_z and H_y in the one-dimensional case.*

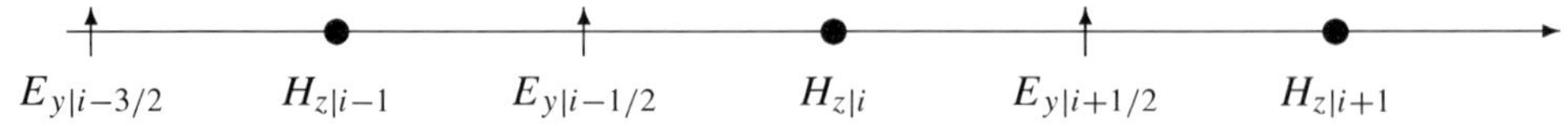

Figure 7.12. *Spatial arrangement for E_y and H_z in the one-dimensional case.*

Using the semi-implicit approximation on E, since its value in $n\Delta t$ is not stored, we have

$$E^{n+1/2}_{z|i-1/2} = \frac{1-\frac{\Delta t\sigma_{i-1/2}}{2\varepsilon_{i-1/2}}}{1+\frac{\Delta t\sigma_{i-1/2}}{2\varepsilon_{i-1/2}}} E^{n-1/2}_{z|i-1/2} + \frac{\frac{\Delta t}{\varepsilon_{i-1/2}\Delta x}}{1+\frac{\Delta t\sigma_{i-1/2}}{2\varepsilon_{i-1/2}}}\left(H^n_{y|i} - H^n_{y|i-1} - \Delta x J^n_{s\,z|i-1/2}\right).$$

The same reasoning is applied to the set of equations (7.69). Therefore we obtain the following discretized set of equations, and their spatial arrangements are given in Figures 7.11 and 7.12

$$\begin{cases} H^{n+1}_{y|i} = \dfrac{1-\frac{\Delta t\sigma^*_i}{2\mu_i}}{1+\frac{\Delta t\sigma^*_i}{2\mu_i}} H^n_{y|i} + \dfrac{\frac{\Delta t}{\mu_i\Delta x}}{1+\frac{\Delta t\sigma^*_i}{2\mu_i}}\left(E^{n+1/2}_{z|i+1/2} - E^{n+1/2}_{z|i-1/2} - \Delta x M^{n+1/2}_{s\,y|i}\right), \\ E^{n+1/2}_{z|i-1/2} = \dfrac{1-\frac{\Delta t\sigma_{i-1/2}}{2\varepsilon_{i-1/2}}}{1+\frac{\Delta t\sigma_{i-1/2}}{2\varepsilon_{i-1/2}}} E^{n-1/2}_{z|i-1/2} + \dfrac{\frac{\Delta t}{\varepsilon_{i-1/2}\Delta x}}{1+\frac{\Delta t\sigma_{i-1/2}}{2\varepsilon_{i-1/2}}}\left(H^n_{y|i} - H^n_{y|i-1} - \Delta x J^n_{s\,z|i-1/2}\right), \end{cases}$$

$$\begin{cases} E^{n+1/2}_{y|i-1/2} = \dfrac{1-\frac{\Delta t\sigma_{i-1/2}}{2\varepsilon_{i-1/2}}}{1+\frac{\Delta t\sigma_{i-1/2}}{2\varepsilon_{i-1/2}}} E^{n-1/2}_{y|i-1/2} + \dfrac{\frac{\Delta t}{\varepsilon_{i-1/2}\Delta x}}{1+\frac{\Delta t\sigma_{i-1/2}}{2\varepsilon_{i-1/2}}}\left(H^n_{z|i-1} - H^n_{z|i} - \Delta x J^n_{s\,y|i-1/2}\right), \\ H^{n+1}_{z|i} = \dfrac{1-\frac{\Delta t\sigma^*_i}{2\mu_i}}{1+\frac{\Delta t\sigma^*_i}{2\mu_i}} H^n_{z|i} + \dfrac{\frac{\Delta t}{\mu_i\Delta x}}{1+\frac{\Delta t\sigma^*_i}{2\mu_i}}\left(E^{n+1/2}_{y|i-1/2} - E^{n+1/2}_{y|i+1/2} - \Delta x M^{n+1/2}_{s\,z|i}\right). \end{cases}$$

Typically, in the implementation of the FDTD algorithm it is not needed to logically offset the fields. Thus, when considering the $i = 1$ position in the arrays which contain the values of the fields E_z and H_y, we have the values $E_{z|1/2}$ and $H_{y|1}$.

While we do not write the equations for the two-dimensional case and refer the reader to [246, 287] for details, we would like to show the spatial arrangements of the electromagnetic fields in the Yee cell as depicted in Figure 7.10. The TE_z mode is obtained if we consider the fields in the plane $k+1/2$; the TM_z mode is obtained in the cut plane k. Thus, in the bidimensional lattice, the two modes are as shown in the Figures 7.13 and 7.14.

The FDTD scheme may result in nonphysical dispersion of the waves in the grid, while the need of a stable scheme requires us to satisfy a relation between the time step and the space increment, thus influencing the numerical accuracy of the solution. Concerning dispersion, one obtains waves with phase velocities different from the phase velocities

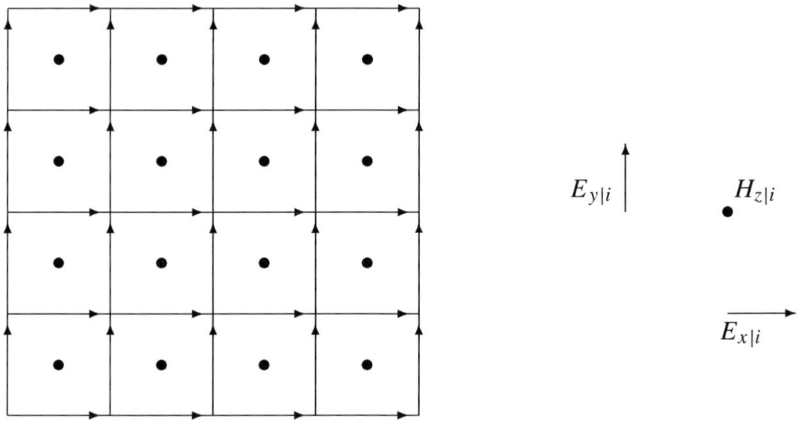

Figure 7.13. *Spatial arrangement for the TE_z mode. In the right-hand picture, the fields of a single cell are depicted.*

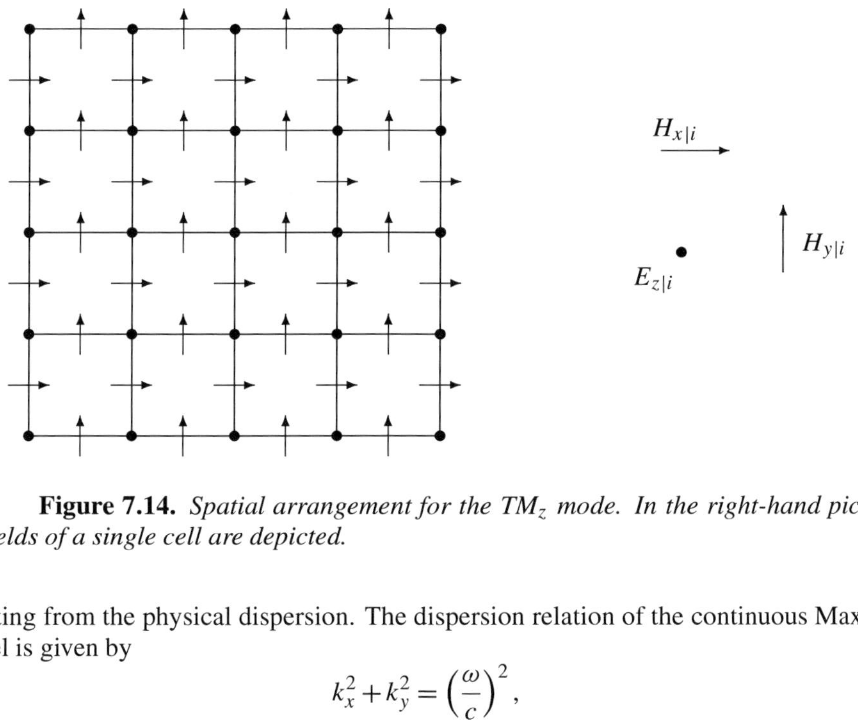

Figure 7.14. *Spatial arrangement for the TM_z mode. In the right-hand picture, the fields of a single cell are depicted.*

resulting from the physical dispersion. The dispersion relation of the continuous Maxwell model is given by

$$k_x^2 + k_y^2 = \left(\frac{\omega}{c}\right)^2,$$

and the numerical dispersion relation is given by

$$\left[\frac{\omega}{2c}\,\frac{\sin(\omega\Delta x/2)}{\omega\Delta x/2}\right]^2 = \left[\frac{k_x}{2}\,\frac{\sin(k_x\Delta x/2)}{k_x\Delta x/2}\right]^2 + \left[\frac{k_y}{2}\,\frac{\sin(k_y\Delta y/2)}{k_y\Delta y/2}\right]^2.$$

Notice that if $\Delta t \to 0$, $\Delta x \to 0$, and $\Delta y \to 0$, the numerical dispersion relation reduces to the ideal dispersion relation.

Concerning stability, stability imposes a bound on the maximum time step relative to the spatial increments as follows (Courant condition)

$$\frac{c\,\Delta t}{\Delta s} \leq \frac{1}{\sqrt{d}}, \tag{7.70}$$

where Δs is the length of a side of a uniform cell and d is the number of space dimensions in the problem.

Concerning accuracy of the FDTD scheme, we report results given in [258], where it is proved that the FDTD scheme is second-order accurate on general nonuniform rectangular grids. Consider a three-dimensional cavity given by a parallelepiped $\Omega = [0, L_x] \times [0, L_y] \times [0, L_z]$. We assume that the electric field satisfies a perfectly conducting (PEC) boundary condition on the boundary of the domain for all $t \in [0, T]$ as

$$\vec{E} \times \vec{n} = 0,$$

and we assume the following initial conditions

$$\vec{E}(\vec{r}, 0) = \vec{E}_0(\vec{r}), \qquad \vec{H}(\vec{r}, 0) = \vec{H}_0(\vec{r}).$$

We also request particular regularity properties of the field solutions of the Maxwell equations (7.62) in the generic three-dimensional case as follows

$$\vec{E} \in C\left([0, T]; [C^3(\Omega)]^3\right), \qquad \vec{H} \in C\left([0, T]; [C^3(\Omega)]^3\right) \cap C^1\left([0, T]; [C^2(\Omega)]^3\right).$$

Let $\vec{E}$ and $\vec{H}$ be the continuous fields and $\vec{E}^h$ and $\vec{H}^h$ be the discretized fields. With these hypotheses, it is possible to prove the following theorem [258].

Theorem 7.3. *Suppose that $\vec{E}$ and $\vec{H}$ are three times continuously differentiable on V, that $\partial \vec{H}/\partial t$ is twice continuously differentiable on V, and that all these derivative are continuous in time. Then, for any fixed $T > 0$, there is a constant C depending on T such that for $0 \leq t \leq T$, we have*

$$\|\vec{E}(t) - \vec{E}^h(t)\|_E + \|\vec{H}(t) - \vec{H}^h(t)\|_H \leq Ch^2, \tag{7.71}$$

where $\|\cdot\|_E$ and $\|\cdot\|_H$ are L^2-norms on the staggered grid.

7.4.5 Perfectly Matched Layer

An FDTD scheme applies on computational domains that are finite, while we need to simulate waves that radiate outward in the unbounded free space. To model this phenomenon with a bounded computational domain, Berenger [33] defined an innovative absorbing material scheme called the perfectly matched layer (PML). The computational domain extends over this medium that generates reflection neither at the interface between the main domain and the PML nor at the end of the PML. This scheme requires matching at the interface of material properties and attenuation of the field propagating in the PML region; see Figure 7.15.

In the one-dimensional case the PML represents a lossy medium, with electric conductivity σ and magnetic conductivity σ^* and without any anisotropy; i.e., the electromagnetic characteristics do not depend on the direction. Consider a TEM mode x-directed and

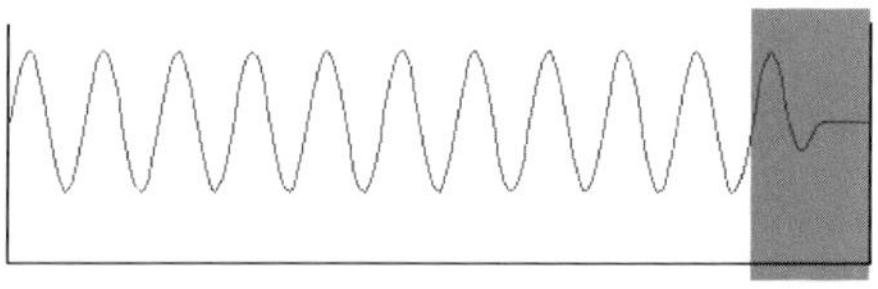

Figure 7.15. *Idea for the truncation of computational domain in the one-dimensional case. At the interface between the main domain (Region* 1*) and the PML region (Region* 2*), the wave is not reflected and it is attenuated.*

z-polarized in a structure similar to that depicted in Figure 7.15. Assume the wave propagating from the lossless Region 1 to Region 2 that corresponds to the PML region. We assume that the PML region ends with a PEC boundary. We assume that Region 2 is made of lossy material. The Maxwell equations that govern the wave propagation in the two regions are as follows

$$\begin{cases} \dfrac{\partial H_y}{\partial t} = \dfrac{1}{\mu_1}\dfrac{\partial E_z}{\partial x}, \\ \dfrac{\partial E_z}{\partial t} = \dfrac{1}{\varepsilon_1}\dfrac{\partial H_y}{\partial x}, \end{cases} \tag{7.72}$$

$$\begin{cases} \dfrac{\partial H_y}{\partial t} = \dfrac{1}{\mu_2}\left(\dfrac{\partial E_z}{\partial x} - \sigma^* H_y\right), \\ \dfrac{\partial E_z}{\partial t} = \dfrac{1}{\varepsilon_2}\left(\dfrac{\partial H_y}{\partial x} - \sigma E_z\right). \end{cases} \tag{7.73}$$

Now, consider the problem of defining different material properties at the interface of the two regions in such a way as to avoid reflection at the interface. For this purpose, consider (7.72) and (7.73) in the frequency domain. We have

$$\begin{cases} j\omega\mu_1 \tilde{H}_y = \dfrac{\partial \tilde{E}_z}{\partial x}, \\ j\omega\varepsilon_1 \tilde{E}_z = \dfrac{\partial \tilde{H}_y}{\partial x}, \end{cases} \tag{7.74}$$

$$\begin{cases} j\omega \tilde{H}_y = \dfrac{1}{\mu_2}\left(\dfrac{\partial \tilde{E}_z}{\partial x} - \sigma^* \tilde{H}_y\right), \\ j\omega \tilde{E}_z = \dfrac{1}{\varepsilon_2}\left(\dfrac{\partial \tilde{H}_y}{\partial x} - \sigma \tilde{E}_z\right). \end{cases} \tag{7.75}$$

In the equations (7.75), corresponding to Region 2, it is possible to group some terms as follows

$$\begin{cases} j\omega\mu_2\left(1 + \dfrac{\sigma^*}{j\omega\mu_2}\right)\tilde{H}_y = \dfrac{\partial \tilde{E}_z}{\partial x}, \\ j\omega\varepsilon_2\left(1 + \dfrac{\sigma}{j\omega\varepsilon_2}\right)\tilde{E}_z = \dfrac{\partial \tilde{H}_y}{\partial x}. \end{cases} \tag{7.76}$$

With these new equations, we can identify the reflectionless conditions. For the normal incidence case the reflection coefficient Γ is given by

$$\Gamma = \frac{\eta_1 - \eta_2}{\eta_1 + \eta_2}, \tag{7.77}$$

where η_1 and η_2 are the impedances in Region 1 and in Region 2, respectively. Note that in the one-dimensional case, there is always normal incidence to the interface.

To obtain the impedance in Region 2 it is sufficient to observe the structure of equations (7.76). We find that, analogously to Region 1, the equivalent electric permittivity ε_2' and the magnetic permeability μ_2' are given by

$$\varepsilon_2' = \varepsilon_2 \left(1 + \frac{\sigma}{j\omega\varepsilon_2}\right), \qquad \mu_2' = \mu_2 \left(1 + \frac{\sigma^*}{j\omega\mu_2}\right).$$

Now, the impedances in Region 1 and in Region 2 are as follows

$$\eta_1 = \sqrt{\frac{\mu_1}{\varepsilon_1}}, \qquad \eta_2 = \sqrt{\frac{\mu_2(1 + \sigma^*/(j\omega\mu_2))}{\varepsilon_2(1 + \sigma/(j\omega\varepsilon_2))}}.$$

Notice from (7.77) that Γ is zero if the impedances are equal. For this reason, we require that $\eta_1 = \eta_2$ and the following

$$\varepsilon_1 = \varepsilon_2, \qquad \mu_1 = \mu_2.$$

Now, we have

$$\sqrt{\frac{\mu_1}{\varepsilon_1}} = \sqrt{\frac{\mu_1}{\varepsilon_1}} \sqrt{\frac{(1 + \sigma^*/(j\omega\mu_1))}{(1 + \sigma/(j\omega\varepsilon_1))}}$$

and we require that

$$\frac{\left(1 + \frac{\sigma^*}{j\omega\mu_1}\right)}{\left(1 + \frac{\sigma}{j\omega\varepsilon_1}\right)} = 1,$$

which implies $\frac{\sigma^*}{\mu_1} = \frac{\sigma}{\varepsilon_1}$, and we obtain

$$\sigma^* = \sigma \frac{\mu_1}{\varepsilon_1} = \sigma(\eta_1)^2. \tag{7.78}$$

The same analysis and properties are obtained for a PML region with a TEM mode x-directed and z-polarized.

Notice that the implementation of a one-dimensional PML region is not completed by the appropriate choice of values σ and σ^* at the interface. In fact in a continuous space, the wave impacts the PML region, crosses this layer, and is reflected by the PEC wall that ends the layer. Then, the wave comes back in the computational domain after a second crossing of the PML layer. Thus we have an effective reflection factor defined as follows

$$R = e^{-2\sigma\,\eta d},$$

where d is the PML thickness with a PEC wall ending.

Now, the straightforward setting of σ and σ^* involves significant numerical reflection. To reduce this reflection error, Berenger proposed to gradually increase the losses in the PML region, starting from 0 at the interface and increasing to a maximum value towards the end of the region. With this approach, the reflection factor becomes

$$R = e^{-2\eta \int_0^d \sigma(x)dx}. \tag{7.79}$$

Typically, we use a polynomial variation for the PML electric loss, like the following

$$\sigma(x) = \left(\frac{x}{d}\right)^m \sigma_{max}, \tag{7.80}$$

where d is the PML layer length, x is the position in the PML layer, varying from 0 to d, σ_{max} is the electric conductivity in $x = d$, and m is the grading exponential. The integral in (7.79) becomes

$$\int_0^d \sigma(x)dx = \int_0^d \sigma_{max}\left(\frac{x}{d}\right)^m dx = \frac{\sigma_{max}}{d^m}\left[\frac{x^{m+1}}{m+1}\right]_0^d = \frac{\sigma_{max}\, d}{m+1}.$$

So, with a polynomial grading the reflection factor is given by

$$R = e^{-2\eta\sigma_{max}\, d/(m+1)}. \tag{7.81}$$

Using this equation with chosen m, d and the desired reflection factor, we can solve the equation for the maximum value of σ at the end of the PML layer. Then, solving (7.81) for σ_{max}, we obtain

$$\sigma_{max} = -\frac{(m+1)\ln(R)}{2\eta d}.$$

With σ_{max}, we can determine σ^* with (7.78). The PML is completely definite.

Typical values of m are chosen to be between 3 and 4, whereas the values of d and R depend on the applications. For example, R may be equal to e^{-8} or e^{-16}; d may correspond to 5 to 10 cells or more.

Exponential Time Differencing in the PML Region

Consider a computational domain surrounded by a PML layer, with conductivities σ and σ^*, related by (7.78). We can see this relation from another point of view: σ^*/μ and σ/ε are in [Hz], so μ/σ^* and ε/σ are in [sec]. Thus, we can write (7.78) in this manner:

$$\tau = \frac{\varepsilon}{\sigma}, \quad \tau^* = \frac{\mu}{\sigma^*} \quad \Rightarrow \quad \tau = \tau^*.$$

Typically, to have a small size of the PML layer, σ is chosen large, so τ is of order $O(10^{-12})$s; i.e., the wave attenuation in the PML region can be very rapid. But in the computational domain, the time scale of typical FDTD simulation is $T_c > 10^{-9}$s. Then, $\tau \ll T_c$. If we set the FDTD parameters to resolve τ, it follows that T_c is overresolved. This implies that, because of the Courant criterion, the spatial cell size Δx overresolves the fields. On the other hand, if we consider only T_c and not τ, the standard FDTD could fail.

Then, we need to introduce an approximation scheme that gradually changes in a way so as to accommodate the variation of the lossy material in the PML region. The resulting scheme is known as the exponential time differencing (ETD) method; see [275]. The ETD scheme is used in the implementation of the FDTD scheme instead of the standard FDTD such that the setting in the computational grid is that of the classic FDTD scheme.

Consider a TEM mode x-directed and z-polarized in a lossy medium. We have

$$\begin{cases} \dfrac{\partial E_z}{\partial t} + \dfrac{\sigma}{\varepsilon} E_z = \dfrac{1}{\varepsilon} \dfrac{\partial H_y}{\partial x}, \\ \dfrac{\partial H_y}{\partial t} + \dfrac{\sigma^*}{\mu} H_y = \dfrac{1}{\mu} \dfrac{\partial E_z}{\partial x}. \end{cases}$$

We focus on the first equation, but the reasoning is the same for the second equation and for the TEM mode x-directed and y-polarized. We can observe that

$$\frac{\partial}{\partial t}\left(e^{(\sigma/\varepsilon)t} E_z\right) = \left(\frac{\partial E_z}{\partial t} + \frac{\sigma}{\varepsilon} E_z\right) e^{(\sigma/\varepsilon)t}.$$

This means that we can rewrite the considered equation, after multiplying both sides by $e^{(\sigma/\varepsilon)t}$ as follows

$$\frac{\partial}{\partial t}\left(e^{(\sigma/\varepsilon)t} E_z\right) = e^{(\sigma/\varepsilon)t}\left(\frac{1}{\varepsilon}\frac{\partial H_y}{\partial x}\right).$$

Using the central difference formula in the point $(i-1/2, n)$, we have

$$\frac{e^{(\sigma_{i-1/2}/\varepsilon)(n+1/2)\Delta t} E_{z|i-1/2}^{n+1/2} - e^{(\sigma_{i-1/2}/\varepsilon)(n-1/2)\Delta t} E_{z|i-1/2}^{n-1/2}}{\Delta t}$$
$$= \frac{1}{\varepsilon} \frac{H_{y|i}^n - H_{y|i-1}^n}{\Delta x} e^{(\sigma_{i-1/2}/\varepsilon)(n)\Delta t},$$

that is,

$$e^{(\sigma_{i-1/2}/\varepsilon)(n+1/2)\Delta t} E_{z|i-1/2}^{n+1/2}$$
$$= e^{(\sigma/\varepsilon)(n-1/2)\Delta t} E_{z|i-1/2}^{n-1/2} + \frac{\Delta t}{\varepsilon} \frac{H_{y|i}^n - H_{y|i-1}^n}{\Delta x} e^{(\sigma_{i-1/2}/\varepsilon)(n)\Delta t}. \tag{7.82}$$

Solving with respect to the next field value, i.e., dividing both sides by $e^{(\sigma_{i-1/2}/\varepsilon)(n+1/2)\Delta t}$, we obtain

$$E_{z|i-1/2}^{n+1/2} = E_{z|i-1/2}^{n-1/2} e^{-\sigma_{i-1/2}\Delta t/\varepsilon} + \frac{\Delta t}{\varepsilon \Delta x}(H_{y|i}^n - H_{y|i-1}^n) e^{-\sigma_{i-1/2}\Delta t/(2\varepsilon)}. \tag{7.83}$$

We can now use a Taylor approximation. Developing e^{-x} and approximating that for a sufficiently small x, we obtain

$$e^{-x} \cong e^{-x}\Big|_0 - e^{-x}\Big|_0 x = 1 - x,$$

that is,

$$x = 1 - e^{-x}.$$

Thus, we have

$$\frac{1}{\sigma}\frac{\Delta t \sigma}{\varepsilon} = \frac{1}{\sigma}(1 - e^{-\Delta t \sigma/\varepsilon}).$$

Applying these considerations in (7.83), we obtain

$$E^{n+1/2}_{z|i-1/2} = E^{n-1/2}_{z|i-1/2} e^{-\sigma_{i-1/2}\Delta t/\varepsilon} + \frac{(1-e^{-\Delta t \sigma_{i-1/2}/\varepsilon})}{\sigma_{i-1/2}\Delta x} e^{-\sigma_{i-1/2}\Delta t/(2\varepsilon)} (H^n_{y|i} - H^n_{y|i-1}).$$

In the same manner we derive the other equations. Thus, the set of equations for a TEM mode x-directed and y-polarized is given by

$$E^{n+1/2}_{z|i-1/2} = E^{n-1/2}_{z|i-1/2} e^{-\sigma_{i-1/2}\Delta t/\varepsilon} + \frac{(1-e^{-\Delta t \sigma_{i-1/2}/\varepsilon})}{\sigma_{i-1/2}\Delta x} e^{-\sigma_{i-1/2}\Delta t/(2\varepsilon)} (H^n_{y|i} - H^n_{y|i-1}),$$

$$H^{n+1}_{y|i} = H^n_{y|i} e^{-\sigma^*_i \Delta t/\mu} + \frac{(1-e^{-\Delta t \sigma^*_i/\mu})}{\sigma^*_i \Delta x} e^{-\sigma^*_i \Delta t/(2\mu)} (E^{n+1/2}_{z|i+1/2} - E^{n+1/2}_{z|i-1/2}),$$

whereas the TEM mode x-directed and y-polarized set of equations is as follows

$$E^{n+1/2}_{y|i-1/2} = E^{n-1/2}_{y|i-1/2} e^{-\sigma_{i-1/2}\Delta t/\varepsilon} + \frac{(1-e^{-\Delta t \sigma_{i-1/2}/\varepsilon})}{\sigma_{i-1/2}\Delta x} e^{-\sigma_{i-1/2}\Delta t/(2\varepsilon)} (H^n_{z|i-1} - H^n_{z|i}),$$

$$H^{n+1}_{z|i} = H^n_{z|i} e^{-\sigma^*_i \Delta t/\mu} + \frac{(1-e^{-\Delta t \sigma^*_i/\mu})}{\sigma^*_i \Delta x} e^{-\sigma^*_i \Delta t/(2\mu)} (E^{n+1/2}_{y|i-1/2} - E^{n+1/2}_{y|i+1/2}).$$

For a detailed discussion on the PML scheme and ETD scheme in two spatial dimensions we refer the reader to [246].

7.4.6 An Inverse Scattering Scenario

In this section, we report results of validation of the inverse scattering solution procedure based on the PDE optimization framework; see [246] for details and additional results. Consider the computational domain is a square whose edges are three times larger than the edges of a scatter square domain D. We take the edges of D corresponding to a length of 20 cells. The number of time steps is set to $n_{max} = 200$ and the time step size is$\Delta t = 10^{-12}$ s. The time domain is $(0, T)$, where $T = n_{max}\Delta t$s. The spatial increment, Δ, is equal in the x- and y-directions and it is set equal to $\Delta = 2c\Delta t = 5.99\,10^{-4}$ m, such that it satisfies the Courant limit discussed in Section 7.4.4. Thus, the length of the computational square domain is 0.036 m. The electromagnetic properties of the free space are given by $\mu_0 = 4\pi\, 10^{-7}$ H/m and $\varepsilon_0 = 8.854212\,10^{-12}$ F/m.

The measurement points and the source points are placed in the region surrounding the domain D. We consider four different incidence points and eight measurement points, i.e., $N = 4$ and $K = 8$. See Figure 7.16.

To generate the incidence fields, hard sources are used. This means a variation of the magnetic field is set directly where the transmitters are placed. For every n the excitation of the magnetic field was taken equal to the following

$$H_z(t) = \begin{cases} \eta_0^{-1}\left[0.3\sin\left(\frac{2\pi t}{T_1}\right) + \sin\left(\frac{2\pi t}{T_2}\right)\right], & 0 \le t \le T_1, \\ 0, & T_1 < t. \end{cases} \tag{7.84}$$

For the evaluation of the inversion algorithm, we consider two hidden scatterers that are two squares with relative electrical permittivities of $\varepsilon_r = 1.5$ and $\varepsilon_r = 2$, respectively.

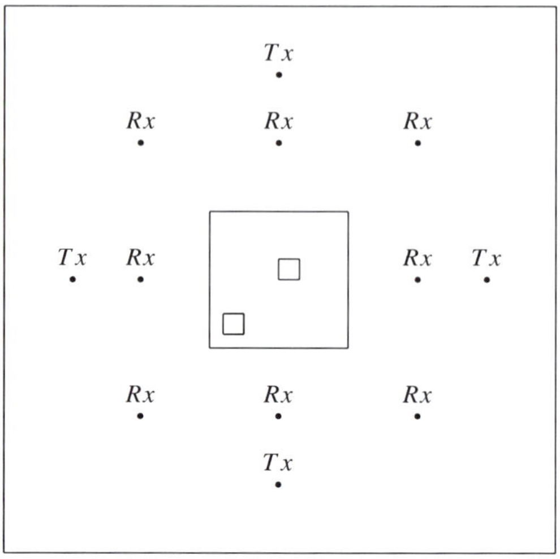

Figure 7.16. *Position of transmitters (Tx) and receivers (Rx) in the domain. Two scatterers are depicted in the domain D.*

With these scatterers, the associated direct scattering problem is solved and the resulting fields are recorded at the receivers for all times in $(0,T)$ and adding noise to avoid inverse crime in the construction of the synthetic data which is used as measurements. We have a white Gaussian noise such that the signal-to-noise ratio is SNR = 10 dB. In Figure 7.17, the two objects used to design the synthetic data are depicted (the smaller square corresponds to $\varepsilon_r = 2$) together with the reconstructed objects (after 20 steps of a steepest descent scheme).

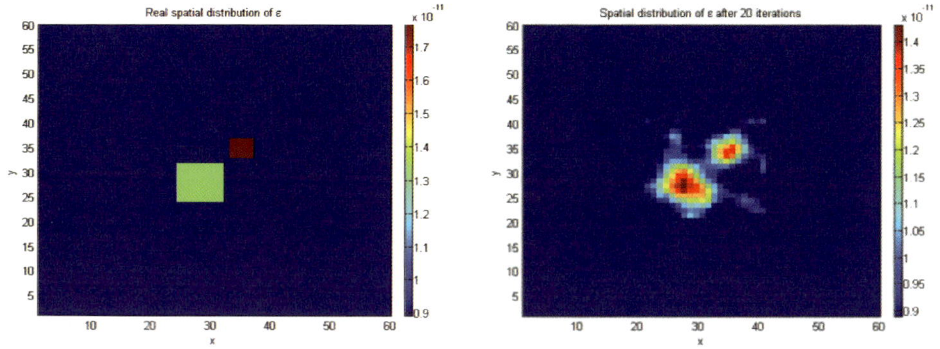

Figure 7.17. *Two-dimensional visualization of the two square scatterer scenario (left) and two-dimensional visualization of the reconstructed objects.*

Bibliography

[1] N.U. Ahmed and K.L. Teo, *Optimal Control of Distributed Parameter Systems*, Elsevier, New York, 1981.

[2] V. Akçelik, G. Biros, A. Draganescu, O. Ghattas, J. Hill, and B. van Bloemen Waanders, *Dynamic data-driven inversion for terascale simulations: Real-time identification of airborne contaminants*, ACM/IEEE SCXY conference series, 2005.

[3] R.R. Aliev and A.V. Panfilov, *A simple two-variable model of cardiac excitation*, Chaos Solitons Fractals, 7 (1996), pp. 293–301.

[4] F. Allgöwer, T.A. Badgwell, J.S. Qin, J.B. Rawlings, and S.J. Wright, *Nonlinear predictive control and moving horizon estimation—An introductory overview*, in P.M. Frank (Ed.), Advances in Control, Highlights of ECC'99, pp. 391–449.

[5] Committee on AMO 2010, Board on Physics and Astronomy, Division on Engineering and Physical Sciences, *Controlling the Quantum World: The Science of Atoms, Molecules, and Photons*, The National Academies Press, Washington, D.C., 2010.

[6] M. Annunziato and A. Borzì, *Fast solvers of Fredholm optimal control problems*, Numer. Math. Theory Methods Appl., 3 (2010), pp. 431–448.

[7] M. Annunziato and A. Borzì, *Optimal control of probability density functions of stochastic processes*, Math. Model. Anal., 15 (2010), pp. 393–407.

[8] X. Antoine, A. Arnold, C. Besse, M. Ehrhardt, and A. Schädle, *A review of transparent and artificial boundary conditions techniques for linear and nonlinear Schrödinger equations*, Commun. Comput. Phys., 4 (2008), pp. 729–796.

[9] N. Arada, E. Casas, and F. Tröltzsch, *Error estimates for a semilinear elliptic control problem*, Comput. Optim. Appl., 23 (2002), pp. 201–229.

[10] E. Arian and S. Ta'asan, *Smoothers for optimization problems*, in Seventh Copper Mountain Conference on Multigrid Methods, Vol. CP3339, NASA Conference Publication, NASA, N.D. Melson, T.A. Manteuffel, S.F. McCormick, and C.C. Douglas, editors, Hampton, VA, 1995, pp. 15–30.

[11] E. Arian and S. Ta'asan, *Multigrid one shot methods for optimal control problems: Infinite dimensional control*, ICASE Report No. 94-52, 1994.

[12] U. Ascher and E. Haber, *A multigrid method for distributed parameter estimation problems*, Electron. Trans. Numer. Anal., 15 (2003), pp. 1–12.

[13] U.M. Ascher and L.R. Petzold, *Computer Methods for Ordinary Differential Equations and Differential-Algebraic Equations*, SIAM, Philadelphia, 1998.

[14] K.E. Atkinson, *The Numerical Solution of Integral Equations of the Second Kind*, Cambridge Monographs on Applied and Computational Mathematics, Cambridge University Press, Cambridge, UK, 2009.

[15] J.A. Atwell, J.T. Borggaard, and B.B. King, *Reduced order controllers for Burgers' equation with a nonlinear observer*, Int. J. Appl. Math. Comput. Sci., 11 (2001), pp. 1311–1330.

[16] A. Auger, A. Ben Haj Yedder, E. Cances, C. Le Bris, C.M. Dion, A. Keller, and O. Atabek, *Optimal laser control of molecular systems: Methodology and results*, Math. Models Methods Appl. Sci., 12 (2002), pp. 1281–1315.

[17] I. Babuška, F. Nobile, and R. Tempone, *A stochastic collocation method for elliptic partial differential equations with random input data*, SIAM J. Numer. Anal., 45 (2007), pp. 1005–1034.

[18] I. Babuška, R. Tempone, and G. E. Zouraris, *Galerkin finite element approximations of stochastic elliptic partial differential equations*, SIAM J. Numer. Anal., 42 (2004), pp. 800–825.

[19] Z.Z. Bai, *Block preconditioners for PDE-constrained optimization problems*, Computing, 91 (2011), pp. 379–395.

[20] J.M. Ball, J.E. Marsden, and M. Slemrod, *Controllability for distributed bilinear systems*, SIAM J. Control Optim., 20 (1982), pp. 575–597.

[21] H.T. Banks and K. Kunisch, *Estimation Techniques for Distributed Parameter Systems*, Birkhäuser, Boston, 1989.

[22] W. Bao and Q. Du, *Computing the ground state solution of Bose–Einstein condensates by a normalized gradient flow*, SIAM J. Sci. Comput., 25 (2004), pp. 1674–1697.

[23] W. Bao, D. Jaksch, and P.A. Markowich, *Numerical solution of the Gross-Pitaevskii Equation for Bose–Einstein condensation*, J. Comput. Phys., 187 (2003), pp. 318–342.

[24] J.L. Barron, D.J. Fleet, and S.S. Beauchemin, *Performance of optical flow techniques*, IJCV, 12 (1994), pp. 43–77.

[25] M. Batista and A.A. Karawia, *The use of the Sherman-Morrison-Woodbury formula to solve cyclic block tri-diagonal and cyclic block penta-diagonal linear systems of equations*, Appl. Math. Comput., 210 (2009), pp. 558–563.

[26] A. Battermann and E. Sachs, *Block preconditioner for KKT systems in PDE-governed optimal control problems*, in Fast Solution of Discretized Optimization Problems, International Series on Numerical Mathematics, Vol. 138, Birkhäuser, Basel, 2001, pp. 1–18.

[27] K. Beauchard and J.-M. Coron, *Controllability of a quantum particle in a moving potential well*, J. Funct. Anal., 232 (2006), pp. 328–389.

[28] R. Becker, M. Braack, D. Meidner, R. Rannacher, and B. Vexler, *Adaptive finite element methods for PDE-constrained optimal control problems*, in Reactive Flows, Diffusion and Transport, Springer, Berlin, 2006, pp. 177–205.

[29] R. Becker and B. Vexler, *Optimal control of the convection-diffusion equation using stabilized finite element methods*, Numer. Math., 106 (2007), pp. 349–367.

[30] M.P. Bendsoe, *Optimization of Structural Topology, Shape, and Material*, Springer-Verlag, Berlin, 1995.

[31] C.L. Bennett, *Time domain inverse scattering*, IEEE Trans. Antennas and Propagation, 29 (1981), pp. 213–219.

[32] M. Benzi, G.H. Golub, and J. Liesen, *Numerical solution of saddle point problems*, Acta Numer., 14 (2005), pp. 1–137.

[33] J.P. Berenger, *A perfectly matched layer for the absorption of electromagnetic waves*, J. Comput. Phys., 114 (1994), pp. 185–200.

[34] M. Bergounioux, K. Ito, and K. Kunisch, *Primal-dual strategy for constrained optimal control problems*, SIAM J. Control Optim., 37 (1999), pp. 1176–1194.

[35] L.D. Berkovitz, *Optimal Control Theory*, Springer-Verlag, New York, Heidelberg, 1974.

[36] D.P. Bertsekas, *Nonlinear Programming*, Athena Scientific, Belmont, MA, 1995.

[37] G. Biros and O. Ghattas, *Parallel Lagrange–Newton–Krylov–Schur methods for PDE-constrained optimization. Part* I: *The Krylov–Schur solver*, SIAM J. Sci. Comput., 27 (2005), pp. 687–713.

[38] G. Biros and O. Ghattas, *Parallel Lagrange–Newton–Krylov–Schur methods for PDE-constrained optimization. Part* II: *The Lagrange–Newton solver and its application to optimal control of steady viscous flows*, SIAM J. Sci. Comput., 27 (2005), pp. 714–739.

[39] H.G. Bock, *Randwertproblemmethoden zur Parameteridentifizierung in Systemen nichtlinearer Differentialgleichungen*, Bonner Mathematische Schriften 183, Bonn, 1987.

[40] H. Bock, M. Diehl, and E. Kostina, *SQP Methods with Inexact Jacobians for Inequality Constrained Optimization*, IWR-Preprint, Universität Heidelberg, Heidelberg, 2004.

[41] H. Bock, M. Diehl, E. Kostina, and J. Schlöder, *Constrained optimal feedback control of systems governed by large differential algebraic equations*, in L. Biegler, O. Ghattas, M. Heinkenschloss, D. Keyes, and B. van Bloemen Waanders (Eds.), Real-Time PDE-Constrained Optimization, Vol. 3, SIAM, Philadelphia, 2007, pp. 3–24.

[42] L. Borcea, *Nonlinear multigrid for imaging electrical conductivity and permittivity at low frequency*, Inverse Problems, 17 (2001), pp. 329–359.

[43] F.A. Bornemann and P. Deuflhard, *The cascadic multigrid method for elliptic problems*, Numer. Math., 75 (1996), pp. 135–152.

[44] A. Borzì, *Multigrid and sparse-grid schemes for elliptic control problems with random coefficients*, Comput. Vis. Sci., 13 (2010), pp. 153–160.

[45] A. Borzì, *On the convergence of the MG/OPT method*, PAMM, Proceedings GAMM Annual Meeting - Luxembourg 2005, 5 (2005), pp. 735–736.

[46] A. Borzì, *Space-time multigrid methods for solving unsteady optimal control problems*, in L. Biegler, O. Ghattas, M. Heinkenschloss, D. Keyes, and B. van Bloemen Waanders (Eds.), Real-Time PDE-Constrained Optimization, Computational Science and Engineering, Vol. 3, SIAM, Philadelphia, 2007, pp. 97–113.

[47] A. Borzì, *Smoothers for control- and state-constrained optimal control problems*, Comput. Vis. Sci., 11 (2008), pp. 59–66.

[48] A. Borzì, *High-order discretization and multigrid solution of elliptic nonlinear constrained optimal control problems*, J. Comput. Appl. Math., 200 (2007), pp. 67–85.

[49] A. Borzì, *Fast multigrid methods for parabolic optimal control problems*, Proceedings of the 18th GAMM-Seminar Leipzig, 2002, pp. 1–10.

[50] A. Borzì, *Multigrid methods for parabolic distributed optimal control problems*, J. Comput. Appl. Math., 157 (2003), pp. 365–382.

[51] A. Borzì and G. Borzì, *An algebraic multigrid method for a class of elliptic differential systems*, SIAM J. Sci. Comput., 25 (2003), pp. 302–323.

[52] A. Borzì and G. Borzì, *An efficient algebraic multigrid method for solving optimality systems*, Comput. Vis. Sci., 7 (2004), pp. 183–188.

[53] A. Borzì and R. Griesse, *Distributed optimal control of lambda-omega systems*, J. Numer. Math., 14 (2006), pp. 17–40.

[54] A. Borzì and U. Hohenester, *Multigrid optimization schemes for solving Bose–Einstein condensate control problems*, SIAM J. Sci. Comput., 30 (2008), pp. 441–462.

[55] A. Borzì, K. Ito, and K. Kunisch, *Optimal control formulation for determining optical flow*, SIAM J. Sci. Comput., 24 (2002), pp. 818-847.

[56] A. Borzì, K. Ito, and K. Kunisch, *An optimal control approach to optical flow computation*, Internat. J. Numer. Methods Fluids., 40 (2002), pp. 231–240.

[57] A. Borzì and K. Kunisch, *The numerical solution of the steady state solid fuel ignition model and its optimal control*, SIAM J. Sci. Comput., 22 (2000), pp. 263–284.

[58] A. Borzì and K. Kunisch, *A multigrid method for optimal control of time-dependent reaction diffusion processes*, in K.H. Hoffmann, R. Hoppe, and V. Schulz (Eds.), Fast Solution of Discretized Optimization Problems, International Series on Numerical Mathematics, Vol. 138, Birkhäuser, Basel, 2001, pp. 50–57.

[59] A. Borzì and K. Kunisch, *A multigrid scheme for elliptic constrained optimal control problems*, Comput. Optim. Appl., 31 (2005), pp. 309–333.

[60] A. Borzì and K. Kunisch, *A globalization strategy for the multigrid solution of elliptic optimal control problems*, Optim. Methods Softw., 21 (2006), pp. 445–459.

[61] A. Borzì, K. Kunisch, and D.Y. Kwak, *Accuracy and convergence properties of the finite difference multigrid solution of an optimal control optimality system*, SIAM J. Control Optim., 41 (2003), pp. 1477–1497.

[62] A. Borzì, K. Kunisch, and M. Vanmaele, *A multigrid approach to the optimal control of solid fuel ignition problems*, in E. Dick, K. Riemslagh, and J. Vierendeels (Eds.), Lecture Notes in Computer Science and Engineering, Vol. 14, European Multigrid Meeting 1999, Springer-Verlag, Berlin, 2000, pp. 59–65.

[63] A. Borzì, J. Salomon, and S. Volkwein, *Formulation and numerical solution of finite-level quantum optimal control problems*, J. Comput. Appl. Math., 216 (2008), pp. 170–197.

[64] A. Borzì and V. Schulz, *Multigrid methods for PDE optimization*, SIAM Rev., 51 (2009), pp. 361–395.

[65] A. Borzì, V. Schulz, C. Schillings, and G. von Winckel, *On the treatment of distributed uncertainties in PDE-constrained optimization.* GAMM-Mitteilungen, Vol. 33, No. 2, 2010, pp. 230–246.

[66] A. Borzì, G. Stadler, and U. Hohenester, *Optimal quantum control in nanostructures: Theory and application to a generic three-level system*, Phys. Rev. A, 66 (2002), 053811.

[67] A. Borzì and G. von Winckel, *Multigrid methods and sparse-grid collocation techniques for parabolic optimal control problems with random coefficients*, SIAM J. Sci. Comput., 31 (2009), pp. 2172–2192.

[68] A. Borzì and G. von Winckel, *A POD framework to determine robust controls in PDE optimization*, Comput. Vis. Sci., to appear.

[69] A. Borzì and G. von Winckel, *SPQUAD: Computes the sparse grid quadrature abscissae and weights on an orthotope/hyperrectangle using the Clenshaw-Curtis rule*, http://www.ing.unisannio.it/borzi/spquad.m, 2008.

[70] U. Boscain, G. Charlot, J.-P. Gauthier, S. Guerin, and H.-R. Jauslin, *Optimal control in laser-induced population transfer for two- and three-level quantum systems*, J. Math. Phys., 43 (2002), pp. 2107–2132.

[71] J. Bourgain, *Global Solutions of Nonlinear Schrödinger Equations*, American Mathematical Society Colloquium Publications, Vol. 46, American Mathematical Society, Providence, RI, 1999.

[72] D. Bouwmeester, A. Ekert, and A. Zeilinger (Eds.), *The Physics of Quantum Information*. Springer, Berlin, 2000.

[73] D. Braess, *Towards algebraic multigrid for elliptic problems of second order*, Computing, 55 (1995), pp. 379–393.

[74] D. Braess and R. Sarazin, *An efficient smoother for the Stokes problem*, Appl. Numer. Math., 23 (1997), pp. 3–19.

[75] J.H. Bramble, *Error estimates for difference methods in forced vibration problems*, SIAM J. Numer. Anal., 3 (1966), pp. 1–12.

[76] J.H. Bramble, *Multigrid Methods*, Pitman Research Notes in Mathematics Series, Vol. 294, Longman Scientific and Technical, Essex, 1993.

[77] J.H. Bramble and B.E. Hubbard, *On a finite difference analogue of an elliptic boundary problem which is neither diagonally dominant nor of non-negative type*, J. Math. Phys., 43 (1964), pp. 117–132.

[78] J.H. Bramble, D.Y. Kwak, and J.E. Pasciak, *Uniform convergence of multigrid V-cycle iterations for indefinite and nonsymmetric problems*, SIAM J. Numer. Anal., 31 (1994), pp. 1746–1763.

[79] J.H. Bramble and J.E. Pasciak, *New convergence estimates for multigrid algorithms*, Math. Comp., 49 (1987), pp. 311–329.

[80] J.H. Bramble and J.E. Pasciak, *The analysis of smoothers for multigrid algorithms*, Math. Comp., 58 (1992), pp. 467–488.

[81] J.H. Bramble, J.E. Pasciak, J. Wang, and J. Xu, *Convergence estimates for multigrid algorithms without regularity assumptions*, Math. Comp., 57 (1992), pp. 23–45.

[82] J.H. Bramble, J.E. Pasciak, and J. Xu, *The analysis of multigrid algorithms with nonnested spaces or noninherited quadratic forms*, Math. Comp., 56 (1991), pp. 1–34.

[83] A. Brandt, *Multi-level adaptive solutions to boundary-value problems*, Math. Comp., 31 (1977), pp. 333–390.

[84] A. Brandt, *Multigrid Techniques:* 1984 *Guide with Applications to Fluid Dynamics*, GMD-Studien Vol. 85, St. Augustin, Germany, 1984.

[85] A. Brandt, *Rigorous quantitative analysis of multigrid,* I: *Constant coefficients two-level cycle with L_2-norm*, SIAM J. Numer. Anal., 31 (1994), pp. 1695–1730.

[86] A. Brandt, *Algebraic multigrid theory: The symmetric case*, Appl. Math. Comput., 19 (1986), pp. 23–56.

[87] A. Brandt, *General highly algebraic coarsening*, Electron. Trans. Numer. Anal., 10 (2000), pp. 1–20.

[88] A. Brandt and C.W. Cryer, *Multigrid algorithms for the solution of linear complementarity problems arising from free boundary problems*, SIAM J. Sci. Statist. Comput., 4 (1983), pp. 655–684.

[89] A. Brandt and J. Greenwald, *Parabolic multigrid revisited*, in W. Hackbusch and U. Trottenberg (Eds.), Multigrid Methods III, International Series on Numerical Mathematics, Vol. 98, Birkhäuser, Basel, 1991.

[90] A. Brandt and D. Ron, *Multigrid solvers and multilevel optimization strategies*, in J. Cong and J.R. Shinnerl (Eds.), Multilevel Optimization in VLSICAD, Series: Combinatorial Optimization , Vol. 14, Kluwer Academic Publishers, Dordrecht, The Netherlands, 2003, pp. 1–69.

[91] M. Brezina, A. J. Cleary, R. D. Falgout, V. E. Henson, J. E. Jones, T. A. Manteuffel, S. F. McCormick, and J. W. Ruge, *Algebraic multigrid based on element interpolation (AMGe)*, SIAM J. Sci. Comput., 22 (2000), pp. 1570–1592.

[92] H. Brézis, *Monotonicity Methods in Hilbert Spaces and Some Applications to Nonlinear Partial Differential Equations*, in E. Zarantonello (Ed.), Contributions to Nonlinear Functional Analysis, Academic Press, New York, 1971, pp. 101–156.

[93] H. Brézis, M-G. Crandall, and A. Pazy, *Perturbations of nonlinear maximal monotone sets in Banach spaces*, Comm. Pure Appl. Math., 23 (1970), pp. 123–144.

[94] W.L. Briggs, V.E. Henson, and S.F. McCormick, *A Multigrid Tutorial*, 2nd ed., SIAM, Philadelphia, 2000.

[95] P.W. Brumer and M. Shapiro, *Principles of the Quantum Control of Molecular Processes*, Wiley-VCH, Berlin, 2003.

[96] R. Bulirsch, A. Miele, J. Stoer, and K.H. Well (Eds.), *Optimal Control*, International Series on Numerical Mathematics, Vol. 111, Birkhäuser, Basel, 1993.

[97] H.-J. Bungartz and S. Dirnstorfer, *Multivariate quadrature on adaptive sparse grids*, Computing, 71 (2003), pp. 89–114.

[98] H.-J. Bungartz and M. Griebel, *Sparse grids*, Acta Numer., 13 (2004), pp. 147–269.

[99] J. Burkardt, M. Gunzburger, and J. Peterson, *Insensitive functionals, inconsistent gradients, spurious minima, and regularized functionals in flow optimization problems*, Int. J. Comput. Fluid Dyn., 16 (2002), pp. 171–185.

[100] A.G. Butkovskiy and Yu.I. Samoilenko, *Control of Quantum-Mechanical Processes and Systems*, Kluwer Academic Publishers, Dordrecht, The Netherlands, 1990.

[101] A.Z. Capri, *Nonrelativistic Quantum Mechanics*, World Scientific, River Edge, NJ, 2002.

[102] E. Catinas, *Inexact Newton methods: The inexact, inexact perturbed, and quasi-Newton methods are equivalent models*, Math. Comp., 74 (2004), pp. 291–301.

[103] T. Chambrion, P. Mason, M. Sigalotti, and U. Boscain, *Controllability of the discrete-spectrum Schrödinger equation driven by an external field*, Ann. Inst. H. Poincaré Anal. Non Linéaire, 26 (2009), pp. 329–349.

[104] F.H. Clarke, *Optimization and Nonsmooth Analysis*, John Wiley and Sons, New York, 1983.

[105] D. Colton, H.W. Engl, A.K. Louis, J.R. McLaughlin, and W. Rundell (Eds.), *Surveys on Solution Methods for Inverse Problems*, Springer-Verlag, Vienna, 2000.

[106] J.-M. Coron, A. Grigoriu, C. Lefter, and G. Turinici, *Quantum control design by Lyapunov trajectory tracking for dipole and polarizability coupling*, New J. Phys., 11 (2009), 105034.

[107] Y.H. Dai and Y. Yuan, *A nonlinear conjugate gradient with a strong global convergence property*, SIAM J. Optim., 10 (1999), pp. 177–182.

[108] D. D'Alessandro, *Introduction to Quantum Control and Dynamics*, Chapman & Hall, Boca Raton, FL, 2008.

[109] F. Dalfovo, G. Stefano, L.P. Pitaevskii, and S. Stringari, *Theory of Bose–Einstein condensation in trapped gases*, Rev. Modern Phys., 71 (1999), pp. 463–512.

[110] M. C. Delfour and J.-P. Zolésio, *Shapes and Geometries: Analysis, Differential Calculus, and Optimization*, Advances in Design and Control, Vol. 4, SIAM, Philadelphia, 2001.

[111] R.S. Dembo, S.C. Eisenstat, and T. Steihaug, *Inexact Newton methods*, SIAM J. Numer. Anal., 19 (1982), pp. 400–408.

[112] W. Desch, F. Kappel, and K. Kunisch (Eds.), *Control and Estimation of Distributed Parameter Systems*, International Series on Numerical Mathematics, Vol. 143, Birkhäuser, Basel, 2003.

[113] M. Diehl, H.G. Bock, and J.P. Schlöder, *A real-time iteration scheme for nonlinear optimization in optimal feedback control*, SIAM J. Control Optim., 43 (2005), pp. 1714–1736.

[114] H.S. Dollar and A.J. Wathen, *Approximate factorization constraint preconditioners for saddle-point matrices*, SIAM J. Sci. Comput., 27 (2006), pp. 1555–1572.

[115] C.C. Douglas, *Madpack: A family of abstract multigrid or multilevel solvers*, Comput. Appl. Math., 14 (1995), pp. 3–20.

[116] C.C. Douglas, *Multi-grid algorithms with applications to elliptic boundary value problems*, SIAM J. Numer. Anal., 21 (1984), pp. 236–254.

[117] A. Draganescu, *Two Investigations in Numerical Analysis: Monotonicity Preserving Finite Element Methods, and Multigrid Methods for Inverse Parabolic Problems*, Ph.D. thesis, University of Chicago, Chicago, IL, 2004.

[118] Th. Dreyer, B. Maar, and V. Schulz, *Multigrid optimization in applications*, J. Comput. Appl. Math., 120 (2000), pp. 67–84.

[119] D. Echeverria and P.W. Hemker, *Space mapping and defect correction*, Comput. Methods Appl. Math., 5 (2005), pp. 107–136.

[120] A.Z. Elsherbeni and V. Demis, *The Finite-Difference Time-Domain Method for Electromagnetics with Matlab Simulations*, Scitech, Raleigh, NC, 2009.

[121] E. Emmrich, *Two-step BDF time discretisation of nonlinear evolution problems governed by monotone operators with strongly continuous perturbations*, Comput. Methods Appl. Math., 9 (2009), pp. 37–62.

[122] M. Engel and M. Griebel, *A multigrid method for constrained optimal control problems*, J. Comput. Appl. Math., 235 (2011), pp. 4368–4388.

[123] K. Eppler, S. Schmidt, V. Schulz, and C. Ilic, *Preconditioning the pressure tracking in fluid dynamics by shape Hessian information*, J. Optim. Theory Appl., 141 (2009), pp. 513–531.

[124] L.C. Evans, *Partial Differential Equations*, Graduate Studies in Mathematics, Vol. 19, American Mathematical Society, Providence, RI, 2002.

[125] H.O. Fattorini, *Infinite Dimensional Optimization and Control Theory*, Cambridge University Press, Cambridge, UK, 1999.

[126] C. Fiolhais, F. Nogueira, and M. Marques (Eds.), *A Primer in Density Functional Theory*, Lectures Notes in Physics, Vol. 620, Springer, Berlin, 2003.

[127] H.P. Flath, L.C. Wilcox, V. Akçelik, J. Hill, B. van Bloemen Waanders, and O. Ghattas, *Fast algorithms for Bayesian uncertainty quantification in large-scale linear inverse problems based on low-rank partial Hessian approximations*, SIAM J. Sci. Comput., 33 (2011), pp. 407–432.

[128] C.A. Floudas and O. Stein, *The adaptive convexification algorithm: A feasible point method for semi-infinite programming*, SIAM J. Optim., 18 (2007), pp. 1187–1208.

[129] M.G. Forbes, M. Guay, and J.F. Forbes, *Control design for first-order processes: Shaping the probability density of the process state*, J. Process Control, 14 (2004), pp. 399–410.

[130] R.W. Freund and N.M. Nachtigal, *A new Krylov-subspace method for symmetric indefinite linear systems*, Proceedings of the 14th IMACS World Congress on Computational and Applied Mathematics (W. F. Ames, Ed.), 1994, pp. 1253–1256.

[131] M. Galle, *Ein Verfahren zur numerischen Simulation kompressibler, reibungsbehafteter Strömungen auf hybriden Netzen*. Technical Report DLR-FB 99-04, Deutsches Zentrum für Luft und Raumfahrt e.V., Cologne, Germany, 1999.

[132] B. Ganapathysubramanian and N. Zabaras, *Sparse grid collocation schemes for stochastic natural convection problems*, J. Comput. Phys., 225 (2007), pp. 652–685.

[133] J. Garcke, *A dimension adaptive sparse grid combination technique for machine learning*, ANZIAM J., 48 (2007), pp. C725–C740.

[134] J. Garcke, M. Griebel, and M. Thess, *Data mining with sparse grids*, Computing, 67 (2001), pp. 225–253.

[135] M. Gasca and T. Sauer, *Polynomial interpolation in several variables*, Adv. Comput. Math., 12 (2000), pp. 377–410.

[136] N. Gauger, *Das Adjungiertenverfahren in der aerodynamischen Formoptimierung*, Ph.D. thesis, TU Braunschweig, Braunschweig, Germany, 2003.

[137] N.R. Gauger, A. Griewank, and J. Riehme, *Extension of fixed point PDE solvers for optimal design by one-shot method - with first applications to aerodynamic shape optimization*, Eur. J. Comput. Mech. (REMN), 17 (2008), pp. 87–102.

[138] K. Geiger and C. Kanzow, *Theorie und Numerik restringierter Optimierungsaufgaben*, Springer, Berlin, 2002.

[139] V. Georgescu, *On the unique continuation property for Schrödinger Hamiltonians*, Helv. Phys. Acta, 52 (1979), pp. 655–670.

[140] T. Gerhold, *Overview of the hybrid RANS TAU-code*, in N. Kroll and J. K. Fassbender (Eds.), MEGAFLOW—Numerical Flow Simulation for Aircraft Design, Notes on Numerical Fluid Mechanics and Multidisciplinary Design, Vol. 89, Springer, Berlin, 2005, pp. 81–92.

[141] T. Gerstner and M. Griebel, *Dimension-adaptive tensor-product quadrature*, Computing, 71 (2003), pp. 65–87.

[142] R.G. Ghanem and P.D. Spanos, *Stochastic Finite Elements: A Spectral Approach*, Springer, New York, 1991.

[143] I. Gherman, *Approximate Partially Reduced SQP Approaches for Aerodynamic Shape Optimization Problems*, Ph.D. thesis, University of Trier, Trier, Germany, 2007.

[144] I. Gihman and A.V. Skorohod, *Stochastic Differential Equations*, Springer-Verlag, New York, 1972.

[145] J.C. Gilbert and J. Nocedal, *Global convergence properties of conjugate gradient methods for optimization*, SIAM J. Optim., 2 (1992), pp. 21–42.

[146] M.B. Giles and N.A. Pierce, *Adjoint equations in CFD: Duality, boundary conditions and solution behaviour*, AIAA Paper 97-1850.

[147] K. Glashoff and E. Sachs, *On theoretical and numerical aspects of the bang-bang principle*, Numer. Math., 29 (1977), pp. 93–113.

[148] H. Goldberg and F. Tröltzsch, *Second-order sufficient optimality conditions for a class of nonlinear parabolic boundary control problems*, SIAM J. Control Optim., 31 (1993), pp. 1007–1025.

[149] H. Goldberg and F. Tröltzsch, *On a SQP–multigrid technique for nonlinear parabolic boundary control problems*, in W.W. Hager and P.M. Pardalos (Eds.), Optimal Control: Theory, Algorithms, and Applications, Kluwer Academic Publishers, Dordrecht, The Netherlands, 1998, pp. 154–177.

[150] S. González Andrade and A. Borzì, *Multigrid second-order accurate solution of parabolic control-constrained problems*, Comput. Optim. Appl., to appear.

[151] I. G. Graham, F. Y. Kuo, D. Nuyens, R. Scheichl, and I. H. Sloan, *Quasi-Monte Carlo Methods for Computing Flow in Random Porous Media*, Preprint 04/10, Bath Institute for Complex Systems, Bath, UK, 2010.

[152] S. Gratton, A. Sartenaer, and Ph.L. Toint, *Second-order convergence properties of trust-region methods using incomplete curvature information, with an application to multigrid optimization*, J. Comput. Math., 24 (2006), pp. 676–692.

[153] A. Griewank, *The local convergence of Broyden-like methods in Lipschitzian problems in Hilbert spaces*, SIAM J. Numer. Anal., 24 (1987), pp. 684–705.

[154] A. Griewank, *Evaluating Derivatives: Principles and Techniques of Algorithmic Differentiation*, Frontiers in Applied Mathematics, Vol. 19, SIAM, Philadelphia, 2000.

[155] L. Grüne and A. Rantzer, *On the infinite horizon performance of receding horizon controllers*, IEEE Trans. Automat. Control, 53 (2008), pp. 2100–2111.

[156] C.R. Gumbert, P.A. Newman, and G.J.-W. Hou, *High-fidelity computational optimization for 3-d flexible wings: Part II—effect of random geometric uncertainty on design*, Optim. Eng., 6 (2005), pp. 139–156.

[157] M.D. Gunzburger, *Perspectives in Flow Control and Optimization*, SIAM, Philadelphia, 1987.

[158] M.D. Gunzburger, L. Hou, and T.P. Svobodny, *Finite element approximations of an optimal control problem associated with the scalar Ginzburg-Landau equation*, Comput. Math. Appl., 21 (1991), pp. 123–131.

[159] M. Gustafsson and S. He, *An optimization approach to two-dimensional time domain electromagnetic inverse problems*, Radio Science, 35 (2000), pp. 525–536.

[160] W. Haack, *Geschoßformen kleinsten Wellenwiderstandes*, Bericht der Lilienthal-Gesellschaft, 136 (1941), pp. 14–28.

[161] E. Haber, *A parallel method for large scale time domain electromagnetic inverse problems*, Appl. Numer. Math., 58 (2008), pp. 422–434.

[162] E. Haber and U. Ascher, *Preconditioned all-at-once methods for large, sparse parameter estimation problems*, Inverse Problems, 17 (2001), pp. 1847–1864.

[163] E. Haber, U. Ascher, and D. Oldenburg, *Inversion of 3D electromagnetic data in frequency and time domain using an inexact all-at-once approach*, Geophysics, 69 (2004), pp. 1216–1228.

[164] E. Haber, D. Oldenburg, and R. Shekhtman, *Inversion of time domain three-dimensional electromagnetic data*, Geophysics, 171 (2007), pp. 550–564.

[165] W. Hackbusch, *On the regularity of difference schemes*, Ark. Mat., 19 (1981), pp. 71–95.

[166] W. Hackbusch, *On the regularity of difference schemes*. II. *Regularity estimates for linear and nonlinear problems*, Ark. Mat., 21 (1983), pp. 3–28.

[167] W. Hackbusch, *On the multi-grid method applied to difference equations*, Computing, 20 (1978), pp. 291–306.

[168] W. Hackbusch, *Fast solution of elliptic control problems*, J. Optim. Theory Appl., 31 (1980), pp. 565–581.

[169] W. Hackbusch, *A numerical method for solving parabolic equations with opposite orientations*, Computing, 20 (1978), pp. 229–240.

[170] W. Hackbusch, *Parabolic multigrid methods*, in R. Glowinski and J.-L. Lions (Eds.), Computing Methods in Applied Sciences and Engineering VI, North–Holland, Amsterdam, 1984, pp. 189–197.

[171] W. Hackbusch, *On the fast solving of parabolic boundary control problems*, SIAM J. Control Optim., 17 (1979), pp. 231–244.

[172] W. Hackbusch, *Numerical Solution of Linear and Nonlinear Parabolic Optimal Control Problems*, Lecture Notes in Control and Information Science, Vol. 30, Springer-Verlag, Berlin, 1981.

[173] W. Hackbusch, *Multi-grid Methods and Applications*, Springer-Verlag, New York, 1985.

[174] W. Hackbusch, *Elliptic Differential Equations*, Springer-Verlag, New York, 1992.

[175] W. Hackbusch, *Iterative Solution of Large Sparse Systems of Equations*, Springer-Verlag, New York, 1994.

[176] W. Hackbusch, *Integral Equations: Theory and Numerical Treatment*, Birkhäuser, Basel, 1995.

[177] W. Hackbusch and M. Griebel (Org.), *Multigrid and related methods for optimization problems*, Proceedings of the 18th GAMM-Seminar Leipzig, 2002.

[178] W. Hackbusch and A. Reusken, *Analysis of a damped nonlinear multilevel method*, Numer. Math., 55 (1989), pp. 225–246.

[179] W. Hackbusch and G. Wittum (Org.), *Seventh European MultiGrid Conference*, Pforzheim, 2002.

[180] A. Hagberg, E. Meron, I. Rubinstein, and B. Zaltzman, *Controlling domain patterns far from equilibrium*, Phys. Rev. Lett., 76 (1996), pp. 427–430.

[181] W.W. Hager and H. Zhang, *A new conjugate gradient method with guaranteed descent and an efficient line search*, SIAM J. Optim., 16 (2005), pp. 170–192.

[182] S.B. Hazra and V. Schulz, *Simultaneous pseudo-timestepping for aerodynamic shape optimization problems with state constraints*, SIAM J. Sci. Comput., 28 (2006), pp. 1078–1099.

[183] S.B. Hazra, V. Schulz, J. Brezillon, and N. Gauger, *Aerodynamic shape optimization using simultaneous pseudo-timestepping*, J. Comput. Phys., 204 (2005), pp. 46–64.

[184] M. Heinkenschloss, *The numerical solution of a control problem governed by a phase field model*, J. Optim. Theory Appl., 94 (1997), pp. 237–309.

[185] M. Heinkenschloss, R.H.W. Hoppe, and V. Schulz (Org.), *Numerical Techniques for Optimization Problems with PDE Constraints*, Mathematisches Forschungsinstitut Oberwolfach, 2003.

[186] M. Heinkenschloss and H. Nguyen, *Neumann–Neumann domain decomposition preconditioners for linear-quadratic elliptic optimal control problems*, SIAM J. Sci. Comput., 28 (2006), pp. 1001–1028.

[187] M. Heinkenschloss and L.N. Vicente, *Analysis of inexact trust-region SQP algorithms*, SIAM J. Optim., 12 (2001), pp. 283–302.

[188] R. Herzog and E. Sachs, *Preconditioned conjugate gradient method for optimal control problems with control and state constraints*, SIAM J. Matrix Anal. Appl., 31 (2010), pp. 2291–2317.

[189] R. M. Hicks and P. A. Henne, *Wing design by numerical optimization*, J. Aircraft, 15 (1978), pp. 407–412.

[190] M. Hintermüller and R.H.W. Hoppe, *Goal-oriented adaptivity in control constrained optimal control of partial differential equations*, SIAM J. Control Optim., 47 (2008), pp. 1721–1743.

[191] M. Hintermüller, K. Ito, and K. Kunisch, *The primal-dual active set strategy as a semismooth Newton method*, SIAM J. Optim., 13 (2003), pp. 865–888.

[192] M. Hintermüller and L.N. Vicente, *Space mapping for optimal control of partial differential equations*, SIAM J. Optim., 15 (2005), pp. 1002–1025.

[193] M. Hinze, *A variational discretization concept in control constrained optimization: The linear-quadratic case*, J. Comput. Optim. Appl., 30 (2005), pp. 45–63.

[194] M. Hinze, R. Pinnau, M. Ulbrich, and S. Ulbrich, *Optimization with PDE constraints*, Mathematical Modelling: Theory and Applications, Vol. 23, Springer Science + Business Media B.V., Dordrecht, The Netherlands, 2009.

[195] M. Hochbruck and C. Lubich, *On Magnus integrators for time-dependent Schrödinger equations*, SIAM J. Numer. Anal., 41 (2003), pp. 945–963.

[196] K.H. Hoffmann, R. Hoppe, and V. Schulz (Eds.), *Fast Solution of Discretized Optimization Problems*, International Series on Numerical Mathematics, Vol. 138, Birkhäuser, Basel, 2001.

[197] K.H. Hoffmann, I. Lasiecka, G. Leugering, J. Sprekels, and F. Tröltzsch (Eds.), *Optimal Control of Complex Structures*, International Series on Numerical Mathematics, Vol. 139, Birkhäuser, Basel, 2002.

[198] U. Hohenester, *Optical properties of semiconductor nanostructures: Decoherence versus quantum control*, in M. Rieth and W. Schommers (Eds.), Handbook of Theoretical and Computational Nanotechnology, American Scientific Publishers, Valencia, CA, 2005.

[199] U. Hohenester, P.-K. Rekdal, A. Borzì, and J. Schmiedmayer, *Optimal quantum control of Bose-Einstein condensates in magnetic microtraps*, Phys. Rev. A, 75 (2007), 023602.

[200] R.H.W. Hoppe, S.I. Petrova, and V.H. Schulz, *A primal-dual Newton-type interior-point method for topology optimization*, J. Optim. Theory Appl., 114 (2002), pp. 545–571.

[201] B.K.P. Horn and B.G. Schunck, *Determining optical flow*, Artificial Intelligence, 17 (1981), pp. 185–204.

[202] G. Horton and S. Vandewalle, *A space-time multigrid method for parabolic partial differential equations*, SIAM J. Sci. Comput., 16 (1995), pp. 848–864.

[203] P.A. Howard, *Multigrid Methods for Elliptic Optimal Control Problems*, CAAM Technical Report TR 06-14, Rice University, Houston, TX, 2006.

[204] S.P. Huang, S. Mahadevan, and R. Rebba, *Collocation-based stochastic finite element analysis for random field problems*, Probabilist. Eng. Mech., 22 (2007), pp. 194–205.

[205] L. Huyse, R. M. Lewis, W. Li, and S. L. Padula, *Probabilistic approach to free-form airfoil shape optimization under uncertainty*, AIAA J., 40 (2002), pp. 1764–1772.

[206] A. Iserles, H.Z. Munthe-Kaas, S. Nørsett, and A. Zanna, *Lie-group methods*, Acta Numer., 9 (2000), pp. 215–365.

[207] K. Ito and K. Kunisch, *Lagrange Multiplier Approach to Variational Problems and Applications*, SIAM, Philadelphia, 2008.

[208] K. Ito and K. Kunisch, *Asymptotic properties of receding horizon optimal control problems*, SIAM J. Control Optim., 40 (2002), pp. 1585–1610.

[209] K. Ito and K. Kunisch, *Optimal control of the solid fuel ignition model with H^1-cost*, SIAM J. Control Optim., 40 (2002), pp. 1455–1472.

[210] K. Ito and K. Kunisch, *BV-type regularization methods for convoluted objects with edge, flat and grey scales*, Inverse Problems, 16 (2000), pp. 909–928.

[211] K. Ito, K. Kunisch, V. Schulz, and I. Gherman, *Approximate nullspace iterations for KKT systems*, SIAM J. Matrix Anal. Appl., 31 (2010), pp. 1835–1847.

[212] G. Jumarie, *Tracking control of nonlinear stochastic systems by using path cross-entropy and Fokker-Planck equation*, Internat. J. Systems Sci., 23 (1992), pp. 1101–1114.

[213] T. Kailat, *Some integral equations with "nonrational" kernels*, IEEE Trans. Inform. Theory, IT-12 (1966), pp. 442–447.

[214] B. Kaltenbacher, *V-cycle convergence of some multigrid methods for ill-posed problems*, Math. Comp., 72 (2003), pp. 1711–1730.

[215] M. Kárný, *Towards fully probabilistic control design*, Automatica, 32 (1996), pp. 1719–1722.

[216] A. Kauffmann, *Optimal Control of the Solid Fuel Ignition Model*, Ph.D. thesis, Technical University of Berlin, Berlin, 1998.

[217] C.T. Kelley, *Iterative Methods for Optimization*, Frontiers in Applied Mathematics, Vol. 18, SIAM, Philadelphia, 1999.

[218] C.T. Kelley and E.W. Sachs, *Fast algorithms for compact fixed point problems with inexact function evaluations*, SIAM J. Sci. Statist. Comput., 12 (1991), pp. 725–742.

[219] C.T. Kelley and E.W. Sachs, *Multilevel algorithms for constrained compact fixed point problems*, SIAM J. Sci. Statist. Comput., 15 (1994), pp. 645–667.

[220] G. Kerschen, J-C. Golival, A.F. Vakakis, and L.A. Bergman, *The method of proper orthogonal decomposition for dynamical characterization and order reduction of mechanical systems: An overview*, Nonlinear Dynam., 41 (2005), pp. 147–169.

[221] J.T. King, *Multilevel algorithms for ill-posed problems*, Numer. Math., 61 (1992), pp. 311–334.

[222] A. Klimke and B. Wohlmuth, *Algorithm* 847: *Spinterp: Piecewise multilinear hierarchical sparse grid interpolation in MATLAB*, ACM Trans. Math. Software, 31 (2005), pp. 561–579.

[223] W. Kohn and L. J. Sham, *Self-consistent equations including exchange and correlation effects*, Phys. Rev. A, 140 (1965), pp. A1133–A1138.

[224] M.L. Krasnov, A.I. Kiselev, and G.I. Makarenko, *Integral Equations*, Mir, Moscow, 1983.

[225] A. Kröner and B. Vexler, *A priori error estimates for elliptic optimal control problems with a bilinear state equation*, J. Comput. Appl. Math., 230 (2009), pp. 781–802.

[226] K. Kunisch, K. Liang, and X. Lu, *Optimal control for an elliptic system with polygonal state constraints*, SIAM J. Control Optim., 48 (2010), pp. 5053–5072.

[227] K. Kunisch and S. Volkwein, *Proper orthogonal decomposition for optimality systems*, M2AN Math. Model. Numer. Anal., 42 (2008), pp. 1–23.

[228] F.S. Kupfer, *An infinite-dimensional convergence theory for reduced SQP methods in Hilbert space*, SIAM J. Optim., 6 (1996), pp. 126–163.

[229] Y. Kuramoto, *Chemical Oscillations, Waves, and Turbulence*, Springer-Verlag, Berlin, 1984.

[230] O. Lass, M. Vallejos, A. Borzì, and C.C. Douglas, *Implementation and analysis of multigrid schemes with finite elements for elliptic optimal control problems*, Computing, 84 (2009), pp. 27–48.

[231] I. Lesanovsky, T. Schumm, S. Hofferberth, L.M. Andersson, P. Krüger, and J. Schmiedmayer, *Adiabatic radio-frequency potentials for the coherent manipulation of matter waves*, Phys. Rev. A, 73 (2006), 033619.

[232] R.M. Lewis and S.G. Nash, *Model problems for the multigrid optimization of systems governed by differential equations*, SIAM J. Sci. Comput., 26 (2005), pp. 1811–1837.

[233] W. Li, L. Huyse, and S. Padula, *Robust airfoil optimization to achieve drag reduction over a range of mach numbers*, Struct. Multidiscip. Optim., 24 (2002), pp. 38–50.

[234] W. Li and S. Padula, *Robust Airfoil Optimization in High Resolution Design Space*, ICASE NASA Langley Research Center, Hampton, VA, 2003.

[235] J.-L. Lions, *Optimal Control of Systems Governed by Partial Differential Equations*, Springer, Berlin, 1971.

[236] J.-L. Lions, *Control of Distributed Singular Systems*, Gauthier-Villars, Paris, 1985.

[237] M. Loeve, *Probability Theory*, Vols. I & II, IV Ed., Springer-Verlag, New York, 1978.

[238] G.J.A. Loeven and H. Bijl, *Airfoil analysis with uncertain geometry using the probabilistic collocation method*, 48th AIAA/ASME/ASCE/AHS/ASC Structures, Structural Dynamics, and Materials, AIAA 2008-2070, 2008.

[239] D.G. Luenberger, *Optimization by Vector Space Methods*, Wiley, New York, 1969.

[240] B. Maar and V. Schulz, *Interior point multigrid methods for topology optimization*, Struct. Optim., 19 (2000), pp. 214–224.

[241] Y. Maday, J. Salomon, and G. Turinici, *Monotonic time-discretized schemes in quantum control*, Numer. Math., 103 (2006), pp. 323–338.

[242] Y. Maday, J. Salomon, and G. Turinici, *Monotonic parareal control for quantum systems*, SIAM J. Numer. Anal., 45 (2007), pp. 2468–2482.

[243] Y. Maday and G. Turinici, *New formulations of monotonically convergent quantum control algorithms*, J. Chem. Phys., 118 (2003), pp. 8191–8196.

[244] L. Magni, D.M. Raimondo, and F. Allgöwer, *Nonlinear Model Predictive Control*, Springer, Berlin, 2009.

[245] K. Malanowski, *Convergence of approximations vs. regularity of solutions for convex, control-constrained optimal-control problems*, Appl. Math. Optim., 8 (1981), pp. 69–95.

[246] R. Mancini, *An Adjoint-Based Optimization Scheme For Solving Time-Domain Electromagnetic Inverse Scattering Problems*, Tesi di Laurea in Computational Electromagnetics, Università degli Studi del Sannio, Facoltà di Ingegneria, Benevento, Italy, 2009.

[247] Y.M. Marzouk, H.N. Najm, and L.A. Rahn, *Stochastic spectral methods for efficient Bayesian solution of inverse problems*, J. Comput. Phys., 224 (2007), pp. 560–586.

[248] H. Matthies and G. Strang, *The solution of nonlinear finite element equations*, Internat. J. Numer. Methods Engrg., 14 (1979), pp. 1613–1626.

[249] H. Maurer and J. Zowe, *First and second order necessary and sufficient optimality conditions for infinite-dimensional programming problems*, Math. Programming, 16 (1979), pp. 98–110.

[250] D.Q. Mayne and H. Michalska, *Receding horizon control for nonlinear systems*, IEEE Trans. Automat. Control, 35 (1990), pp. 814–824.

[251] S.F. McCormick and J. Wade, *Multigrid solution of a linearized, regularized least-squares problem in electrical impedance tomography*, Inverse Problems, 9 (1993), pp. 697–713.

[252] S. McLoone and G. Irwin, *A variable memory quasi-Newton training algorithm*, Neural Process. Lett., 9 (1999), pp. 77–89.

[253] D. Meidner and B. Vexler, *Adaptive space-time finite element methods for parabolic optimization problems*, SIAM J. Control Optim., 46 (2007), pp. 116–142.

[254] C. Meyer, A. Rösch, and F. Tröltzsch, *Optimal control of PDEs with regularized pointwise state constraints*, Comput. Optim. Appl., 33 (2006), pp. 209–228.

[255] C. Meyer, A. Rösch, and F. Tröltzsch, *On an elliptic optimal control problem with pointwise mixed control-state constraints*, Recent Advances in Optimization. Proceedings of the 12th French-German-Spanish Conference on Optimization held in Avignon, September 20–24, 2004, Edited by A. Seeger, Lectures Notes in Economics and Mathematical Systems, Vol. 563, Springer-Verlag, Berlin, Heidelberg, 2006, pp. 187–204.

[256] A. Migdalas, P.M. Pardalos, and P. Varbrand (Eds.), *Multilevel Optimization: Algorithms and Applications (Nonconvex Optimization & Its Applications)*, Kluwer Academic Publishers, Dordrecht, The Netherlands, 1998.

[257] P. Monk, *Finite Element Methods for Maxwell's Equations*, Clarendon Press, Oxford, UK, 2003.

[258] P. Monk and E. Süli, *A convergence analysis of Yee's scheme on nonuniform grids*, SIAM J. Numer. Anal., 31 (1994), pp. 393–412.

[259] S.G. Nash, *Newton-type minimization via the Lanczos method*, SIAM J. Numer. Anal., 21 (1984), pp. 770–788.

[260] S.G. Nash, *A multigrid approach to discretized optimization problems*, Optim. Methods Softw., 14 (2000), pp. 99–116.

[261] S.G. Nash and R.M. Lewis, *Assessing the performance of an optimization-based multilevel method*, Optim. Methods Softw., 26 (2011), pp. 693–717.

[262] P. Neittaanmäki and D. Tiba, *Optimal Control of Nonlinear Parabolic Systems*, Marcel Dekker, New York, 1994.

[263] F. Nobile, R. Tempone, and C.G. Webster, *A sparse grid stochastic collocation method for partial differential equations with random input data*, SIAM J. Numer. Anal., 46 (2008), pp. 2309–2345.

[264] J. Nocedal, *Updating quasi-Newton matrices with limited storage*, Math. Comp., 35 (1980), pp. 773–782.

[265] J. Nocedal and S. Wright, *Numerical Optimization*, Springer, New York, 2000.

[266] D. Oeltz, *An algebraic multigrid method for linear elasticity*, 7th Copper Mountain Conference on Iterative Methods, Copper Mountain, CO, 2002.

[267] S. Oh, C.A. Bouman, and K.J. Webb, *Multigrid tomographic inversion with variable resolution data and image spaces*, IEEE Trans. Image Processing, 15 (2006), pp. 2805–2819.

[268] B. Øksendal, *Optimal control of stochastic partial differential equations*, Stochastic Anal. Appl., 23 (2005), pp. 165–179.

[269] M. Oristaglio and H. Blok, *Wavefield Imaging and Inversion in Electromagnetics and Acoustics*, Notes Study, 1995.

[270] S. Padula, C. Gumbert, and W. Li, *Aerospace applications of optimization under uncertainty*, ISUMA '03: Proceedings of the 4th International Symposium on Uncertainty Modelling and Analysis, IEEE Computer Society, Los Alamitos, CA, 2003.

[271] C.C. Paige and M.A. Saunders, *Solution of sparse indefinite systems of linear equations*, SIAM J. Numer. Anal., 12 (1975), pp. 617–629.

[272] A.V. Panfilov, *Spiral breakup in an array of coupled cells: The role of intercellular conductance*, Phys. Rev. Lett., 88 (2002), pp. 1–4.

[273] A.P. Peirce, M.A. Dahleh, and H. Rabitz, *Optimal control of quantum-mechanical systems: Existence, numerical approximation, and applications*, Phys. Rev. A, 37 (1988), pp. 4950–4964.

[274] H.J. Pesch and M. Plail, *The maximum principle of optimal control: A history of ingenious ideas and missed opportunities*, Control Cybernet., 38 (2009), pp. 973–995.

[275] P.G. Petropoulos, *Analysis of exponential time-differencing for FDTD in lossy dielectrics*, IEEE Trans. Antennas and Propagation, 45 (1997), pp. 1054–1057.

[276] O. Pironneau, *On optimum profiles in Stokes flow*, J. Fluid Mech., 59 (1973), pp. 117–128.

[277] O. Pironneau, *Optimal Shape Design for Elliptic Systems*, Springer, New York, 1984.

[278] M. Plail, *Die Entwicklung der optimalen Steuerungen*, Vandenhoeck & Ruprecht, Göttingen, Germany, 1998.

[279] S. Primak, V. Kontorovitch, and V. Lyandres, *Stochastic Methods and Their Applications to Communications: Stochastic Differential Equations Approach*, Wiley, New York, 2004.

[280] M.M. Putko, P.A. Newman, A.C. Taylor III, and L.L. Green, *Approach for uncertainty propagation and robust design in CFD using sensitivity derivatives*, J. Fluid Eng., 124 (2002), pp. 60–69.

[281] L. Qi and J. Sun, *A nonsmooth version of Newton's method*, Math. Programming, 58 (1993), pp. 353–368.

[282] H. Rabitz, M. Hsieh, and C. Rosenthal, *Quantum optimally controlled transition landscapes*, Science, 303 (2004), p. 998.

[283] H. Rabitz, G. Turinici, and E. Brown, *Control of Molecular Motion: Concepts, Procedures, and Future Prospects*, in P. Ciarlet and J. Lions (Eds.), Handbook of Numerical Analysis, Volume X, Elsevier, Amsterdam, 2003.

[284] V. Ramakrishna, H. Rabitz, M.V. Salapaka, M. Dahleh, and A. Peirce, *Controllability of molecular systems*, Phys. Rev. A, 62 (1995), pp. 960–966.

[285] T. Rees, H.S. Dollar, and A.J. Wathen, *Optimal solvers for PDE-constrained optimization*, SIAM J. Sci. Comput., 32 (2010), pp. 271–298.

[286] H. Regler and U. Ruede, *Layout optimization with algebraic multigrid methods (AMG)*, in Proceedings of the Sixth Copper Mountain Conference on Multigrid Methods, Copper Mountain, April 4–9, 1993, Conference Publication, NASA, 1993.

[287] I.T. Rekanos, *Time-domain inverse scattering using Lagrange multipliers: An iterative FDTD-based optimization technique*, J. Electromagn. Waves Appl., 17 (2003), pp. 271–289.

[288] H. Risken, *The Fokker-Planck Equation: Methods of Solution and Applications*, 2nd ed., Springer, Berlin, 1989.

[289] A. Rösch, *Error estimates for linear-quadratic control problems with control constraints*, Optim. Methods Softw., 21 (2006), pp. 121–134.

[290] E. Rosseel, T. Boonen, and S. Vandewalle, *Algebraic multigrid for stationary and time-dependent partial differential equations with stochastic coefficients*, Numer. Linear Algebra Appl., 15 (2008), pp. 141–163.

[291] J.W. Ruge and K. Stüben, *Algebraic multigrid (AMG)*, in S. McCormick (Ed.), Multigrid Methods, Frontiers in Applied Mathematics, Vol. 3, SIAM, Philadelphia, 1987, pp. 73–130.

[292] S. Rusinkiewicz, *Estimating curvatures and their derivatives on triangle meshes*, in Symposium on 3D Data Processing, Visualization, and Transmission, 2004.

[293] Y. Saad, *SPARSKIT: A Basic Tool Kit for Sparse Matrix Computations*, Rep. No. 90-20, Research Institute for Advanced Computer Science, NASA Ames Research Center, Moffett Field, CA, 1990.

[294] Y. Saad, J.R. Chelikowsky, and S.M. Shontz, *Numerical methods for electronic structure calculations of materials*, SIAM Rev., 52 (2010), pp. 3–54.

[295] Y. Saad and M.H. Schultz, *GMRES: A generalized minimal residual algorithm for solving nonsymmetric linear systems*, SIAM J. Sci. Statist. Comput., 7 (1986), pp. 856–869.

[296] A. Sacchetti, *Nonlinear time-dependent one-dimensional Schrödinger equation with double-well potential*, SIAM J. Math. Anal., 35 (2004), pp. 1160–1176.

[297] E. Sachs, *Broyden's method in Hilbert space*, Math. Programming, 35 (1986), pp. 71–82.

[298] M. Sadiku, *Numerical Techniques in Electromagnetics*, CRC Press, Boca Raton, FL, 2001.

[299] S. Schaffer, *Higher order multi-grid methods*, Math. Comp., 43 (1984), pp. 89–115.

[300] C. Schillings, S. Schmidt, and V. Schulz: *Efficient shape optimization for certain and uncertain aerodynamic design*, Comput. Fluids, 46 (2011), pp. 78–87.

[301] S.G. Schirmer, H. Fu, and A. Solomon, *Complete controllability of quantum systems*, Phys. Rev. A, 63 (2001), 063410.

[302] S. Schmidt, *Efficient Large Scale Aerodynamic Design Based on Shape Calculus*, Ph.D. thesis, University of Trier, Trier, Germany, 2010.

[303] S. Schmidt, C. Ilic, V. Schulz, and N. Gauger, *Airfoil design for compressible inviscid flow based on shape calculus*, Optim. Eng., 12 (2011), pp. 349–369.

[304] S. Schmidt and V. Schulz, *Shape derivatives for general objective functions and the incompressible Navier–Stokes equations*, Control Cybernet., 39 (2010), pp. 677–713.

[305] S. Schmidt and V. Schulz, *Impulse response approximations of discrete shape Hessians with application in CFD*, SIAM J. Control Optim., 48 (2009), pp. 2562–2580.

[306] S. Schmidt, V. Schulz, C. Ilic, and N. Gauger, *Three dimensional large scale aerodynamic shape optimization based on shape calculus*, in 41st AIAA Fluid Dynamics Conference and Exhibit, number AIAA 2011-3718, Honolulu, HI, June 27–30, 2011.

[307] S. Schmidt, V. Schulz, C. Ilic, and N. R. Gauger, *Three dimensional large scale aerodynamic shape optimization based on shape calculus*, AIAA J., submitted.

[308] J. Schöberl and W. Zulehner, *Symmetric indefinite preconditioners for saddle point problems with applications to PDE-constrained optimization problems*, SIAM J. Matrix Anal. Appl., 29 (2007), pp. 752–773.

[309] V. Schulz, *Numerical optimization of the cross-sectional shape of turbine blades*, Z. Angew. Math. Mech., 76 (1996), Suppl. 1, pp. 207–210.

[310] V. Schulz, *Solving discretized optimization problems by partially reduced SQP methods*, Comput. Vis. Sci., 1 (1998), pp. 83–96.

[311] V. Schulz, A. Bardossy, and R. Helmig, *Conditional statistical inverse modeling in groundwater flow by multigrid methods*, Comput. Geosci., 3 (1999), pp. 49–68.

[312] V. Schulz and I. Gherman, *One-Shot Methods for Aerodynamic Shape Optimization*, in "MEGADESIGN and MegaOpt - German Initiatives for Aerodynamic Simulation and Optimization in Aircraft Design," Eds. N. Kroll, D. Schwamborn, K. Becker, H. Rieger, and F. Thiele, "Notes on Numerical Fluid Mechanics and Multidisciplinary Design," 107 (2009), pp. 207–220.

[313] V. Schulz and C. Schillings, *On the nature and treatment of uncertainties in aerodynamic design*, AIAA J., 47 (2009), pp. 646–654.

[314] V. Schulz and G. Wittum, *Multigrid optimization methods for stationary parameter identification problems in groundwater flow*, in W. Hackbusch and G. Wittum (Eds.), Multigrid Methods V, Lecture Notes in Computational Science and Engineering, Vol. 3, Springer, Berlin, 1998, pp. 276–288.

[315] V. Schulz and G. Wittum, *Transforming smoothers for PDE constrained optimization problems*, Comput. Vis. Sci., 11 (2008), pp. 207–219.

[316] Ch. Schwab and R.A. Todor, *Sparse finite elements for stochastic elliptic problems higher order moments*, Computing, 71 (2003), pp. 43–63.

[317] D. Schwamborn and T. Gerhold, *The DLR TAU-code: Recent application in research and industry*, in P. Wessling, E. Onate, and J. Périaux (Eds.), ECCOMAS CFD 2006, Egmond aan Zee, The Netherlands, September 5–8, 2006.

[318] V.V. Shaidurov, *Some estimates of the rate of convergence of the cascadic conjugate gradient method*, Comput. Math. Appl., 31 (1996), pp. 161–171.

[319] D.F. Shanno, *Conjugate gradient methods with inexact searches*, Math. Oper. Res., 3 (1978), pp. 244–256.

[320] R. E. Showalter, *Monotone Operators in Banach Space and Nonlinear Partial Differential Equations*, Mathematical Surveys and Monographs, Vol. 49, AMS, Providence, RI, 1997.

[321] J. Sokolowski and J.-P. Zolésio, *Introduction to Shape Optimization: Shape Sensitivity Analysis*, Springer, Berlin, 1992.

[322] G. Stadler, *Semismooth Newton and augmented Lagrangian methods for a simplified friction problem*, SIAM J. Optim., 15 (2004), pp. 39–62.

[323] G. Stadler, *Elliptic optimal control problems with L^1-control cost and applications for the placement of control devices*, Comput. Optim. Appl., 44 (2009), pp. 159–181.

[324] K. Stüben, *Algebraic multigrid (AMG): Experiences and comparisons*, Appl. Math. Comput., 13 (1983), pp. 419–452.

[325] K. Stüben, *Algebraic Multigrid (AMG): An Introduction with Applications*, GMD Report 53, March 1999. Appeared as an appendix in the book: U. Trottenberg, C. Oosterlee, and A. Schüller, *Multigrid*, Academic Press, London, 2001.

[326] E. Süli, *Finite Elements Methods for Partial Differential Equations*, Lecture Notes, Oxford University Computing Laboratory, Oxford, UK, 2001.

[327] E. Süli, *An Introduction to the Numerical Analysis of Partial Differential Equations*, Lecture Notes, Oxford University Computing Laboratory, Oxford, UK, 2005.

[328] E. Süli, *Convergence of finite volume schemes for Poisson's equation on nonuniform meshes*, SIAM J. Numer. Anal., 28 (1991), pp. 1419–1430.

[329] B.J. Sussman, D. Townsend, M.Y. Ivanov, and A. Stolow, *Dynamic Stark control of photochemical processes*, Science, 13 (2006), pp. 278–281.

[330] S. Ta'asan, *Optimal multigrid method for inviscid flows*. In P.W. Hemker and P. Wesseling (Eds.), Multigrid Methods, IV (Amsterdam, 1993), Internat. Ser. Numer. Math., 116, Birkhäuser, Basel, 1994, pp. 309–320.

[331] S. Ta'asan, *Introduction to Shape Design and Control, Inverse Design and Optimisation Methods*, von Karman Institute Lecture Notes, VKI LS 1997-05.

[332] S. Ta'asan, G. Kuruvila, and M.D. Salas, *Aerodynamic design and optimization in one shot*, 30th Aerospace Sciences Meeting, Reno, NV, AIAA Paper 92-0025, 1992.

[333] A. Taflove and S.C. Hagness, *Computational Electrodynamics: The Finite-Difference Time-Domain Method*, Artech House, Boston, MA, 2000.

[334] X.-C. Tai and J. Xu, *Global and uniform convergence of subspace correction methods for some convex optimization problems*, Math. Comp., 71 (2002), pp. 105–124.

[335] T. Tanaka, T. Takenaka, and S. He, *An FDTD approach to the time-domain inverse scattering problem for an inhomogeneous cylindrical object*, Microw. Opt. Tech. Lett., 16 (1999), pp. 292–296.

[336] A. Tarantola, *Inverse Problem Theory and Methods for Model Parameter Estimation*, SIAM, Philadelphia, 2005.

[337] J. W. Thomas, *Numerical Partial Differential Equations: Finite Difference Methods*, Springer-Verlag, New York, 1995.

[338] F. Tröltzsch, *Optimality Conditions for Parabolic Control Problems and Applications*, Teubner, Leipzig, 1984.

[339] F. Tröltzsch, *Optimal Control of Partial Differential Equations*, AMS, Providence, RI, 2010.

[340] U. Trottenberg, C. Oosterlee, and A. Schüller, *Multigrid*, Academic Press, London, 2001.

[341] P. Tseng, *Convergence of block coordinate descent method for nondifferentiable minimization*, J. Optim. Theory Appl., 109 (2001), pp. 475–494.

[342] G. Turinici, *On the controllability of bilinear quantum systems*, in M. Defranceschi and C. Le Bris (Eds.), Mathematical Models and Methods for Ab Initio Quantum Chemistry, Lecture Notes in Chemistry, Vol. 74, Springer, Berlin, pp. 75–92.

[343] M. Ulbrich, *Semismooth Newton methods for operator equations in function spaces*, SIAM J. Optim., 13 (2003), pp. 805–841.

[344] M. Vallejos, *MGOPT with gradient projection method for solving bilinear elliptic optimal control problems*, Computing, 87 (2010), pp. 21–33.

[345] M. Vallejos and A. Borzì, *Multigrid optimization methods for linear and bilinear elliptic optimal control problems*, Computing, 82 (2008), pp. 31–52.

[346] A. van den Bos, *Complex gradient and Hessian*, IEEE Proc. Vis. Image Signal Process., 141 (1994), pp. 380–382.

[347] S. Vandewalle, *Parallel Multigrid Waveform Relaxation for Parabolic Problems*, B.G. Teubner, Stuttgart, 1993.

[348] S. Vandewalle and G. Horton, *Fourier mode analysis of the multigrid waveform relaxation and time-parallel multigrid methods*, Computing, 54 (1995), pp. 317–330.

[349] S. Vandewalle and R. Piessens, *Efficient parallel algorithms for solving initial-boundary value and time-periodic parabolic partial differential equations*, SIAM J. Sci. Statist. Comput., 13 (1992), pp. 1330–1346.

[350] P. Vanek, M. Brezina, and J. Mandel, *Convergence of algebraic multigrid based on smoothed aggregation*, Numer. Math., 88 (2001), pp. 559–579.

[351] S. Vanka, *Block-implicit multigrid calculation of two-dimensional recirculating flows*, Comput. Methods Appl. Mech. Engrg., 59 (1986), pp. 29–48.

[352] E. Vanmarcke, *Random Fields: Analysis and Synthesis*, MIT Press, Cambridge, MA, 1983.

[353] R.S. Varga, *Matrix Iterative Analysis*, Springer, Berlin, 2000.

[354] G. von Winckel and A. Borzì, *Computational techniques for a quantum control problem with H^1-cost*, Inverse Problems, 24 (2008), 034007.

[355] G. von Winckel and A. Borzì, *QUCON: A fast Krylov-Newton code for dipole quantum control problems*, Comput. Phys. Comm., 181 (2010), pp. 2158–2163.

[356] G. von Winckel, A. Borzì, and S. Volkwein, *A globalized Newton method for the accurate solution of a dipole quantum control problem*, SIAM J. Sci. Comput., 31 (2009), pp. 4176–4203.

[357] D. Wachsmuth, *Sufficient second-order optimality conditions for convex control constraints*, J. Math. Anal. Appl., 319 (2006), pp. 228–247.

[358] H. Wang, *Robust control of the output probability density functions for multivariable stochastic systems with guaranteed stability*, IEEE Trans. Automat. Control, 44 (1999), pp. 2103–2107.

[359] J. Weickert and C. Schnörr, *Variational optic flow computation with a spatio-temporal smoothness constraint*, J. Math. Imaging Vision, 14 (2001), pp. 245–255.

[360] J. Werschnik and E.K.U. Gross, *Quantum optimal control theory*, J. Phys. B: Atom. Mol. Opt. Phys., 40 (2007), pp. R175–R211.

[361] R. Wienands and C.W. Oosterlee, *On three-grid Fourier analysis for multigrid*, SIAM J. Sci. Comput., 23 (2001), pp. 651–671.

[362] G. Wittum, *Multigrid methods for Stokes and Navier-Stokes equations. Transforming smoothers: Algorithms and numerical results*, Numer. Math., 54 (1988), pp. 543–563.

[363] D. Xiu and J.S. Hesthaven, *High-order collocation methods for differential equations with random inputs*, SIAM J. Sci. Comput., 27 (2005), pp. 1118–1139.

[364] D. Xiu and G.E. Karniadakis, *The Wiener–Askey polynomial chaos for stochastic differential equations*, SIAM J. Sci. Comput., 24 (2002), pp. 619–644.

[365] I. Yavneh and D. Dardyk, *A multilevel nonlinear method*, SIAM J. Sci. Comput., 28 (2006), pp. 24–46.

[366] K.S. Yee, *Numerical solution of initial boundary value problems involving Maxwell's equations in isotropic media*, IEEE Trans. Antennas and Propagation, 14 (1966), pp. 302–307.

[367] D. Young, *Iterative Solution of Large Linear Systems*, Academic Press, New York, 1971. (Reprinted by Dover, 2003.)

[368] H. Yserentant, *Regularity and Approximability of Electronic Wave Functions*, Lecture Notes in Mathematics, Vol. 2000, Springer-Verlag, Berlin, 2010.

[369] N. Zabaras and B. Ganapathysubramanian, *A scalable framework for the solution of stochastic inverse problems using a sparse grid collocation approach*, J. Comput. Phys., 227 (2008), pp. 4697–4735.

[370] E. Zuazua, *Switching control*, J. Eur. Math. Soc. (JEMS), 13 (2011), pp. 85–117.

Index